Advanced Coating and Cladding Technologies

This book explores advancements in coating and cladding technologies designed for high-temperature resistance and wear protection, including laser cladding of nickel-based and high-entropy alloy coatings, cold metal transfer (CMT) technology, and integration of machine learning for predictive modeling and optimization. It addresses joining dissimilar materials, enhancing tool performance through advanced coatings and 3D-printed microstructures, mechanical testing in various conditions, and eco-friendly corrosion protection for aerospace and industrial applications.

Features:

- Explores cutting-edge coating and cladding technologies such as laser cladding and CMT for high-temperature and wear-resistant applications.
- Blends foundational knowledge with cutting-edge advancements like laser cladding, CMT, and AI-driven predictive modeling.
- Provides practical insights into the integration of machine learning and AI for predictive modeling and optimization of coatings and cladding performance.
- Includes introductory sections, illustrative diagrams, and case studies.
- Addresses applications in aerospace, automotive, energy, and biomedical sectors

This book is aimed at graduate students and researchers in materials science, mechanical, industrial, and manufacturing engineering.

Advanced Coating and Cladding Technologies

Edited by

Pankaj Kumar,
Dipen Kumar Rajak
and Manowar Hussain

CRC Press
Taylor & Francis Group
Boca Raton London New York

CRC Press is an imprint of the
Taylor & Francis Group, an **informa** business

First edition published 2026
by CRC Press
2385 NW Executive Center Drive, Suite 320, Boca Raton FL 33431

and by CRC Press
4 Park Square, Milton Park, Abingdon, Oxon, OX14 4RN

CRC Press is an imprint of Taylor & Francis Group, LLC

ISBN: 9781041100171 (hbk)
ISBN: 9781041100737 (pbk)
ISBN: 9781003653226 (ebk)

DOI: 10.1201/9781003653226

Typeset in Times
by SPi Technologies India Pvt Ltd (Straive)

Contents

Preface

The rapid evolution of industrial, aerospace, automotive, energy, and marine sectors demands materials with exceptional performance, durability, and sustainability. Coating and cladding technologies have emerged as vital enablers in meeting these demands, providing critical protection against corrosion, wear, thermal degradation, and environmental impact while enhancing component functionality.

This edited volume, *Advanced Coating and Cladding Technologies*, presents a comprehensive exploration of state-of-the-art protective surface engineering solutions, spanning thermal spray, cold spray, laser cladding, sol–gel systems, nanocoatings, high-entropy alloys, and ultra-high-temperature ceramics. The chapters highlight innovations that align with modern industrial priorities—efficiency, regulatory compliance, environmental stewardship, and cost-effectiveness.

A unique emphasis is placed on emerging trends such as machine learning-driven process optimization, additive manufacturing of functional coatings, bio-based and chrome-free alternatives, and integration of computational modeling with high-throughput experimentation.

The book also underscores the critical intersection of sustainability and performance, demonstrating how eco-friendly solutions can match or surpass traditional toxic systems while delivering significant life-cycle and economic benefits.

By combining fundamental principles, advanced characterization, and industrial case studies, this work serves as a bridge between cutting-edge research and practical application. It is intended for researchers, engineers, materials scientists, and industry professionals seeking to understand and implement next-generation coating and cladding solutions in the era of Industry 4.0.

About the Editors

Pankaj Kumar is the professor and head of the Mechanical Engineering Department, SR University, Warangal. In 2011, he received a Teaching Assistant Fellowship from the Ministry of Human Resource Development of India. He received his PhD from IIT (ISM) Dhanbad. Dr. Pankaj has more than 15 years of experience in the design, fabrication, and characterization of micro- to nanoscale components using unconventional techniques. His current research includes additive manufacturing, micromachining, composite materials processing technology, nano-materials, and sustainable manufacturing, biotechnology. He has published over 50 research articles in various international journals and conferences—5 books, 20 book chapters, and 10 patents. He won the best paper award at the International Conference on Research Trends in Science, Technology, Engineering & Management 2019. He is on the Editorial Board of members of the *PLOS ONE* Journal. He is a reviewer of many international journals. He has also given many invited talks at universities, workshops, and faculty development programs.

Dipen Kumar Rajak is a scientist in the Light Weight Materials Division at CSIR-Advanced Materials and Processes Research Institute (CSIR-AMPRI), Bhopal. He earned his PhD in Mechanical Engineering from IIT (ISM) Dhanbad. His expertise spans materials science, composites, metal foams, high-entropy alloys, additive manufacturing, and fiber-reinforced polymers. With over 200 publications, including journal articles, books, and book chapters, he has contributed extensively to advanced materials research, securing more than 4500 citations and an h-index of 19. Dr. Rajak has been granted patents in fiber-reinforced polymer composites and innovative engineering designs. He serves on the editorial boards of reputed journals such as *Materials Genome Engineering and Advances in Polymer Technology*. His ongoing CSIR-funded project focuses on Cu/graphene-based hybrid coatings for thermal conductivity and corrosion resistance in Mg alloys, underscoring his role in developing strategic, sustainable material solutions.

Manowar Hussain is an assistant professor in the Department of Production & Industrial Engineering at B.I.T. Sindri, Dhanbad. He holds a PhD and MTech in Mechanical (Manufacturing) Engineering from IIT (ISM) Dhanbad and a BTech in Production Engineering from B.I.T. Sindri. His research expertise includes manufacturing, laser material processing, and non-conventional machining processes. He has published extensively in reputed journals such as the *Arabian Journal for Science and Engineering*, *Optics and Lasers in Engineering*, and the *Journal of Mechanical Science and Technology*, contributing significantly to additive manufacturing, laser sintering, and micro-EDM studies. Dr. Hussain holds two published patents in biomedical implants and magnetic nanofluid manufacturing. He has served in academic roles at SR University and Chaitanya Bharathi Institute of Technology, Hyderabad. A life member of ISTE and the Indian Laser Association, he is also experienced in teaching machine tool engineering, metrology, production technology, and modern manufacturing processes.

Contributors

R Abhishek
Department of Mechanical Engineering
RV College of Engineering
Bengaluru, India

Kumar Jain Amit
Department of Mechanical Engineering
Birla Institute of Technology and Science
Pilani, Rajasthan, India

Ghosh Anima
Department of Physics, School of Sciences and Humanities
SR University
Warangal, India

Ghosh Arka
Department of Mechanical Engineering
C.V. Raman Global University
Bhubaneswar, India

Yelamasetti Balram
Department of Mechanical Engineering
MLR Institute of Technology
Hyderabad, India

Das Bappa
Department of Materials Engineering
Indian Institute of Science Bangalore
Bengaluru, India

A Bharatish
Department of Mechanical Engineering
RV College of Engineering
Bengaluru, Karnataka, India

Choudhury Bishub
Department of Materials Engineering
Indian Institute of Science Bangalore
Bengaluru, India

Oraon Buddhadeb
Department of Mechanical Engineering
Jadavpur University
Kolkata, India

Bommana Divakar
Department of Mechanical Engineering
GST, GITAM
Visakhapatnam, India

Tejonidhi G. K.
Department of Mechanical Engineering
RV College of Engineering,
Bengaluru, India

Joshi G. R.
Department of Mechanical Engineering
Marwadi University
Rajkot, India

Rajkumar G. R.
Department of Mechanical Engineering
RV College of Engineering
Bengaluru, India

Surja Kiran Ghorui
Department of Physics
Vaagdevi Engineering College
Warangal, India

Suresh Gudala
Department of Mechanics and Mechanical Engineering
Siberian State Industrial University
Novokuznetsk, Russia

Varshith K
Department of Mechanical Engineering
RV College of Engineering
Bengaluru, India

Chandu K. Pavan
Applied Materials Bangalore
Indian Institute of Science Bangalore
Bengaluru, India

Chouhan Koustubh
Department of Production & Industrial Engineering
BIT Sindri
Dhanbad, India

Kiran Kumar Kupireddi
Department of Mechanical Engineering
National Institute of Technology Warangal
Warangal, India

Khan MD. F.
Department of Mechanical Engineering
King Faisal University
Al Hufūf, Al-Ahsa, Saudi Arabia

Hussain Manowar
Department of Production and Industrial Engineering
BIT Sindri
Jharkhand, India

Subbarao Medabalimi
Department of Mechanical Engineering
SR University
Warangal, India

Shahid Raza Mohammad
Department of Mechanical Engineering
Indian Institute of Engineering Science and Technology
Shibpur, Howrah, WB, India

Albalawi Mona O.
Department of Chemistry, Faculty of Science
University of Tabuk
Tabuk, Saudi Arabia

Sumanta Mukherjee
Department of Production & Industrial Engineering
BIT Sindri
Dhanbad, Jharkhand, India

Pallavi Senapati N
Department of Mechanical Engineering
Nadimpalli Satyanarayana Raju Institute of Technology
Visakhapatnam, India

Shanmukha N
Department of Mechanical Engineering
RV College of Engineering
Bengaluru, India

Vijay Kumar P. R.
Department of Mechanical Engineering
RV College of Engineering
Bengaluru, India

Kumar Pankaj
Department of Mechanical Engineering
SR University
Warangal, India

Kumar Prakash
Department of Metallurgical and Materials Engineering
IIT Roorkee
Haridwar, India

Kumar Prakash
Faculty of Engineering and Applied Sciences
Mechanical Engineering
Usha Martin University
Ranchi, India

Nagireddy Praveena Devi
Department of Mechanical Engineering
SR University
Warangal, India

Kesari Priya
Department of Applied Sciences
NIMS Institute of Engineering and Technology
NIMS University
Jaipur, India

Ravi Kumar R.
Indira Gandhi Centre for Atomic Research
Kalpakkam, India

Moatassim Raoof
Department of Materials Science
Institute of Graduate Studies and Research
Alexandria University, Egypt

Ande Rohan
Department of Mechanical Engineering
SR University
Warangal, India

Uzwalkiran Rokkala
Department of Mechanical Engineering
Aditya University
Surampalem, India

Madan Royal
Department of Mechanical Engineering
Graphic Era University
Dehradun, India

Sivakumar S
Department of Mechanical Engineering
RV College of Engineering
Bengaluru, India

Harisha S. K.
Department of Mechanical Engineering
RV College of Engineering
Bengaluru, India

Pulla Sammaiah
Department of Mechanical Engineering
SR University
Warangal, India

Begum SD Ruksar
Department of Mechanical Engineering
SR University
Warangal, India

Tamboli Shahid
Department of Mechanical Engineering
RV College of Engineering
Bengaluru, India

Kumar Shakti
Department of Materials Engineering
Indian Institute of Science Bangalore
Bengaluru, India

Salman Basha Sheik
Department of Mechanical Engineering
Sasi Institute of Technology & Engineering
Tadepalligudem, India

Lal Shiv
Department of Mechanical Engineering
University Department
Rajasthan Technical University
Kota, India

Sinha Sumit Kumar
Department of Materials Engineering
Indian Institute of Science Bangalore
Bengaluru, India

Pratap Suresh
Department of Mechanical Engineering
GL Bajaj Institute of Technology
Greater Noida, India

Dora T. Rajesh Kumar
Samant Chandrasekhar Institute of Technology & Management
Semiliguda, India

Addanki Rama Swamy
Department of Mechanical Engineering
Sasi Institute of Technology & Engineering
Tadepalligudem, India

Kumar Vikas
Department of Mechanical Engineering
University Department
Rajasthan Technical University
Kota, India

Kumar Vikash
Department of Mechanical Engineering
Jadavpur University
Kolkata, India

Mohammed Zubairuddin
Department of Mechanical Engineering
Aditya University
Surampalem, India

1 Introduction to Advanced Coating and Cladding Technologies

Pankaj Kumar, Manowar Hussain, Mohammad Shahid Raza, and Amit Kumar Jain

1.1 INTRODUCTION

In modern engineering and material science, coating and cladding technologies have become indispensable for enhancing the surface properties of materials to meet the demanding requirements of industrial applications. These technologies are primarily employed to improve wear resistance, corrosion protection, thermal insulation, oxidation resistance, and biocompatibility, depending on the application. With the emergence of advanced manufacturing methods, the scope and precision of surface engineering have evolved significantly. Coating typically involves the application of a thin layer of a protective material on a substrate surface using methods such as thermal spraying, chemical vapor deposition (CVD), physical vapor deposition (PVD), laser surface modification, and electroplating. In contrast, cladding refers to the process of bonding a thick layer of a different material onto a substrate, often through welding, laser cladding, hot rolling, or explosive bonding. The integration of nanotechnology, robotics, AI-driven process control, and additive manufacturing has accelerated innovations in coating and cladding. These advancements enable microstructural tailoring, enhanced adherence, and graded coatings, offering multifunctionality and improved service life for components.

1.2 ADVANCED COATING PROCESSES

Advanced coating processes are essential for imparting enhanced surface characteristics to materials, enabling their use in aggressive environments and specialized applications. Unlike conventional coating techniques, advanced processes focus on precision, material compatibility, and controlled microstructural evolution to achieve superior performance.

1.2.1 Plasma Spraying

The plasma spray process offers several advantages that make it a versatile and widely adopted surface modification technique for industrial applications. This

DOI: 10.1201/9781003653226-1

enables the deposition of a broad spectrum of materials, including metals, ceramics, and composites, onto a variety of substrates with complex geometries. The process generates extremely high flame temperatures, ranging from 6000°C to over 15,000°C, enabling the melting of high-melting-point materials, such as zirconia, alumina, and tungsten carbide. One of the key properties is its ability to produce coatings with controllable porosity, microstructure, and thickness (ranging from a few micrometers to several millimeters), tailored to specific functional requirements, such as wear resistance, thermal insulation, and corrosion protection. The high particle velocities achieved in this process result in strong mechanical interlocking and good adhesion strength, often exceeding 20 MPa. Plasma spraying is also characterized by a low thermal input to the substrate, minimizing distortion and making it suitable for heat-sensitive components. This process allows for atmospheric, low-pressure, or inert gas environments, enabling both conventional and advanced variants such as vacuum plasma spraying (VPS) and suspension plasma spraying (SPS). These properties collectively contribute to the widespread adoption of plasma spray coatings in aerospace, biomedical, energy, and automotive industries. The process flow of the plasma spray coating is illustrated in Figure 1.1, which shows key components, such as the plasma torch, powder feed system, and substrate orientation.

The principle of the plasma spray process is based on the generation and utilization of a high-temperature plasma jet to heat and propel the coating material onto a substrate surface, where it rapidly cools and solidifies to form a coating. This was achieved by creating a plasma arc between the tungsten cathode and copper anode within a specially designed plasma torch. When an inert gas such as argon, nitrogen, or helium is introduced into the torch, it becomes ionized owing to the high-energy

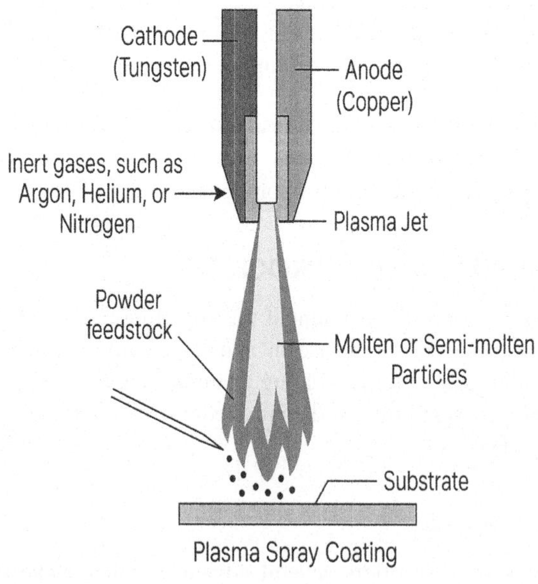

FIGURE 1.1 Schematic diagram for plasma spray coating.

arc, forming a plasma state characterized by extremely high temperatures (typically 6000–15,000°C). In this plasma jet, fine powder particles of the desired coating material were injected either axially or radially. These particles are rapidly heated to a molten or semi-molten state and then accelerated by the plasma stream toward the substrate. Upon impact, the molten particles flatten and solidify almost instantaneously, forming thin lamellae known as splats. Successive splat layering builds up a coating with the desired thickness and surface characteristics. The process is governed by parameters such as the plasma gas composition, arc current, powder feed rate, and spray distance, which influence the adhesion strength, microstructure, and performance of the final coating.

The plasma spray process parameters include the plasma gas composition and flow rate, arc voltage and current, powder feed rate, particle size, stand-off distance, and substrate surface temperature. These parameters influence the coating thickness, adhesion, porosity, microstructure, and overall performance and require precise control of the coating quality and functional properties.

Commonly used coating materials in plasma spray coating include ceramics such as alumina (Al_2O_3), zirconia (ZrO_2), and titania (TiO_2), and carbides such as tungsten carbide (WC) and chromium carbide (Cr_3C_2). Metallic coatings such as nickel-based alloys, cobalt-based alloys, and molybdenum are also frequently applied. Typical substrate materials include steels (carbon and stainless steel), titanium alloys, aluminum alloys, and superalloys such as Inconel. Compatibility between the coating and substrate materials is critical to ensure proper adhesion, thermal expansion matches, and desired functional performance in service environments.

1.2.2 High-Velocity Oxy-Fuel (HVOF)

High-velocity oxy-fuel (HVOF) is an advanced thermal spray process that is used to apply dense, wear-resistant, and well-adhered coatings. In the HVOF, a mixture of fuel gas (commonly hydrogen, propane, or kerosene) and oxygen is combusted in a chamber to generate a high-temperature, high-pressure gas stream. The resulting supersonic jet propels powder particles of the coating material at velocities exceeding 800 m/s toward the substrate. Although the flame temperature in HVOF (≈ 2500–3000°C) is lower than that in plasma spray, extremely high particle velocities result in strong mechanical interlocking and minimal oxidation, producing coatings with low porosity (<1%) and high bond strength (>70 MPa) [1, 2].

HVOF is particularly effective for spraying carbide-based materials such as tungsten carbide-cobalt (WC-Co) and chromium carbide, as well as metallic alloys such as Inconel, Hastelloy, and stainless steels. This process yields coatings with excellent hardness, wear resistance, and corrosion protection, making it ideal for aerospace, petrochemical, power generation, and automotive applications. Because of the lower thermal input compared to plasma spraying, HVOF also minimizes residual stresses and distortions in heat-sensitive components. The working principle of the HVOF process is illustrated in Figure 1.2, showing key elements such as the combustion chamber, nozzle, and particle trajectory.

A critical review of the experimental procedures for the deposition and characterization of partially oxidized coatings reveals the importance of systematic

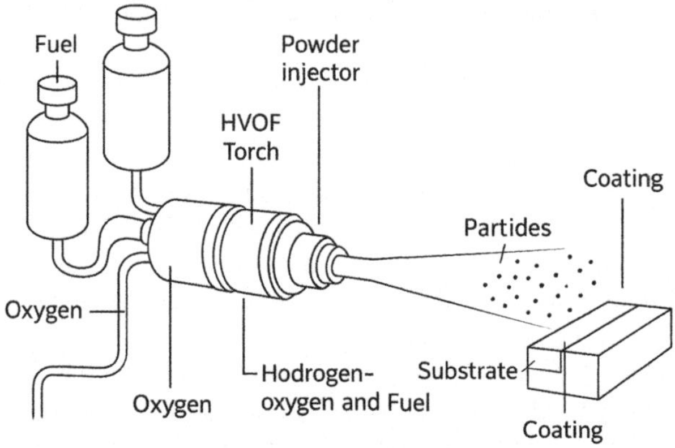

FIGURE 1.2 HVOF process parameters and their relationships.

methodologies to ensure reliable performance assessments. Deposition via thermal spray techniques, particularly HVOF and plasma spray, allows for controlled partial oxidation during in-flight particle exposure to the surrounding atmosphere, which can tailor oxide dispersion to improve wear and corrosion resistance [3]. Post-deposition, cross-sectional sample preparation using metallographic polishing and etching is crucial for revealing microstructural features such as lamellae, interplay boundaries, and oxide inclusions [4]. Surface roughness is typically measured using contact or optical profilometry, offering insights into tribological behavior and coating adhesion.

Porosity analysis was conducted using image analysis software on scanning electron microscopy (SEM) micrographs or via Archimedes' principle, both of which influence mechanical and thermal properties. The bond strength evaluation followed ASTM C633 pull-off tests to determine the coating adherence under tensile loading. Micro-hardness, measured using the Vickers or Knoop methods, reflects localized mechanical resistance. The density and erosion resistance further characterize the coating durability. Studies emphasize that integrating these techniques provides comprehensive insight into the structure–property relationships of partially oxidized coatings, directly impacting their application in the aerospace and power generation industries.

Critical literature highlights that the bond strength, microhardness, and density of partially oxidized coatings are directly influenced by the oxide dispersion and inter-splat bonding formed during thermal spraying. High-velocity processes, such as HVOF, yield coatings with improved density and hardness owing to reduced oxidation and better particle cohesion [5]. X-ray diffraction (XRD) was used to identify the phase composition and quantify oxide formation, while SEM provided insights into the coating morphology, lamellar structures, and porosity. Energy dispersive spectroscopy (EDS) complements SEM by mapping the elemental distribution, revealing oxidation gradients across the splats [3].

A review of recent studies on MDN steel and partially oxidized coatings such as NiCr, NiCr+Al, and NiCrBSiFe revealed significant differences in wear performance and surface behavior. The wear rate and coefficient of friction (COF) were primarily governed by the hardness, oxide content, and splat bonding of the coatings. Among these, NiCrBSiFe coatings, owing to their self-fluxing nature and high hardness, consistently exhibit the lowest wear rate and COF under dry sliding conditions [6]. XRD analysis revealed the formation of stable oxides and intermetallic phases, which enhanced the surface integrity and wear resistance. SEM studies of the worn morphologies show that NiCr and NiCr+Al coatings suffer from abrasive and oxidative wear, whereas NiCrBSiFe coatings exhibit smoother wear tracks and minimal delamination, indicating better mechanical integrity. Comparative analysis confirmed that NiCrBSiFe is the most effective coating among those evaluated for wear-critical applications in the aerospace and tooling sectors.

1.2.2.1 Process Parameters and Their Effects

The HVOF process involves several critical parameters that directly influence coating quality and performance. Key parameters include the type of fuel (hydrogen, propane, and kerosene), fuel and oxygen flow rates (typically 15–40 L/min), and oxygen-to-fuel ratio, which is usually slightly oxidizing to enhance combustion efficiency. The combustion pressure ranges from 0.7 to 1.5 MPa, producing flame temperatures of approximately 2500–3000°C and particle velocities exceeding 800 m/s. The powder feed rate (10–50 g/min) and particle size (commonly 15–30 μm) influenced the deposition efficiency and coating morphology. The spray distance was typically maintained between 150 and 300 mm to balance the thermal input and particle impact. Substrates are usually preheated to 100–200°C and grit-blasted to achieve a roughness (Ra) of 3–5 μm for better adhesion. Carrier gases such as nitrogen or argon transport the powder to the flame. Other influential parameters include the number of torch passes and the cooling method (air or water). Proper optimization of these parameters results in dense low-porosity coatings with excellent bond strength and wear resistance, which are suitable for industrial applications. As illustrated in Figure 1.3, the HVOF process parameters are closely interrelated, influencing the coating quality and performance.

In the plasma spraying process, several key parameters—temperature, chamber pressure, coating thickness, deposition rate, and power input—play critical roles in determining the quality, structure, and performance of the resulting coatings. The plasma temperature, typically in the range of 6000–15,000°C, directly affects the degree of particle melting, thereby influencing the coating microstructure, hardness, and inter-splat bonding. Higher temperatures generally enhance coating densification but can also cause the decomposition or oxidation of sensitive materials [7, 8]. Chamber pressure, particularly in low-pressure or VPS, controls the particle velocity and heat transfer efficiency. Lower pressures yield higher kinetic energy, producing denser coatings with improved adhesion and lower porosity [9]. The coating thickness is governed by the deposition rate, which depends on the powder feed rate and spray distance. An optimal balance is essential to avoid excessive residual stresses or poor interlamellar cohesion. Power input—a function of arc current and voltage—affects energy transfer to the gas and ultimately to the particles. Higher power increases the melting efficiency but may introduce thermal stresses if not managed properly. Collectively, these parameters influence the porosity, surface roughness,

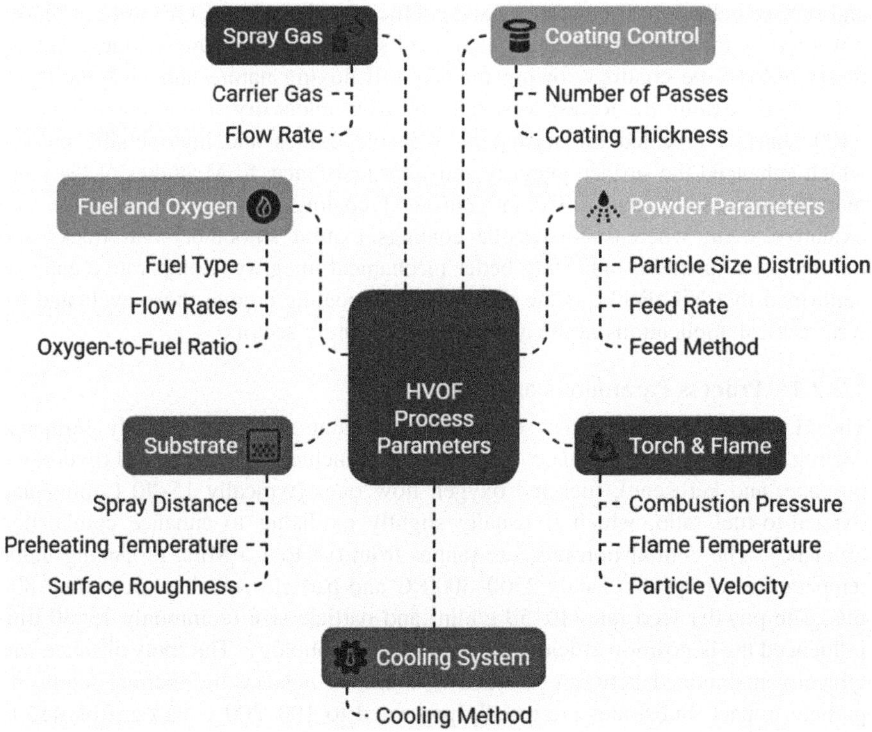

FIGURE 1.3 Schematic diagram for HVOF process.

adhesion strength, and hardness of the coating, which are critical for functional applications, such as wear resistance and thermal insulation [10, 11].

1.2.2.2 Porosity and Density of Coating

Porosity refers to the presence of voids or unfilled spaces within a coating, which affect its mechanical strength and barrier properties. The density indicated the compactness of the coating structure. Lower porosity and higher density enhance coating durability, corrosion resistance, and wear performance, making them critical for high-performance applications.

1.2.2.3 Surface Roughness, Microhardness, and Adhesion Strength Test

The surface roughness, microhardness, and adhesion strength are the key parameters used to evaluate the quality of plasma-sprayed coatings. The surface roughness measures the texture and irregularities on the coating surface, influencing the tribological behavior and coating-substrate interaction. The microhardness assesses the localized hardness of the coating material, reflecting its resistance to indentation and wear. It is typically measured by using a Vickers or Knoop hardness tester. The adhesion strength determines the bond integrity between the coating and substrate, which is commonly evaluated using a tensile pull-off test (ASTM C633). The high values in these tests indicate well-bonded, dense, and mechanically robust coatings suitable for industrial applications.

1.2.2.4 Microstructure and Phase Analysis

Microstructure and phase analyses are essential for understanding the performance and integrity of plasma-sprayed coatings. Microstructural evaluations, typically conducted using optical microscopy and SEM, reveal features such as lamellar splats, porosity, microcracks, unmelted particles, and interface quality. These features influence the mechanical strength, wear resistance, and thermal behavior. Phase analysis, usually performed using XRD, identifies the crystalline phases present in the coating and detects any phase transformations or oxidation during spraying. The distribution and stability of phases are critical for achieving the desired properties, such as hardness, thermal stability, and corrosion resistance, in functional coatings for high-performance applications.

1.2.2.5 Solid Particle Erosion Test

The solid particle erosion test evaluates the resistance of a coating or material to wear caused by high-velocity solid particles, typically under controlled laboratory conditions. In this test, abrasive particles, such as alumina or silica, were propelled at the coated surface using a gas jet at specific angles and velocities. The erosion rate was calculated by measuring the mass loss or depth of the material removed after exposure. Parameters such as the particle size, impact angle, velocity, and test duration significantly affected the results. This test is critical for assessing the durability of coatings used in aerospace, turbines, and offshore environments.

1.2.2.6 Oxidation and Hot Corrosion Tests

Oxidation and hot corrosion tests were conducted to assess the high-temperature stability and chemical resistance of the coatings and materials in aggressive environments. Oxidation tests exposed the samples to elevated temperatures in an oxygen-rich atmosphere to evaluate scale formation, weight gain, and oxide layer integrity over time. In contrast, hot corrosion tests simulate more severe conditions involving molten salts (e.g., Na_2SO_3 and V_2O_3) at high temperatures, which accelerate material degradation through chemical attack. These tests help determine the suitability of a coating for applications in gas turbines, boilers, and aerospace engines, where thermal and corrosive stresses are simultaneously present, affecting long-term performance.

1.2.2.7 Wear Test

The wear test evaluates the resistance of a material or coating to material loss caused by mechanical interaction with another surface under controlled conditions. Common wear tests include the pin-on-disk, ball-on-disk, and reciprocating wear methods, which simulate sliding or rolling contact. Test parameters, such as load, sliding speed, duration, and counter body material, were carefully selected to replicate real-world conditions. The wear rate is typically determined by measuring the volume loss or weight loss after the test. This test is crucial for assessing the durability and lifespan of coatings in applications involving friction, abrasion, and mechanical contact, such as tools, bearings, and engine parts.

1.2.3 Cold Spray

Cold spraying is a solid-state coating process that uses a high-velocity gas stream to accelerate powder particles to supersonic speeds, allowing them to plastically deform and bond to the substrate without melting. This process relies on kinetic energy rather than thermal energy, making it ideal for temperature-sensitive materials.

In cold spraying, compressed gas (typically nitrogen or helium) is heated moderately and passed through a converging–diverging nozzle. Powder particles were injected into the gas stream and propelled at velocities of up to 1200 m/s [5]. Upon impact with the substrate, the particles underwent severe plastic deformation, leading to solid-state bonding through mechanical interlocking and metallurgical adhesion. The cold-spray process utilizes supersonic gas flow to accelerate powder particles toward the substrate, achieving bonding without melting (see Figure 1.4).

1.2.3.1 Properties of Cold Spray Process

The cold-spray process produces dense, oxide-free coatings through solid-state bonding, avoiding melting and minimizing thermal distortion. It preserves the original properties of the feedstock material, resulting in a high adhesion strength, low residual stress, and excellent fatigue resistance. This process enables thick coatings, is ideal for temperature-sensitive substrates, and maintains high electrical and thermal conductivities, making it suitable for structural repairs and functional surface enhancements. These properties make cold-spray coatings ideal for corrosion resistance, electrical conductivity, and structural repair. The cold spray process offers key advantages, such as low residual stress, oxide-free deposition, and excellent adhesion strength, as illustrated in Figure 1.5.

1.2.3.2 Process Parameters

The key process parameters in the cold-spray process include the gas temperature (typically 200–800°C) and gas pressure (2–4 MPa), which control the particle

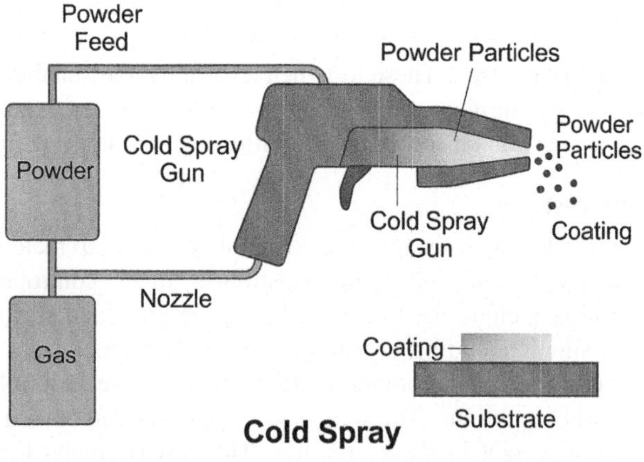

FIGURE 1.4 Schematic representation of the cold spray process showing high-pressure gas supply, converging-diverging nozzle, particle injection, and solid-state deposition on the substrate.

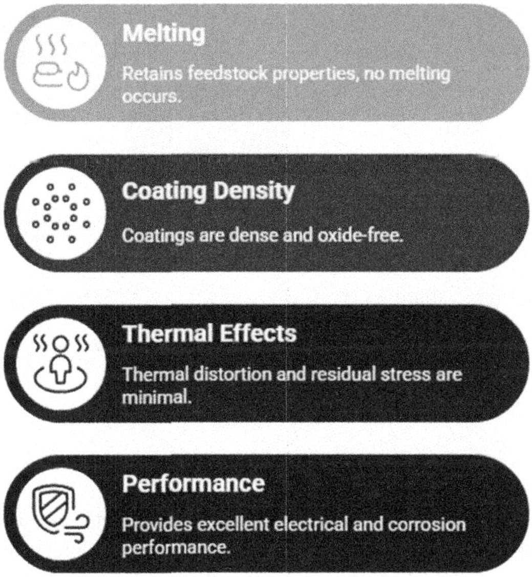

FIGURE 1.5 Key properties of the cold spray process, including solid-state bonding, minimal oxidation, high deposition efficiency, and preservation of feedstock microstructure.

velocity essential for bonding. The gas type (helium or nitrogen) influences cost and efficiency. The particle size, nozzle design, standoff distance, and traverse speed affect the deposition quality, layer thickness, and coating uniformity. Optimizing these parameters ensures dense, adherent coatings with minimal oxidation and thermal distortion. Proper control of these parameters ensures strong bonding, minimal porosity, and desired coating thickness. The influence of various process parameters, such as gas temperature, pressure, and particle velocity, on the cold-spray coating quality is summarized in Figure 1.6.

Commonly used coating materials in surface engineering include metals, such as aluminum, copper, titanium, stainless steel, and Inconel, along with alloys and composites, such as nickel-based superalloys, aluminum–silicon (Al–Si), and tungsten carbide–cobalt (WC–Co). These materials were selected for their corrosion protection, wear resistance, and electrical conductivity. Typical substrate materials include aluminum and magnesium alloys, steels, and titanium, with the selective use of polymers and ceramics depending on their application.

1.2.4 Physical Vapor Deposition (PVD)

In PVD, the coating material is vaporized in a vacuum environment and condensed on the substrate.

PVD is a widely used vacuum-based coating technique for producing thin, high-purity, and strongly adherent films on a variety of substrates [12, 13]. It is especially important in semiconductors, aerospace, biomedical devices, and cutting tools, where precise control of microstructure, composition, and surface properties is essential. In the PVD process, a solid source material (target) is vaporized through thermal

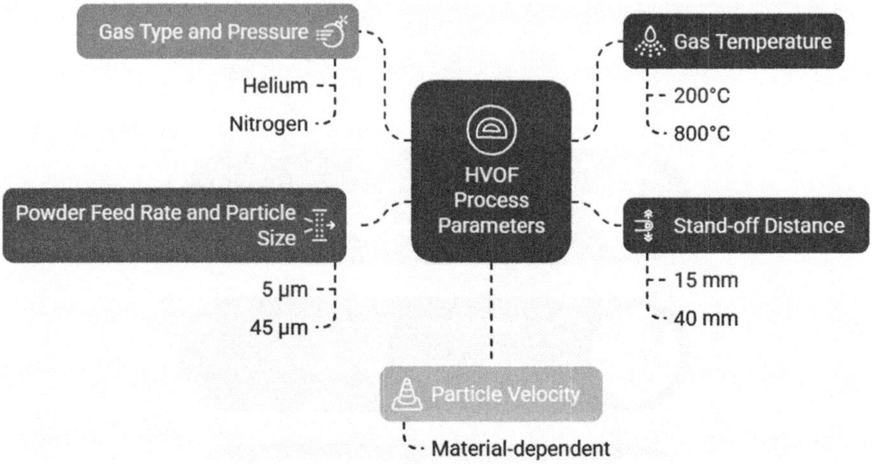

FIGURE 1.6 Cold spray process parameters and their effects on coating characteristics, showing relationships between gas temperature, pressure, particle velocity, and resulting adhesion strength, deposition efficiency, and microstructure.

evaporation, an electron beam, or sputtering in a high-vacuum chamber. The resulting vapor is transported through a vacuum and condenses onto the substrate to form a thin film. The process relies on the physical transport of atoms, rather than chemical reactions, ensuring high purity and controllable deposition rates. The vacuum environment minimizes contamination and allows for fine control over film thickness, crystallinity, and stress. The PVD process, involving vaporization and condensation of the coating material under vacuum, is illustrated schematically in Figure 1.7.

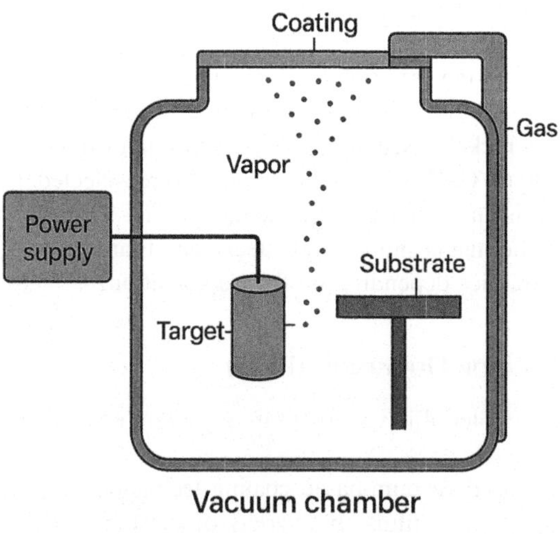

FIGURE 1.7 Schematic diagram of the physical vapor deposition (PVD) process, showing the vacuum chamber, target material, vapor transport, and substrate for thin film deposition.

1.2.4.1 Properties of PVD Coatings

PVD coatings exhibit high hardness, excellent wear and corrosion resistance, and good adhesion owing to their dense and uniform structure. They provide low friction coefficients and enhance the performance of the cutting tools and mechanical components. This process enables precise thickness control and strong bonding of thin films to various substrates. In addition, PVD coatings offer thermal stability and can be tailored for optical, decorative, or electronic applications, making them highly versatile across industries. These features render PVD coatings ideal for tools, MEMS devices, optics, and biomedical implants. As shown in Figure 1.8, PVD coatings offer a combination of high hardness, wear resistance, and strong adhesion, making them suitable for a wide range of industrial applications.

1.2.4.2 PVD Coating Process Parameters

Key PVD coating process parameters include the power source (DC, RF, or pulsed), which controls plasma generation, and chamber pressure (10^{-3} to 10^{-6} Torr), which is essential for vapor transport. The substrate temperature (100–600°C) influences coating adhesion and crystallinity. The deposition rate (0.1–10 μm/h) and target–substrate distance (5–30 cm) affect the film thickness and uniformity. Precise control of these variables ensures optimal coating properties such as density, adhesion, and microstructure. Precise control of these parameters influences the film adhesion, grain size, residual stress, and composition uniformity. The influence of key parameters, such as the power source, chamber pressure, and substrate temperature, on the PVD coating characteristics is illustrated in Figure 1.9.

Commonly used coating materials include metals such as titanium (Ti), chromium (Cr), aluminum (Al), and copper (Cu), along with ceramics and nitrides such as TiN,

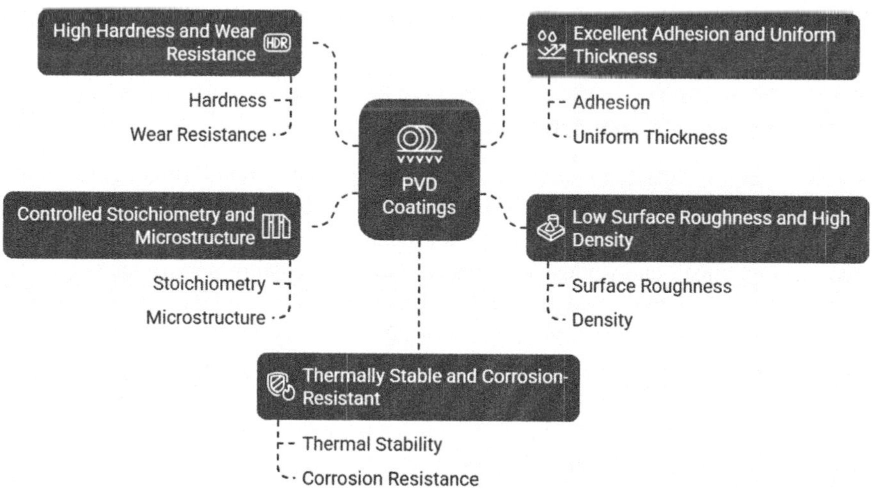

FIGURE 1.8 Key properties of physical vapor deposition (PVD) coatings, including high hardness, excellent wear and corrosion resistance, good adhesion, low friction, and thermal stability.

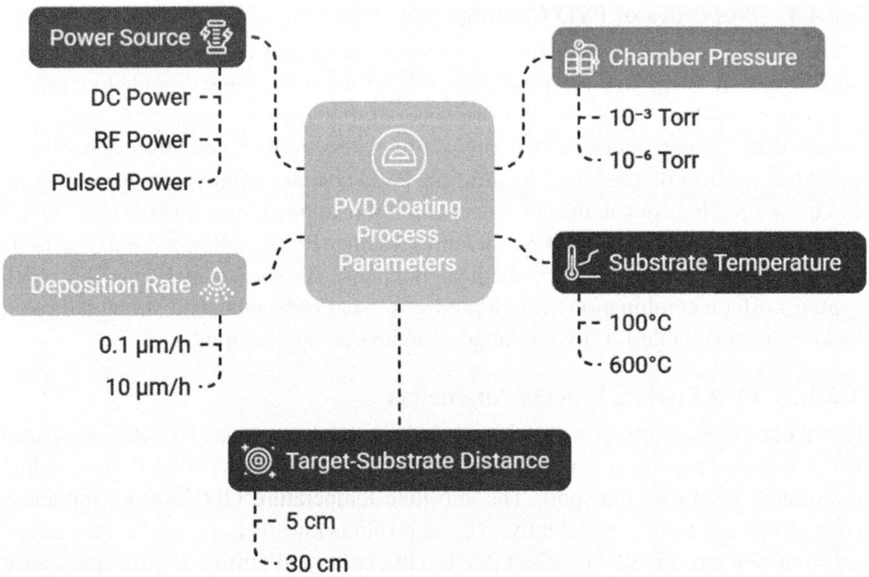

FIGURE 1.9 Key process parameters for physical vapor deposition (PVD), including power source, chamber pressure, substrate temperature, deposition rate, and target–substrate distance, which influence coating uniformity, adhesion, and microstructure.

TiAlN, AlCrN, ZrN, SiC, and DLC. Multilayers and nanocomposites are also used for advanced applications that require high hardness, corrosion resistance, and specific optical and electrical properties. Substrate materials range from stainless steel, aluminum, and tool steels to polymers, ceramics, and semiconductors such as silicon wafers. Appropriate pretreatment, such as biasing or plasma cleaning, is essential for strong adhesion and stress management.

1.2.5 Chemical Vapor Deposition (CVD)

CVD is a highly versatile and widely used thin-film deposition technique for producing dense, uniform, and adherent coatings with excellent conformality, even for complex geometries [14, 15]. It plays a crucial role in the microelectronics, aerospace, biomedical, and optical industries, where the coating performance depends heavily on the chemical composition, purity, and microstructure. This summary addresses the working principles, essential properties, key process parameters, and commonly used coating and substrate materials, highlighting their roles in achieving high-quality coatings. A schematic representation of the CVD process, showing the precursor gas flow, reaction chamber, and deposition onto the heated substrate, is illustrated in Figure 1.10.

CVD involves the chemical reaction of volatile precursor gases at elevated temperatures (typically 300–1000°C) in a reaction chamber [16, 17]. These gases decompose or react near or on the heated substrate surface, forming a solid coating layer, while the gaseous by-products are removed. This process allows for excellent

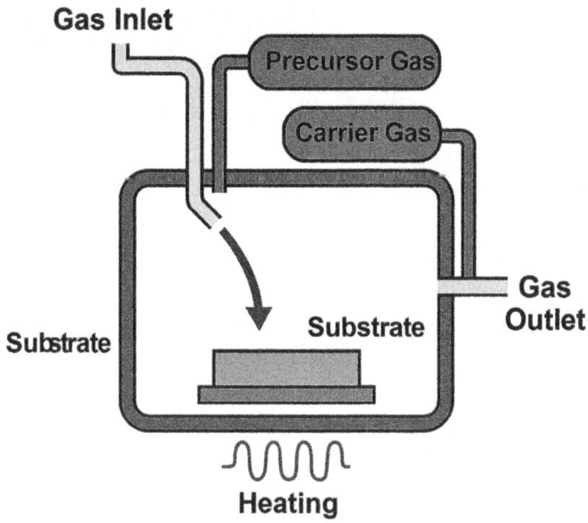

FIGURE 1.10 Schematic diagram of the chemical vapor deposition (CVD) process, illustrating precursor delivery, reaction chamber environment, and thin-film deposition on the heated substrate.

control of the composition, phase, and thickness. Variants for specific applications include low-pressure CVD (LPCVD), plasma-enhanced CVD (PECVD), and metal organic CVD (MOCVD).

1.2.5.1 Properties of CVD Coatings

The PVD coatings exhibit high density and purity, ensuring defect-free surfaces. They offer strong chemical bonding and adhesion and improve durability under mechanical and thermal stresses. With excellent step coverage and conformality, they uniformly coated complex geometries. These coatings also provide high thermal and chemical stabilities, making them ideal for harsh environments. Additionally, their tailorable microstructure and stoichiometry allow precise tuning of their properties for applications in electronics, optics, and advanced manufacturing. These properties make them ideal for protective, dielectric, semiconducting, and optical coatings. The key attributes of PVD coatings, such as high purity, chemical stability, and microstructural tunability, are shown in Figure 1.11.

1.2.5.2 Process Parameters of CVD

CVD process parameters include precursor type (halides, metal-organics, and hydrides), which determines the film composition and reactivity. Carrier gases such as hydrogen, argon, or nitrogen transport the precursors and influence the deposition uniformity. Gas flow rates and pressure influence film thickness and quality, ranging from low pressure (0.1–10 Torr) to atmospheric pressure. The substrate temperature (300–1000°C) and reaction time (minutes to hours) controlled the crystallinity, uniformity, and growth rate of the deposited layer. These parameters affected the deposition rate, grain structure, stress, and film composition. The influence of various CVD process parameters such as precursor type, carrier gas, substrate temperature, and gas pressure is illustrated in Figure 1.12.

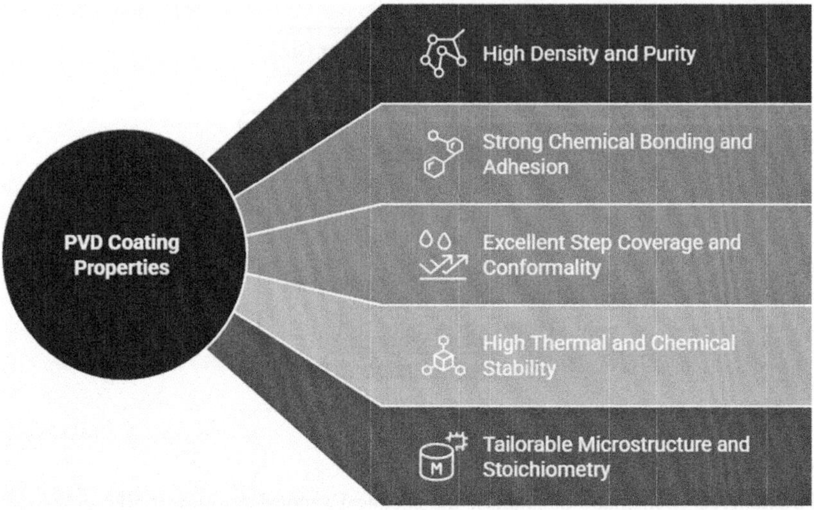

FIGURE 1.11 Core properties of physical vapor deposition (PVD) coatings, including high density and purity, strong adhesion, excellent step coverage, thermal and chemical stability, and customizable microstructure and stoichiometry.

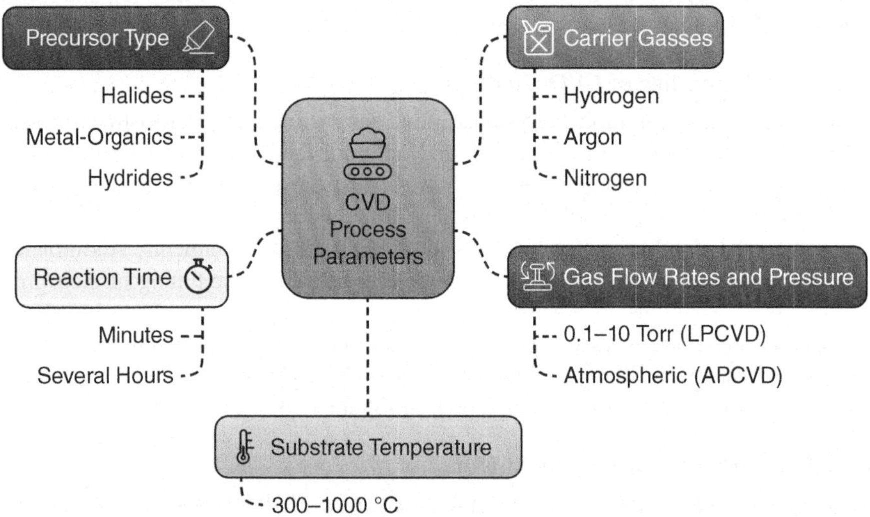

FIGURE 1.12 Schematic representation of chemical vapor deposition (CVD) process parameters and their influence, including precursor type, carrier gases, reaction time, gas flow rate and pressure, and substrate temperature.

Commonly used coating materials for advanced thin film applications include carbides and nitrides such as TiC, SiC, TiN, and BN; oxides such as SiO_2, Al_2O_3, and ZrO_2; and metals and semiconductors such as tungsten (W), polysilicon, GaN, and ZnO. These were selected for their electrical, thermal, and mechanical performances. Substrate materials include semiconductors such as silicon wafers and GaAs; metals such as stainless steel, titanium, and nickel alloys; and ceramics or glasses for optical or wear-resistant applications. Surface preparation is critical to achieve uniform adherent coatings.

1.2.6 Laser Surface Coating

Laser-based coatings utilize a high-energy laser beam to melt the feedstock powder or wire along with a portion of the substrate surface to form metallurgical bonds [18, 19]. Laser cladding is ideal for localized deposition and controlled dilution. Laser surface coating is an advanced surface-engineering technique that uses a high-energy laser beam to deposit, melt, or alloy coating materials onto a substrate surface with precision and localized heat input. This is highly significant for applications that demand tailored microstructures, minimal dilution, and superior metallurgical bonding, such as aerospace, automotive, power generation, and tooling. Understanding the working principle, key properties, critical process parameters, and common materials is essential for optimizing coating performance and extending component life. The working mechanism of the laser surface coating process, involving a focused laser beam and powder feed, is illustrated in Figure 1.13.

In laser surface coating, a focused laser beam irradiates the substrate while simultaneously melting an injected coating material, typically in the form of a powder or wire. The intense localized heat from the laser forms a molten pool on the substrate surface, where the feedstock material fuses metallurgically upon solidification. This process achieves strong adhesion with minimal dilution of the base material owing to the precise energy input and rapid cooling rates [20, 21].

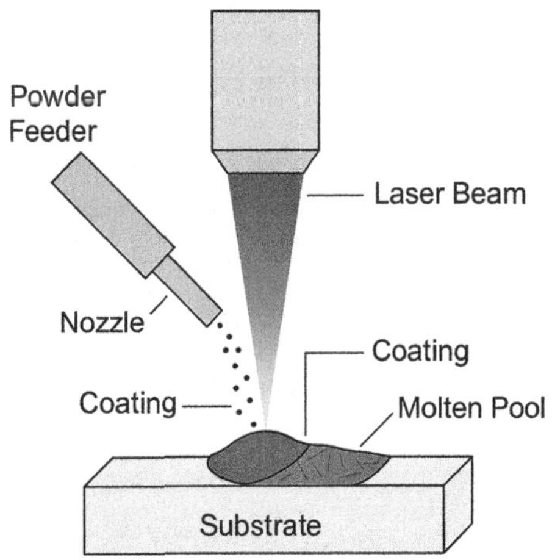

FIGURE 1.13 Schematic diagram of the laser surface coating process showing the interaction between the laser beam, powder feeder, nozzle, and substrate, leading to the formation of a molten pool and solidified coating layer.

1.2.6.1 Properties of Laser-Coated Surfaces

Laser-coated surfaces exhibit high bond strengths owing to metallurgical fusion, providing excellent adhesion. These coatings demonstrated improved hardness and wear resistance, enhancing surface durability. The refined grains resulted in a dense microstructure, whereas minimal porosity and cracking reduced the number of defects. Additionally, these coatings offer enhanced corrosion and oxidation resistance, ensuring environmental protection and extended component life. These properties make laser coatings ideal for industrial and high-performance applications. Figure 1.14 illustrates the properties of laser-coated surfaces, highlighting the bond strength, wear resistance, microstructural density, and corrosion resistance.

These properties render laser coatings ideal for high-stress and high-temperature environments.

Commonly used coating materials in thermal spraying and high-performance cladding include nickel-based alloys such as NiCr and Inconel, cobalt-based alloys such as Stellite, and various stainless steels. Carbides and ceramics, such as WC, TiC, and Al_2O_3, along with self-fluxing alloys, such as NiCrBSi, are also widely used to enhance wear, corrosion, and thermal barrier properties. When the process parameters are tightly controlled, substrate materials include carbon steel, stainless steel, tool steel, and high-performance superalloys such as Inconel and Hastelloy, as well as titanium and aluminum alloys. Surface preparation, particularly grit blasting, is crucial to promote wettability and strong metallurgical bonding.

1.2.6.2 Process Parameters

The laser power (typically 500–5000 W) determines the melt pool size and depth during laser cladding. The scanning speed (5–50 mm/s) controls the coating thickness and dilution, whereas the powder feed rate (5–30 g/min) or wire feed rate affects the deposition rate. The beam spot size and focal length influenced the energy density and coating width. The shielding gas type (Ar and He) and flow rate prevent oxidation. Optimizing these parameters is essential for balancing the microstructure, HAZ, and overall coating quality. The laser surface coating process is influenced by

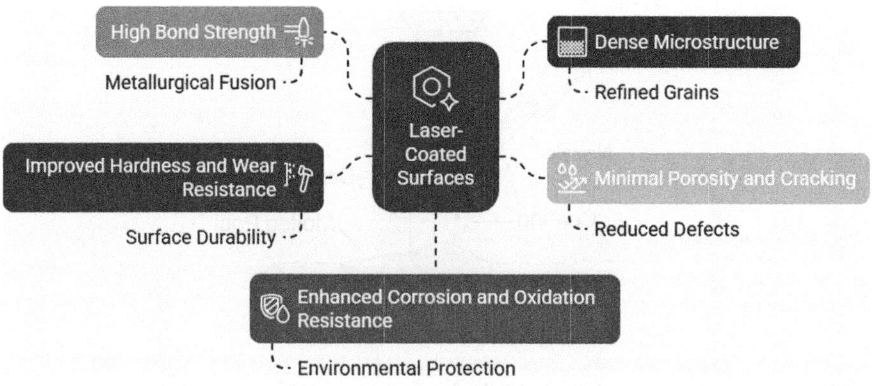

FIGURE 1.14 Properties of laser-coated surfaces, highlighting bond strength, wear resistance, microstructural density, and corrosion resistance.

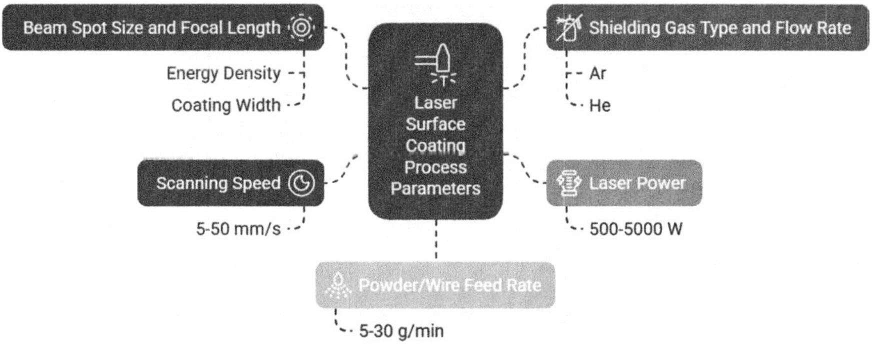

FIGURE 1.15 Laser surface coating process parameters, illustrating the influence of beam spot size, scanning speed, feed rate, laser power, and shielding gas on coating characteristics.

various parameters, such as laser power, scanning speed, and shielding gas type, which directly affect the uniformity and quality of the coating (Figure 1.15).

A critical review of laser surface coatings highlights that laser power, scanning speed, powder feed rate, beam spot size, and shielding gas flow are decisive factors governing the coating integrity, microstructure, and heat-affected zone (HAZ). Laser power, typically in the range of 500–5000 W, directly influences the melt pool depth and width, which can cause deeper substrate melting, increased dilution, and residual stresses, whereas insufficient power leads to poor metallurgical bonding and lack of fusion defects [22, 23]. The scanning speed is equally vital: higher speeds reduce the heat input per unit length, resulting in shallower melt pools and finer microstructures, but may cause incomplete bonding if too fast. Conversely, slower speeds increase the coating thickness but enlarge the HAZ, which can cause distortion [24]. The powder feed rate (or wire feed) determines the deposition efficiency and layer uniformity; mismatched feed rates cause porosity, nonuniform layers, or excessive spatter [25]. The beam spot size and focal length affect the energy density and interaction area; a smaller spot yields higher precision and deeper penetration but may increase thermal gradients, whereas a larger spot improves the surface finish but reduces depth control. Finally, employing an appropriate shielding gas (typically argon or helium) at an optimized flow rate minimizes the oxidation and contamination in the molten pool. Improper shielding can result in oxide inclusions and a poor surface quality [26, 27]. The literature consistently emphasizes that synergistic optimization of these parameters is crucial for producing coatings with a dense microstructure, minimal dilution, controlled residual stresses, and superior wear and corrosion resistance.

1.2.7 Electrodeposition (Electroplating)

This electrochemical process involves depositing a metal coating from a solution onto a conductive substrate [28, 29]. Common coatings include Ni, Cr, Zn, and Au. Electrodeposition, commonly known as electroplating, is a well-established surface-finishing process used to deposit a thin, uniform layer of metal onto a conductive substrate by electrochemical reduction [30, 31]. It plays a crucial role in industries

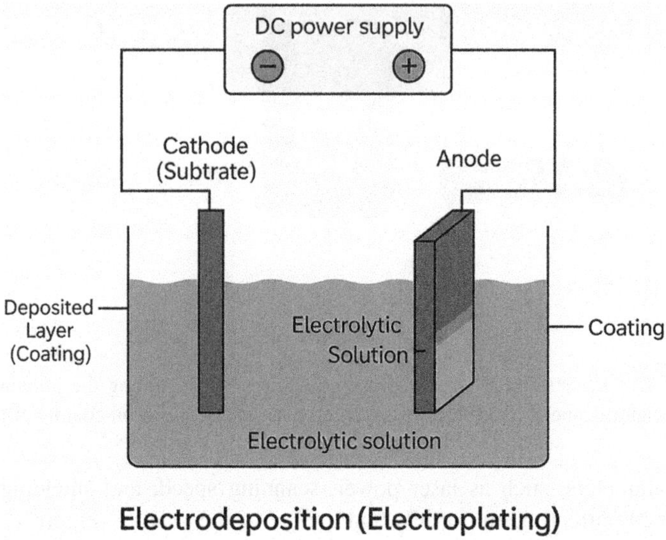

FIGURE 1.16 Schematic representation of the electrodeposition (electroplating) process showing the cathode, anode, electrolyte bath, and DC power supply used to deposit metal ions onto the substrate surface.

such as automotive, electronics, aerospace, jewelry, and corrosion protection, where enhanced surface properties, such as wear resistance, corrosion resistance, decorative appeal, and electrical conductivity, are vital. Understanding the working principle, coating properties, critical process parameters, and commonly used materials is essential for achieving optimal coating performance and reliability. Electrodeposition, also known as electroplating, involves the reduction of metal ions on a conductive substrate under an applied electric field, as illustrated in Figure 1.16.

Electrodeposition is based on the principle of passing an electric current through an electrolytic solution containing metal ions. The workpiece to be coated acts as the cathode, whereas a suitable metal or inert material serves as the anode. When current flows, metal cations from the electrolyte migrate to the cathode and are reduced to form a solid metallic layer, whereas the anode material may dissolve to replenish ions or remain inert depending on the system.

1.2.7.1 Properties of Electroplated Coatings

Electroplated coatings offer a uniform and smooth surface finish with precise thickness control, typically ranging from a few micrometers to several tens of micrometers, thereby ensuring consistent performance and appearance [32, 33]. These coatings significantly enhance corrosion resistance and wear resistance, thereby extending the lifespan of components in harsh environments. Additionally, electroplating improves electrical conductivity and solderability, making it indispensable for electronic and electrical applications. Beyond their functional benefits, electroplated layers provide an attractive decorative appeal, available in bright,

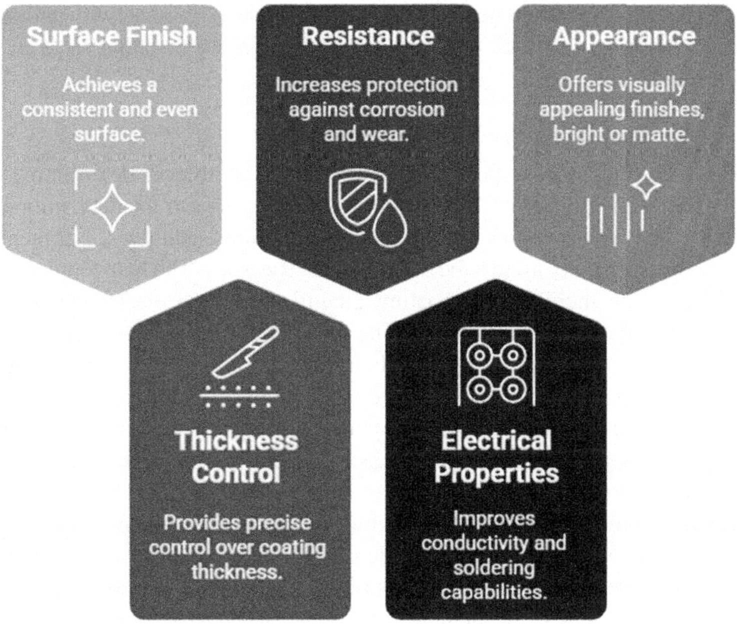

FIGURE 1.17 Key properties of electroplated coatings, including smooth finish, precise thickness, enhanced corrosion and wear resistance, and improved electrical conductivity.

glossy, or matte finishes, which adds aesthetic value to consumer goods, automotive parts, and jewelry. These properties make the electroplated layers suitable for functional and aesthetic applications. Electroplated coatings are valued for their surface uniformity, controlled thickness, and enhanced functional properties, such as wear resistance and electrical conductivity (Figure 1.17).

1.2.7.2 Commonly Used Coating Materials

Commonly used electroplating materials include single metals, such as nickel, chromium, copper, zinc, gold, and silver, which are selected for their excellent corrosion resistance, conductivity, and decorative qualities [34, 35]. Additionally, various alloys such as nickel–phosphorus (Ni–P), zinc–nickel (Zn–Ni), and tin–lead (Sn–Pb) have been employed to impart specialized properties, such as enhanced wear resistance, improved solderability, and superior corrosion protection. The choice of coating material depends on the specific functional or aesthetic requirements of the component, which ensures optimal performance in diverse industrial and consumer applications.

1.2.7.3 Substrate Materials

Typical substrates for electroplating include metals such as mild steel, stainless steel, copper alloys, and aluminum (which require pretreatment to ensure adhesion). Additionally, conductive polymers can be electroplated after metallization, thereby expanding their application possibilities. Electronic components and connectors commonly receive electroplated coatings to enhance their conductivity and corrosion resistance. Thorough surface cleaning and activation are essential steps to achieve strong bonding and defect-free, high-quality plated layers.

1.2.7.4 Electroplating Process Parameters

The electroplating process parameters play a critical role in determining coating quality. The current density (0.1–10 A/dm^2) influences both the deposition rate and grain size. The bath composition, including metal ions, additives, brighteners, and leveling agents, directly affects the deposit characteristics. The bath temperature (25–75°C) affects ion mobility and coating quality, while the electrolyte pH governs metal ion stability and uniformity. The deposition time controlled the coating thickness, and agitation or stirring enhanced the ion transport. Optimizing these variables ensures uniform deposition, refined microstructure, and strong adhesion. The quality and performance of electroplated coatings are significantly affected by process parameters, such as current density, electrolyte composition, temperature, and deposition time, as depicted in Figure 1.18.

A review of the current literature highlights that precise control of electroplating process parameters, including current density, bath composition, bath temperature, electrolyte pH, deposition time, and agitation, is crucial for achieving coatings with optimal microstructures, mechanical properties, and surface finishes. The current

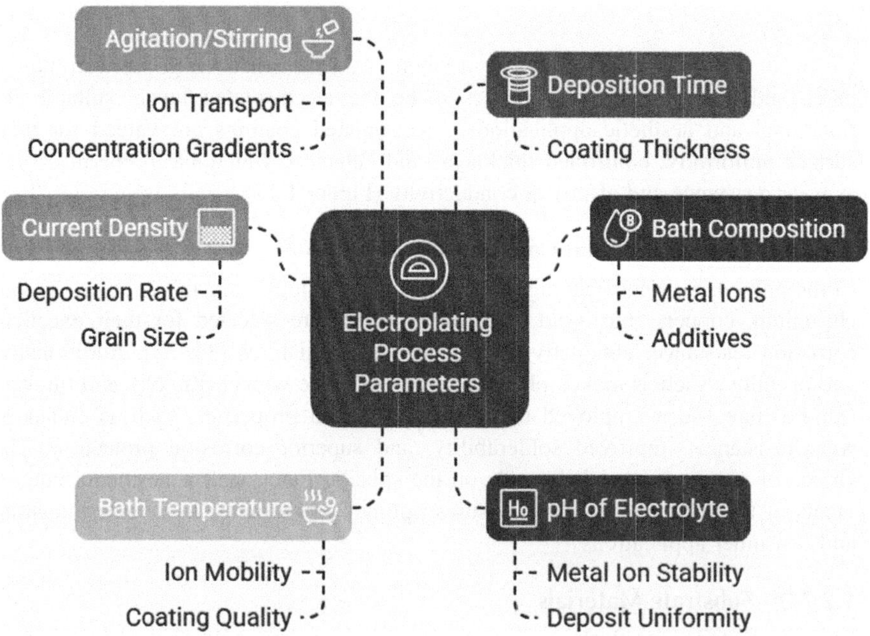

FIGURE 1.18 Key electroplating process parameters, including current density, bath composition, temperature, pH, deposition time, and agitation, which influence coating quality and uniformity.

density (0.1–10 A/dm^2) directly determines the rate of metal ion reduction and significantly affects grain size, internal stress, and texture. High current densities can cause coarse, nodular, or burnt deposits, whereas low densities may produce overly fine but slow-to-build films [36]. The bath composition, including the type and concentration of metal ions, organic additives, brighteners, and leveling agents, governs the deposit brightness, smoothness, and stress relief [37]. Bath temperature (25–75°C) influences ion mobility and deposition kinetics; higher temperatures generally increase deposition rates and promote smoother deposits, but may accelerate unwanted side reactions or degrade additives [38]. The pH of the electrolyte critically affects the metal ion speciation and cathodic reaction pathways; improper pH can lead to hydrogen evolution, porosity, or poor adhesion [39]. The deposition time linearly correlates with the coating thickness, but prolonged times can cause stress accumulation or roughness if the other parameters are not balanced. Finally, agitation or stirring ensures uniform ion distribution near the cathode, reduces concentration gradients, and minimizes dendritic growth, which is vital for achieving a consistent thickness and microstructure [40]. Collectively, the literature underscores that multiparameter optimization is essential for producing high-quality electroplated coatings with the desired functional and decorative attributes.

1.2.8 Atomic Layer Deposition (ALD)

A precise technique allows for the layer-by-layer growth of films through self-limiting chemical reactions, providing atomic-level control over thickness. Atomic layer deposition (ALD) is a cutting-edge thin-film deposition technique renowned for its atomic-scale thickness control, exceptional conformality, and uniformity, even in complex 3D nanostructures [15, 41]. This makes ALD indispensable in semiconductor device fabrication, nanotechnology, photovoltaics, and protective barrier coatings, where ultrathin, pinhole-free layers are critical for performance and reliability. Understanding the working principle, unique properties, precise process parameters, and appropriate materials is essential for harnessing ALD's full potential of ALD. ALD enables precise control over film thickness and uniformity through self-limiting surface reactions, as illustrated in Figure 1.19.

ALD operates via a sequential self-limiting surface-reaction mechanism. The process involves alternately exposing the substrate to vapor-phase chemical precursors in separate pulses. Each precursor chemisorbs onto the available reactive sites on the substrate surface, forming a monolayer. Excess precursor and reaction by-products were purged with an inert gas (e.g., nitrogen or argon) before the next precursor pulse. This cyclic sequence ensures controlled layer-by-layer growth, achieving precise thickness control at the atomic scale (typically 0.1–0.3 nm per cycle) [42].

Coating materials commonly used in ALD include oxides such as Al_2O_3, TiO_2, HfO_2, SiO_2, and ZnO, which offer excellent dielectric, barrier, and optical properties. Nitrides such as TiN and AlN provide conductive and diffusion barrier functionalities, whereas metals such as platinum (Pt) and ruthenium (Ru) are deposited using plasma-assisted ALD for catalytic and electronic applications. Substrate materials range from semiconductors, such as silicon wafers, GaN, and SiC, to metals, such as

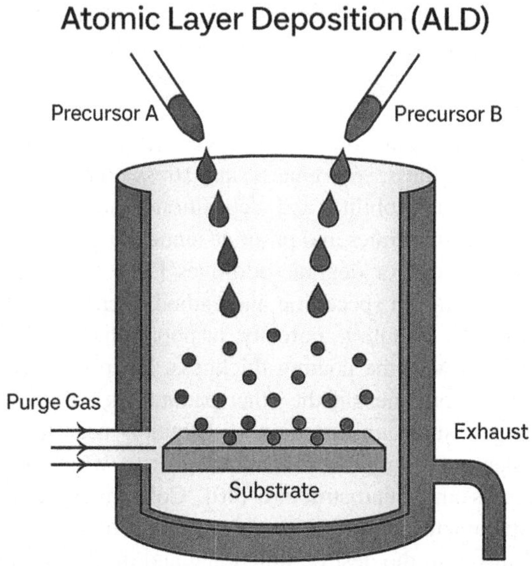

FIGURE 1.19 Schematic diagram of the atomic layer deposition (ALD) process.

copper and aluminum interconnects. Polymers are also used as substrates in flexible electronics, utilizing low-temperature ALD to avoid thermal damage.

1.2.8.1 Properties of ALD Coatings

ALD offers exceptional conformality to high aspect ratios and porous structures, making it ideal for coating complex 3D geometries. This provides atomic-level thickness precision, enabling highly controlled film growth. ALD coatings exhibit high uniformity with low pinhole density, ensuring defect-free layers. These films possess excellent barrier and dielectric properties, which are essential for microelectronic and protective applications. Depending on the deposition temperature and material type, the resulting structure can be amorphous or crystalline and tailored to specific functional needs. ALD coatings are known for their superior conformality, precise thickness control, and excellent barrier properties, as shown in Figure 1.20.

1.2.8.2 Process Parameters of ALD Process

ALD relies on the precise control of the process parameters to ensure high-quality films. The substrate temperature (50–350°C) must be within the ALD window for ideal surface reactions. Precursor pulse and purge times (0.1–10 s) manage the surface saturation and by-product removal. The chamber pressure (mTorr–Torr) affects the gas flow dynamics, whereas the number of cycles determines the film thickness. Proper optimization ensures uniformity, conformality, and defect-free coating on complex geometries. The performance and precision of ALD coatings are governed by parameters such as precursor pulse duration, purge timing, and substrate temperature, as shown in Figure 1.21.

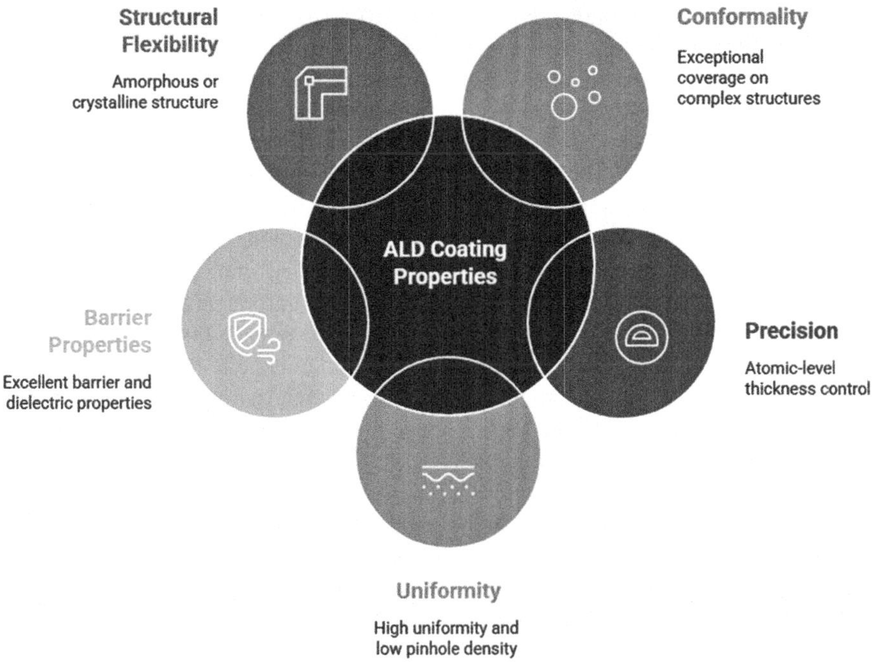

FIGURE 1.20 Key properties of atomic layer deposition (ALD) coatings.

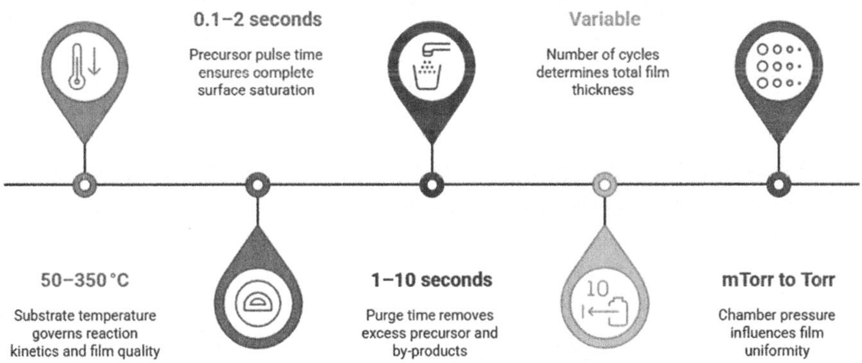

FIGURE 1.21 Key process parameters of atomic layer deposition (ALD).

1.2.8.3 Critical Literature Review on Effects of Key Process Parameters in Atomic Layer Deposition (ALD)

A focused review of recent literature shows that careful control of the substrate temperature, precursor pulse time, purge time, cycle number, and chamber pressure is fundamental to realizing the exceptional conformality and atomic-level thickness precision that define ALD.

The substrate temperature (typically 50–350°C) is crucial for maintaining the ALD temperature window, where surface reactions remain self-limiting and precursor chemisorption is maximized without causing decomposition [43, 44]. Operating outside this window can cause incomplete reactions, precursor condensation, or thermal decomposition, leading to poor film uniformity and defects [45]. The precursor pulse time must be long enough (0.1–2 s) to ensure full surface saturation; underdosing results in non-uniform coverage and incomplete monolayers, while excessively long pulses waste precursors without improving film quality [46]. The purge time between precursor pulses (1–10 s) is equally vital: insufficient purging can cause gas-phase reactions (CVD-like behavior), introducing unwanted nucleation and pinholes, while overly long purges reduce the process throughput [47]. The number of ALD cycles directly dictates the total film thickness in a linear, cycle-by-cycle manner; however, the literature emphasizes that nucleation delays can occur on some substrates, especially for high-k dielectrics or noble metals, requiring cycle optimization for uniform growth initiation. Lastly, chamber pressure (mTorr to Torr) affects precursor diffusion and residence time: lower pressures enhance the mean free path and film conformality in high-aspect-ratio features but may reduce precursor flux, whereas higher pressures increase flux but risk gas-phase reactions if the purge steps are insufficient [48]. Recent studies underline that precise tuning of these parameters is central to ALD's success in producing defect-free, ultrathin films for semiconductor gates, diffusion barriers, nanolaminates, and flexible electronics, driving ongoing process innovation for next-generation applications.

1.2.9 Suspension Plasma Spraying (SPS) and Solution Precursor Plasma Spraying (SPPS)

Emerging variants of thermal spraying techniques using submicron/nanosized particles suspended in a liquid. Fine, porous, or columnar coatings with high functionality were produced. SPS and solution precursor plasma spraying (SPPS) are advanced variants of the traditional plasma spray process developed to deposit fine-structured and nanostructured coatings with enhanced functionality [49–51]. They address the limitations of conventional thermal spraying, which struggles to handle submicron and nanosized feedstocks owing to poor flowability. These techniques are highly significant for producing thermal barrier coatings (TBCs), catalytic layers, and bioactive surfaces, where microstructure control at the nanoscale is crucial for the performance. Advanced thermal spray techniques such as SPS and SPPS allow the deposition of fine-structured and functional coatings using liquid feedstocks, as illustrated in Figure 1.22.

In SPS, a suspension of fine powders (typically nanoparticles or submicron particles dispersed in a liquid such as water or alcohol) is injected directly into the plasma plume. The liquid evaporated rapidly, and the solid particles melted and propelled toward the substrate to form a coating with a refined microstructure and unique columnar or porous architectures. In the SPPS, a solution containing dissolved precursor salts was injected into the plasma jet. The solvent evaporates, and the precursor decomposes in flight, forming oxide particles that melt and deposit on

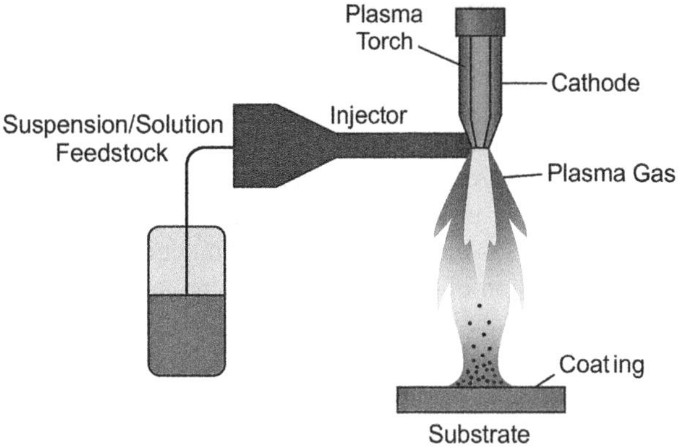

FIGURE 1.22 Schematic representation of suspension plasma spraying (SPS) and solution precursor plasma spraying (SPPS).

the substrate. This allows for the direct synthesis and deposition of desired oxides without the need for a pre-synthesized powder feedstock, offering greater chemical homogeneity and phase purity.

Commonly used coating materials in SPS include yttria-stabilized zirconia (YSZ), gadolinium zirconate, alumina, and titania, while SPPS employs directly synthesized oxides like YSZ, doped ceria, and $LaMnO_3$. These materials were selected for thermal insulation, ionic conductivity, or catalytic activity. Substrate materials include nickel-based superalloys (e.g., turbine components), steels, stainless steels (e.g., industrial equipment), and ceramics (for sensors/electronics). Proper pre-treatment is essential for promoting strong bonding and reducing thermal mismatch stresses.

1.2.9.1 Properties of SPS and SPPS Coatings

The coatings produced by SPS and SPPS exhibit fine and nanostructured grains, controlled porosity, and columnar architectures, enabling low thermal conductivity, which is ideal for thermal barrier applications. Their columnar microstructure enhances strain tolerance, while good adhesion and low residual stress ensure durability. These properties make them highly suitable for next-generation gas-turbine coatings and functional thin films used in advanced energy systems and high-temperature environments.

1.2.9.2 Process Parameters

In the SPS and SPPS processes, the suspension or solution concentration influences the solid content and the resulting particle size post-evaporation. The solvent type (e.g., water or ethanol) affects the droplet breakup and evaporation rate. The plasma power and gas flow rates regulate the plume temperature and velocity, thereby affecting melting and deposition. The injection position and feed rate control the droplet trajectory and coating uniformity, whereas the spray distance is optimized to ensure proper drying, melting, and impact. Fine-tuning of these parameters is essential to prevent incomplete evaporation, nozzle clogging, and poor coating morphology.

1.3 ADVANCED CLADDING PROCESSES

Cladding is a surface modification technique in which a different material is bonded to a substrate to improve surface properties, such as corrosion resistance, wear resistance, oxidation resistance, and thermal stability. Unlike coatings, cladding typically involves thicker layers and stronger metallurgical bonding, making it ideal for structural reinforcement in harsh service environments. With advancements in materials science and process engineering, several modern cladding techniques now enable the precise deposition of complex materials on a wide range of substrates.

1.3.1 Laser Cladding

This is a highly precise technique in which a laser beam is used to melt and fuse a cladding material (in powder or wire form) onto a substrate. This process offers minimal dilution, excellent metallurgical bonding, and localized deposition. Laser cladding is an advanced surface modification and additive manufacturing technique widely used to improve surface properties, repair damaged components, and fabricate wear- or corrosion-resistant layers on critical parts in industries such as aerospace, energy, and tooling [52–54]. Its significance lies in its ability to produce defect-free, metallurgically bonded cladding layers with precise control over dilution, heat input, and geometry, thus outperforming conventional welding and thermal spraying for high-value applications. Laser cladding is a surface-modification technique that achieves strong metallurgical bonds and minimal HAZs, as illustrated in Figure 1.23.

In laser cladding, a high-power focused laser beam is used to melt a thin layer of the substrate surface while simultaneously melting a cladding material, typically delivered as a powder or wire feedstock. The molten cladding material fuses with the substrate and rapidly solidifies, forming a dense metallurgically bonded coating with minimal dilution of the base metal. Localized heat input, precise laser control, and rapid cooling rates enable high-quality deposits with excellent microstructural characteristics.

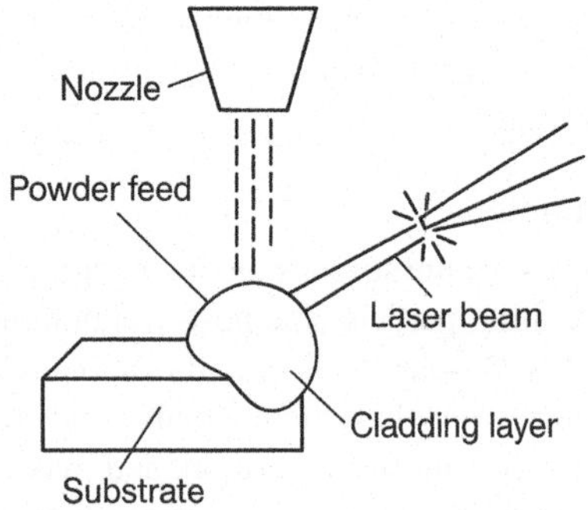

FIGURE 1.23 Schematic diagram of the laser cladding process, illustrating the laser beam interaction with cladding material.

1.3.1.1 Properties of Laser Cladded Layers

Laser cladding offers an excellent metallurgical bond with low dilution, typically below 5–10%, ensuring strong adhesion and compositional integrity. This produces a dense microstructure with minimal porosity, resulting in enhanced hardness, wear resistance, and corrosion resistance. This process allows for a tailored microstructure with refined grains, and its precise energy input minimizes distortion and the HAZ. These superior properties render laser cladding ideal for extending the service life and performance of high-value components. Laser-cladded layers are characterized by refined grain structures, excellent adhesion, and improved mechanical and corrosion properties, as depicted in Figure 1.24.

Commonly used cladding materials include nickel-based alloys, such as Inconel 625, and self-fluxing NiCrBSi, which are known for their excellent corrosion and wear resistance. Cobalt-based alloys such as the Stellite series are used in harsh wear environments, whereas iron-based alloys (Fe–Cr and Fe–Ni) offer cost-effective wear protection. Carbide-reinforced materials (e.g., WC and TiC with metallic binders) and tool steels exhibit high hardness and abrasion resistance. Substrate materials include carbon and low-alloy steels, stainless steels, nickel-based superalloys, and cast iron, which are commonly used in aerospace, energy, and manufacturing industries. Proper surface preparation, including preheating and grit blasting, is essential for achieving strong metallurgical bonding and reducing thermal mismatch stresses.

1.3.1.2 Key Process Parameters

The laser-cladding process parameters significantly influenced the coating quality and performance. The laser power (typically 500–5000 W) controls the melt pool size and penetration depth. The scanning speed (5–50 mm/s) affected the cladding thickness, heat input, and dilution level. The feedstock feed rate, whether powder (5–30 g/min) or wire, regulates the deposition rate and layer uniformity. The laser spot size and beam focal length determine the energy density and cladding bead

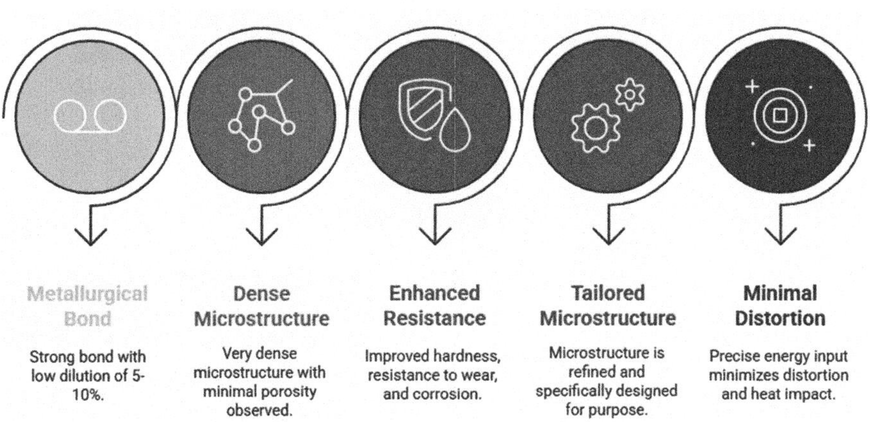

FIGURE 1.24 Key properties of laser-cladded layers.

geometry, respectively. The shielding gas (argon or helium) and its flow rate prevent oxidation and contamination. Additionally, the overlap ratio and number of passes influence the surface finish and multilayer build quality. Optimizing these parameters is essential for achieving consistent cladding quality, controlled dilution, and the desired microstructure for high-performance applications. The quality and performance of laser clads are significantly influenced by process parameters, such as laser power, scanning speed, and powder feed rate, as illustrated in Figure 1.25.

Laser cladding is a highly controllable surface engineering technique, and its process parameters (laser power, scanning speed, powder feed rate, laser spot size, and shielding gas flow) are fundamental to achieving coatings with optimal geometry, microstructure, and mechanical performance. Recent studies have reinforced that laser power (commonly 500–5000 W) is the primary driver of energy input, melt-pool depth, and clad dilution. Excessive power can cause deep substrate melting and cracking, whereas insufficient power leads to poor bonding and a lack of fusion defects [55, 56]. Scanning speed (5–50 mm/s) inversely affects heat input: higher speeds reduce the HAZ and produce finer microstructures, but excessively high speeds may result in incomplete wetting or bead discontinuities [57, 58]. Slower speeds, while enhancing melt pool stability, can increase the residual stress and distortion [59]. The powder feed rate (typically 5–30 g/min) influences the deposition rate and layer uniformity. A mismatch between the feed rate and laser power can lead to porosity, unmelted particles, or excessive spatter [60]. The laser spot size and focal length control the energy density and interaction area: a smaller spot size provides deeper penetration and refined grain structure, but increases thermal

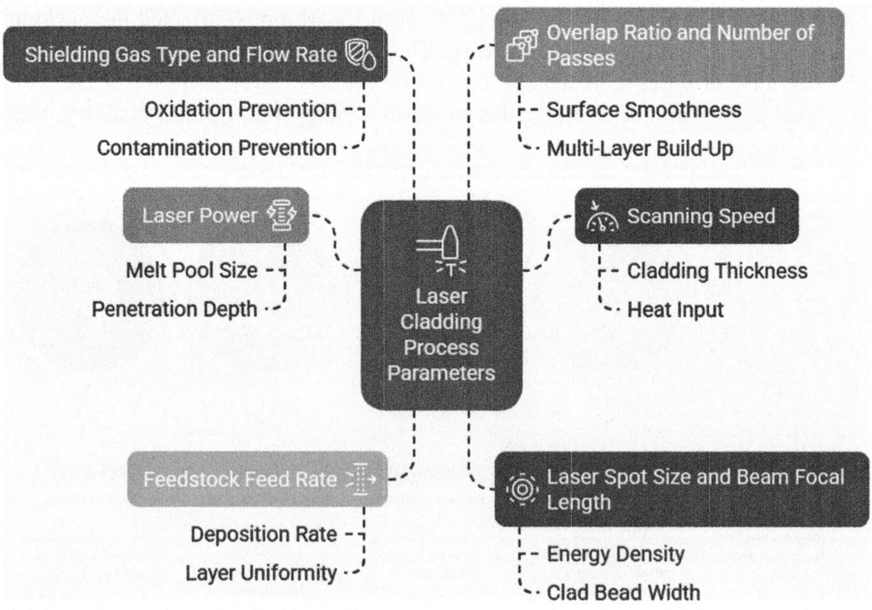

FIGURE 1.25 Key laser cladding process parameters.

gradients, which can lead to cracking. A larger spot promotes smoother surfaces, but may decrease the bonding strength owing to shallow penetration. The shielding gas type (argon or helium) and flow rate are vital for protecting the melt pool from oxidation. Inadequate shielding promotes the inclusion of oxides and leads to poor metallurgical bonding. Modern studies emphasize that the synergistic optimization of these parameters, often supported by real-time monitoring and machine learning, significantly improves clad quality, reduces defects, and tailors microstructural gradients for demanding applications in the aerospace, power generation, and tooling industries.

1.3.2 Cold Spray Cladding

In this technique, metal powders are accelerated at supersonic speeds and plastically deformed upon impact with the substrate to form a dense adherent layer without melting. Cold spray cladding is an innovative solid-state surface engineering process increasingly adopted in the aerospace, defense, marine, and energy sectors for the repair, restoration, and corrosion protection of high-value components. Its significance lies in its unique ability to deposit dense, oxide-free, and well-bonded metallic layers without exposing the materials to high temperatures, thus preserving the base material properties and avoiding thermal distortion or phase changes. In cold-spray cladding, a high-pressure gas (typically helium or nitrogen) accelerates metal or alloy powder particles to supersonic velocities (300–1200 m/s) through a de Laval nozzle [61–63]. Upon impact with the substrate, these solid particles underwent severe plastic deformation, leading to mechanical interlocking and metallurgical bonding at the contact surface without melting. This fundamental difference from thermal spraying or welding ensures the superior retention of feedstock properties and minimal residual stress in the substrate.

1.3.2.1 Properties of Cold Spray Cladding

Cold spray cladding produces dense, oxide-free coatings with high adhesion strength owing to the solid-state bonding. It maintains the original properties of feedstock materials because no melting occurs, thereby preserving the microstructure and avoiding thermal degradation. This process results in minimal residual stresses, excellent fatigue resistance, and high electrical and thermal conductivity. Its low-temperature nature makes it ideal for coating temperature-sensitive substrates and achieving thick, durable layers without distortion or phase transformation. Cold spray cladding offers dense oxide-free coatings with strong adhesion and minimal thermal distortion, as shown in Figure 1.26. These properties make cold spray particularly attractive for repairing heat-sensitive materials, such as high-strength aluminum, titanium, and nickel alloys.

1.3.2.2 Key Process Parameters

The key process parameters in cold-spray cladding include gas temperature (200–800°C), which influences particle acceleration and bonding; gas type, typically helium or nitrogen, which influences flow dynamics and cost efficiency; and gas pressure (2–4 MPa), which determines the particle velocity. Particle size (5–45 μm)

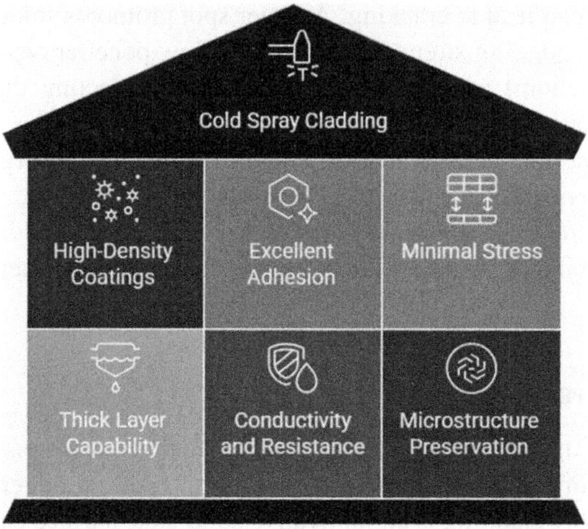

FIGURE 1.26 Properties of cold spray cladding.

is crucial for determining the deposition efficiency and coating density. Additional parameters, such as nozzle design and standoff distance, control the spray focus and efficiency, while the particle velocity is material-dependent. Finally, the traverse speed and number of passes influence the layer thickness and uniformity, ensuring consistent, high-quality coatings. Careful parameter optimization ensures strong bonding, high deposition efficiency, and the desired layer build-up. The performance of cold-spray cladding is governed by process parameters such as gas pressure, particle velocity, and nozzle design, as illustrated in Figure 1.27.

Commonly used cladding materials in cold spraying include aluminum alloys (e.g., Al and Al–Si) for lightweight, corrosion-resistant repairs; copper and its alloys for high electrical conductivity; and titanium alloys for aerospace structural components. Nickel-based superalloys are used for their oxidation and corrosion resistance, while molybdenum and tantalum are used in high-temperature or chemically aggressive environments. Suitable substrate materials include aluminum, magnesium, titanium, nickel-based alloys, high-strength steels, and heat-sensitive materials or composites (with careful control). Proper surface preparation, such as blasting or roughening, is essential for strong mechanical interlocking and adhesion.

1.3.3 Explosive Cladding

A high-energy process in which controlled detonation drives one metal plate against another creates a strong metallurgical bond through plastic deformation and jetting [64, 65]. Explosive cladding, also known as Explosive Welding (EXW), is a unique solid-state bonding process used to join two dissimilar metals or to apply a corrosion-resistant or wear-resistant cladding layer to a base material. Its key advantage lies in producing strong metallurgical bonds between metals that are otherwise difficult or impossible to weld by conventional fusion methods without melting, preserving the individual properties of both materials. It is widely used in the oil and gas, petrochemical, shipbuilding, and nuclear industries for producing bimetallic plates,

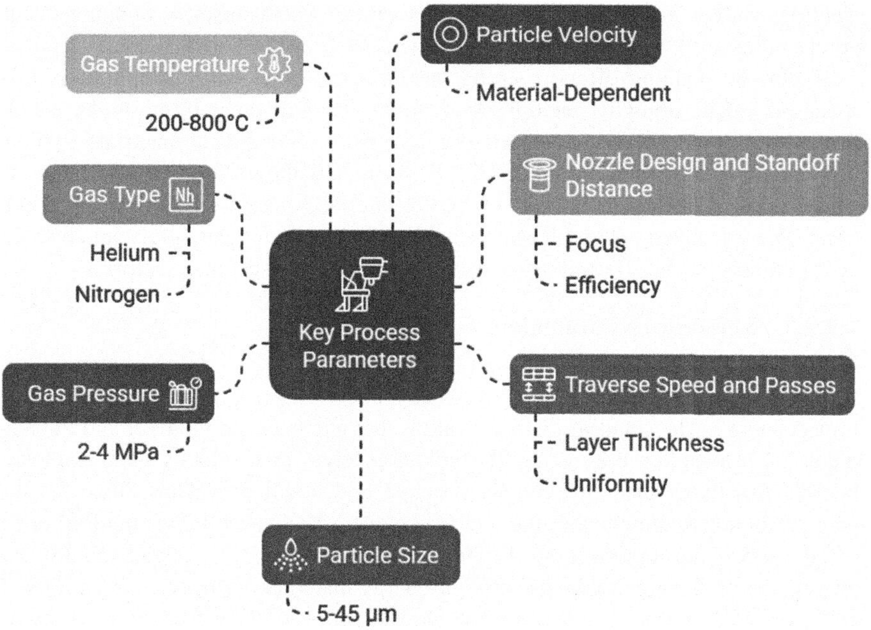

FIGURE 1.27 Key process parameters of cold spray cladding.

pressure vessels, heat exchanger tube sheets, and clad pipes. Explosive cladding utilizes a high-velocity impact between the metal plates driven by controlled detonation, as illustrated in Figure 1.28.

In explosive cladding, the controlled detonation of high-energy explosives propels a thin metal plate (flyer plate) at a high velocity toward the surface of a stationary base plate (parent plate). The flyer plate impacts the base plate at a precise collision angle and speed, generating extreme localized plastic deformation at the interface. This results in a jetting phenomenon that removes surface contaminants and oxide layers, thereby allowing direct metallic bonding through solid-state

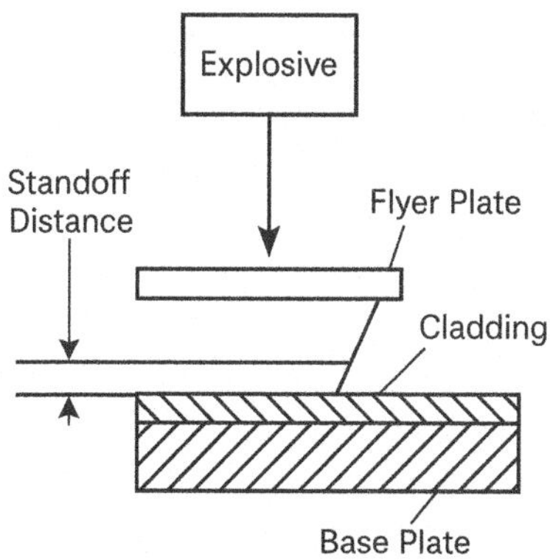

FIGURE 1.28 Schematic diagram of the explosive cladding process.

welding. Unlike fusion welding, no bulk melting occurs; instead, bonding occurs purely because of the kinetic energy and pressure from the shock wave.

Explosive cladding offers a strong metallurgical bond with exceptionally high shear and tensile strengths, enabling the joining of a wide range of dissimilar metals. It preserves the mechanical and corrosion properties of the parent materials owing to its solid-state nature and minimal HAZ. This process allows large-area bonding in a single step, producing distinct wavy or flat interfaces based on the process conditions. These features make it highly suitable for extreme temperatures and corrosive environments, such as in the petrochemical, nuclear, and defense sectors.

1.3.3.1 Key Process Parameters

Key process parameters include the type of explosive (e.g., ANFO, dynamite, and slurry) and its detonation velocity (typically 2000–4000 m/s), which influence the impact energy. The standoff distance, that is, the gap between the flyer and the base plates, is critical for optimizing the collision angle and velocity. The explosive loading density controls the impact pressure and bonding efficacy. The collision parameters determine whether the interface is wavy (enhanced bond strength) or flat (used for delicate applications). Proper plate alignment and boundary control can prevent defects. Owing to the use of high-energy materials, stringent safety protocols and precise control are essential for operational success. The effectiveness of explosive cladding is determined by factors such as detonation velocity, standoff distance, and collision angle, as illustrated in Figure 1.29.

1.3.3.2 Commonly Used Cladding Materials

Commonly used cladding materials in explosive cladding include corrosion-resistant metals, such as stainless steel (SS316, SS304), titanium, and nickel-based alloys, such as Inconel and Hastelloy. These are typically bonded to base materials, such as carbon

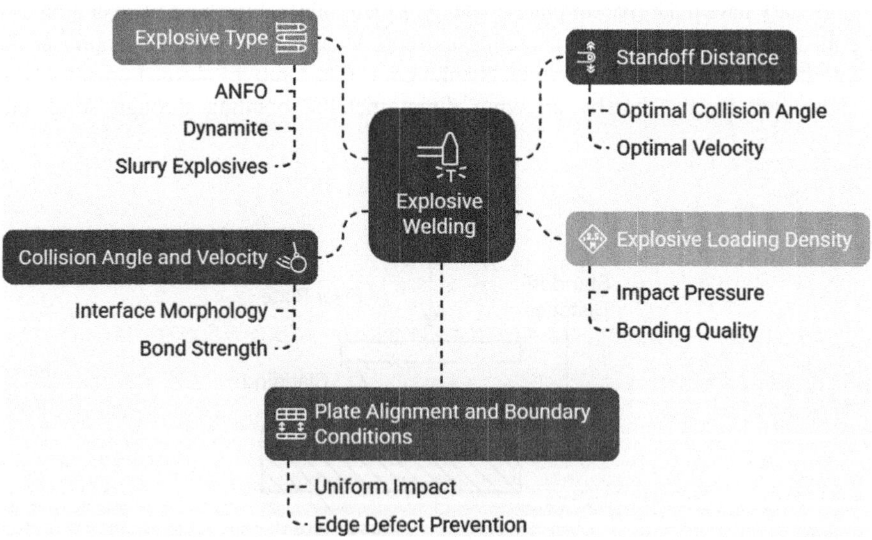

FIGURE 1.29 Key process parameters of explosive cladding.

steels, low-alloy steels, and structural steels. Common clad combinations include stainless steel on carbon steel for chemical and petrochemical vessels, titanium on steel for seawater or aggressive chemical environments, and nickel alloys on steel for high-temperature and high-corrosion conditions. Substrate materials often include large carbon steel plates, pressure vessel shells and heads, pipe sections, and tube sheets used in heat exchangers, ensuring structural strength with enhanced surface properties.

1.3.4 Arc-Based Advanced Cladding

Arc-based advanced cladding refers to a group of surfacing techniques that use an electric arc as a heat source to melt and deposit a cladding layer onto a substrate. These processes are widely used in the repair, refurbishment, and surface enhancement of critical industrial components, particularly where cost-effective, thick, and metallurgically bonded layers are required. Modern advancements in arc cladding include automation, hybridization, and process control, enabling higher deposition rates, better microstructure control, and reduced dilution compared with traditional welding overlays. Arc-based advanced cladding employs an electric arc to fuse the cladding material onto the substrates, enabling surface enhancement and repair, as illustrated in Figure 1.30.

1.3.4.1 Arc Welding-Based Cladding Techniques

In arc-based cladding, an electric arc is struck between a consumable electrode (wire or powder) and the workpiece. The arc heat melted the electrode material and formed a thin surface layer on the substrate. The molten droplets solidify to form a fusion-bonded cladding layer with desired thickness and surface properties. The process can be performed manually, semi-automatically, or fully using robotic arms and CNC systems to ensure consistent quality and efficiency. Arc-welding-based cladding methods such as SMAW, GMAW, GTAW, and PTA are widely employed for surface enhancement and wear protection, as shown in Figure 1.31.

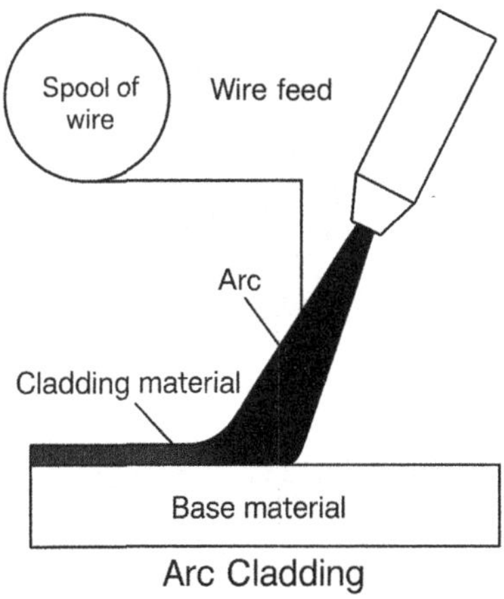

FIGURE 1.30 Schematic representation of arc-based advanced cladding.

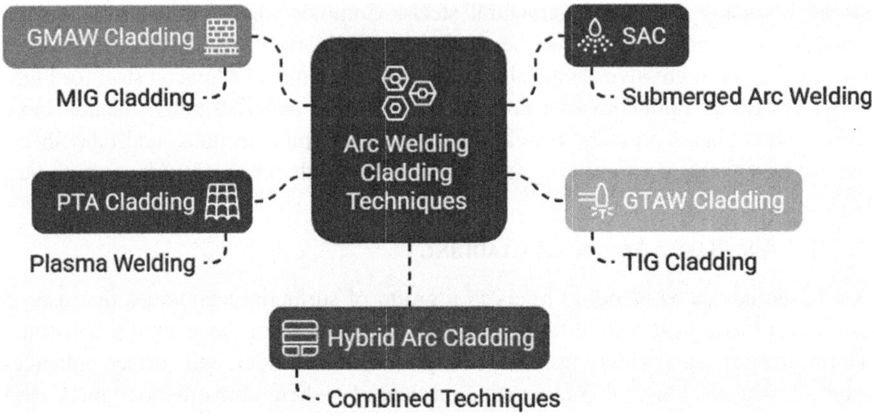

FIGURE 1.31 Arc welding-based cladding techniques.

1.3.4.2 Common Cladding Materials

Common cladding materials for arc-based processes include hardfacing alloys, such as cobalt-based (Stellite), nickel-based, and iron-based systems reinforced with carbides (e.g., Fe–Cr–C), which provide excellent wear and abrasion resistance. Corrosion-resistant alloys, such as Inconel, stainless steels, and Ni–Cr alloys, are also widely used to protect components in harsh environments. For extreme service conditions, special hard materials such as tungsten carbide and chromium carbide blends have been applied. Typical substrates include low-alloy and carbon steels, stainless steels, cast iron, and nickel- or cobalt-based superalloys, selected for their mechanical strength and compatibility with cladding materials. Arc-based cladding processes are valued for their strong bonding, flexible material compatibility, and efficient coverage of large surfaces, as shown in Figure 1.32.

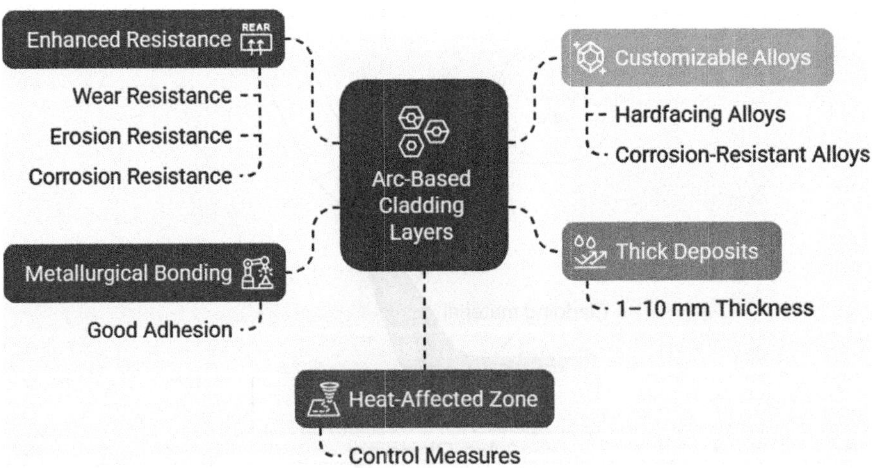

FIGURE 1.32 Properties of arc-based cladding processes.

1.3.5 Additive Friction Stir Deposition (AFSD)

An emerging solid-state additive cladding process, in which a material is deposited layer-by-layer through a rotating feedstock rod, produces dense, forged-like structures. Additive Friction Stir Deposition (AFSD) is a solid-state additive manufacturing process derived from friction stir welding (FSW) principles [66–68]. Unlike traditional fusion-based additive manufacturing, AFSD uses frictional heat and severe plastic deformation to deposit materials layer-by-layer without melting.

AFSD is a solid-state additive manufacturing and cladding process derived from FSW. A rotating non-consumable tool contacts and plasticizes the feedstock rod by generating frictional heat. This softened material flows plastically under the action of a rotating tool and is deposited onto a substrate or previous layers without melting. The tool traverses along a programmed path to deposit material layer by layer, enabling near-net-shape fabrication or surface cladding. The process temperature remains below the melting point of the material, which minimizes defects, such as porosity or hot cracking. AFSD is a solid-state additive manufacturing technique that enables high-integrity, low-distortion metal builds, as illustrated in Figure 1.33.

AFSD produces clads and builds while offering several key advantages. Its solid-state nature eliminates melting, resulting in minimal residual stress and distortions. This process ensures strong metallurgical bonding between the deposited layers and substrate. Through dynamic recrystallization, AFSD created a refined microstructure with fine, uniform grains. This leads to a high mechanical performance, often matching or exceeding that of wrought materials in terms of tensile strength and fatigue resistance. In addition, AFSD achieves low porosity, outperforming many traditional melt-based additive manufacturing techniques.

1.3.5.1 Commonly Used Cladding/Feedstock Materials

Commonly used cladding/feedstock materials in AFSD include a wide range of metals and alloys. Aluminum alloys, such as AA6061, AA2024, and AA7075, are extensively used in aerospace and automotive applications. Titanium alloys, such as Ti-6Al-4V, are employed because of their high strength-to-weight ratio. Copper and its alloys are used for their excellent electrical and thermal properties, whereas nickel-based superalloys,

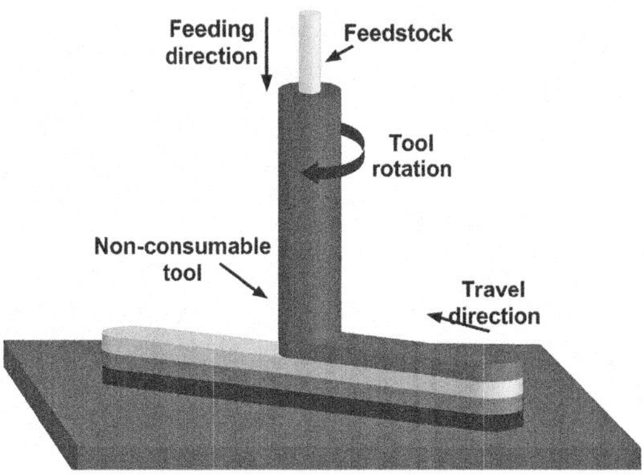

FIGURE 1.33 Schematic representation of the additive friction stir deposition process.

such as Inconel, are suitable for high-temperature environments. Magnesium alloys offer ultra-lightweight solutions for structural components. Typical substrate materials were selected to match or complement the feedstock. The use of the same alloy enables seamless repair or buildup, such as aluminum on aluminum. Dissimilar substrates, such as copper on steel, are used as functional coatings for conductivity. Other structural metals, such as steel, titanium, and nickel alloys, provide strong and stable bases, enabling high-performance builds with tailored functionality. The key process parameters in AFSD include tool rotational speed, traverse speed, axial force, feedstock material properties, and substrate temperature, which influence the deposition rate, layer bonding, microstructure, and mechanical properties (Figure 1.34).

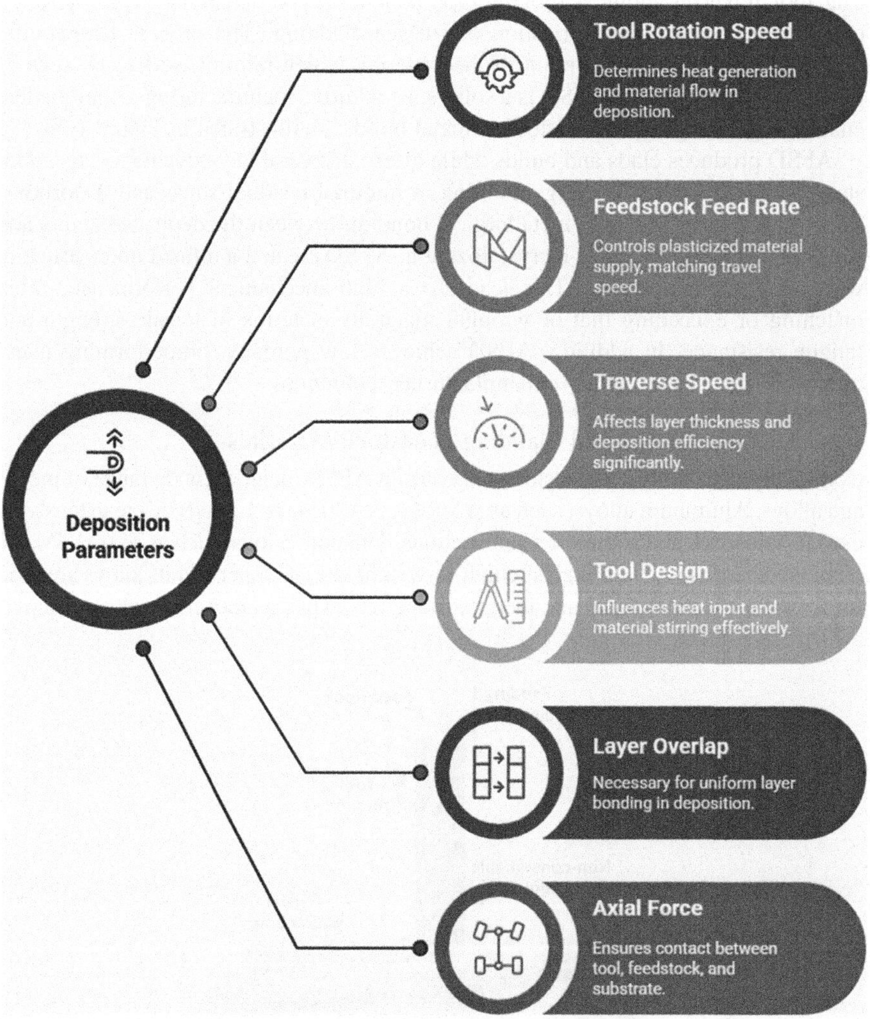

FIGURE 1.34 Key process parameters in additive friction stir deposition.

REFERENCES

[1] Anusha, K., Routara, B. C., & Guha, S. (2023). A review on high-velocity oxy-fuel (HVOF) coating technique. *Journal of The Institution of Engineers (India): Series D*, *104*(2), 831–848.

[2] Lombardi, A. N., Casteletti, L. C., & Totten, G. E. (2013). Thermal spray technologies: An overview. *Encyclopedia of Tribology*, *6*, 3607–3617.

[3] Medabalimi, S., Rokkala, U., Gudala, S. *et al.* (2025). Wear and frictional behaviour of partially oxidized and plasma sprayed NiCr and NiCrBSiFe coatings. *Scientific Reports*, *15*, 19717. https://doi.org/10.1038/s41598-025-99567-8

[4] Pawlowski, L. (2020). *The Science and Engineering of Thermal Spray Coatings*. Hoboken, NJ: Wiley.

[5] Ning, X. J., Jang, J. H., & Kim, H. J. (2007). The effects of powder properties on in-flight particle velocity and deposition process during low pressure cold spray process. *Applied Surface Science*, *253*(18), 7449–7455.

[6] Gupta, G., Tyagi, R. K., Rajput, S. K., Maan, R., Jacob, S., & Verma, S. (2021). Review on thermal spray coating methods and property of different types of metal-based coatings. In Bhupendra Prakash Sharma, G. Srinivasa Rao, Sumit Gupta, Pallav Gupta, and Anamika Prasad (eds.), *Advances in Engineering Materials: Select Proceedings of FLAME 2020* (pp. 427–439). Singapore: Springer Singapore.

[7] Fauchais, P., Vardelle, M., & Vardelle, A. (2001). Thermal spraying: Advances in modeling and control of the spray process. *Surface and Coatings Technology*, *201*(5), 1603–1612.

[8] Fauchais, P., & Vardelle, A. (2012). Thermal spraying: Process evolution, current status and future trends. *Surface and Coatings Technology*, *206*(8–9), 1244–1258.

[9] Kim, Ho Seok, Kang, Bo Ram, & Choi, Seong Man. (2019). Microstructure and mechanical properties of vacuum plasma sprayed HfC, TiC, and HfC/TiC ultra-high-temperature ceramic coatings. *Materials*, *13*(1), 124.

[10] Vardelle, A., Vardelle, M., Fauchais, P., & Moreau, C. (2016). Development of advanced plasma spray processes: The example of suspension and solution precursor spraying. *Journal of Thermal Spray Technology*, *25*(6), 1119–1130.

[11] Zhang, Q., Zhao, Y., Xu, Y., & Zhang, H. (2023). Suspension plasma sprayed thermal barrier coatings: Microstructure design, performance and applications. *Journal of the European Ceramic Society*, *43*(2), 700–717.

[12] Awan, T. I., Afsheen, S., & Kausar, S. (2025). Physical vapor deposition techniques. In Tahir Iqbal Awan, Sumera Afsheen, and Sabah Kausar (eds.), *Thin Film Deposition Techniques: Thin Film Deposition Techniques and Its Applications in Different Fields* (pp. 31–64). Singapore: Springer Nature Singapore.

[13] Wetegrove, M., Duarte, M. J., Taube, K., Rohloff, M., Gopalan, H., Scheu, C., ... & Kruth, A. (2023). Preventing hydrogen embrittlement: The role of barrier coatings for the hydrogen economy. *Hydrogen*, *4*(2), 307–322.

[14] Adeoye, A. E., Adeaga, O. A., & Ukoba, K. (2024). Chemical vapour deposition (CVD) and physical vapour deposition (PVD) techniques: Advances in thin film solar cells. *Nigerian Journal of Technology*, *43*(3), 479–489.

[15] Toma, F. T. Z., Rahman, S., Hussain, K. M., & Ahmed, S. (2024). Thin film deposition techniques: A comprehensive review. *Journal of Modern Nanotechnology*, *4*(6), 1–16.

[16] Jones, A. C., & Hitchman, M. L. (2009). Overview of chemical vapour deposition. *Chemical Vapour Deposition: Precursors, Processes and Applications*, *1*, 1–36.

[17] Carlsson, J. O., & Martin, P. M. (2010). Chemical vapor deposition. In Peter M. Martin (ed.), *Handbook of Deposition Technologies for Films and Coatings* (pp. 314–363). Norwich, NY: William Andrew Publishing.

[18] Lim, W. Y. S., Cao, J., Suwardi, A., Meng, T. L., Tan, C. K. I., & Liu, H. (2022). Recent advances in laser-cladding of metal alloys for protective coating and additive manufacturing. *Journal of Adhesion Science and Technology, 36*(23–24), 2482–2504.
[19] Arulvel, S., Jain, A., Kandasamy, J., & Singhal, M. (2023). Laser processing techniques for surface property enhancement: Focus on material advancement. *Surfaces and Interfaces, 42*, 103293.
[20] Brockmann, W., Geiß, P. L., Klingen, J., & Schröder, K. B. (2009). *Adhesive Bonding: Materials, Applications and Technology*. John Wiley & Sons.
[21] Vilar, R. (1999). Laser cladding. *Journal of Laser Applications, 11*(2), 64–79.
[22] Toyserkani, E., Khajepour, A., & Corbin, S. F. (2004). *Laser Cladding*. CRC Press.
[23] Steen, W. M., & Mazumder, J. (2010). *Laser Material Processing*. Springer Science & Business Media.
[24] Liu, X., Meng, L., Zeng, X., Zhu, B., Wei, K., Cao, J., & Hu, Q. (2024). Studies on high power laser cladding Stellite 6 alloy coatings: Metallurgical quality and mechanical performances. *Surface and Coatings Technology, 481*, 130647.
[25] Kylchbekov, S., Allamyradov, Y., Khuzhakulov, Z., Majidov, I., Banga, S., ben Yosef, J., Duta, L., & Er, A. O. (2023). Bioactivity and structural properties of hydroxyapatite on Ti6Al4V and Si (100) surfaces by pulsed laser deposition. *Coatings, 13*(10), 1681.
[26] Liu, L., Lin, Y., Peng, L., Kang, X., & Wang, X. (2024). Progress in microstructure design and control of high-hardness fe-based alloy coatings via laser cladding. *Coatings, 14*(11), 1351.
[27] Bai, Q., Chen, C., Li, Q., Zhao, C., & Zhang, J. (2024). Status of research on assisted laser cladding and laser cladding posttreatment: A review. *Physics of Metals and Metallography, 125*, 1–16.
[28] Naghdi, S., Rhee, K. Y., Hui, D., & Park, S. J. (2018). A review of conductive metal nanomaterials as conductive, transparent, and flexible coatings, thin films, and conductive fillers: Different deposition methods and applications. *Coatings, 8*(8), 278.
[29] Walsh, F. C., de León, C. P., Kerr, C., Court, S., & Barker, B. D. (2008). Electrochemical characterisation of the porosity and corrosion resistance of electrochemically deposited metal coatings. *Surface and Coatings Technology, 202*(21), 5092–5102.
[30] Landolt, D. (2002). Electrodeposition science and technology in the last quarter of the twentieth century. *Journal of the Electrochemical Society, 149*(3), S9.
[31] Popov, K. I., Djokić, S. S., & Grgur, B. N. (2002). Electroplating and surface finishing. *Fundamental Aspects of Electrometallurgy*, 197–247.
[32] Jagadeesha, T., Soundarrajan, M., Kaliappan, S., & Raj Kamal, M. D. (2024). Advancements in electrochemical surface coatings: Innovations, applications, and future prospects. In R. Thanigaivelan, Pradeep Kumar Krishnan, Kamalakanta Muduli, and Santosh Kumar Tamang (eds.), *New Materials, Processing and Manufacturability: Fabrication and Processing of Advanced Materials* (299–316). New York: Wiley-Scrivener / John Wiley & Sons.
[33] Korka, Z. I. (2017). Precision coatings. In Kapil Gupta (ed.), *Micro and Precision Manufacturing* (pp. 165–193). Cham: Springer International Publishing.
[34] Giurlani, W., Zangari, G., Gambinossi, F., Passaponti, M., Salvietti, E., Di Benedetto, F., . . . & Innocenti, M. (2018). Electroplating for decorative applications: Recent trends in research and development. *Coatings, 8*(8), 260.
[35] Hamza, N. A., Majeed, A. S., & Jawd, S. M. (2021). Review on types and methods of electroplating on metals. *Journal of Power Electronics and Devices, 7*, 44–51.
[36] Low, C. T. J., Wills, R. G. A., & Walsh, F. C. (2006). Electrodeposition of composite coatings containing nanoparticles in a metal deposit. *Surface and Coatings Technology, 201*(1–2), 371–383.

[37] Mohanty, U. S., Tripathy, B. C., Singh, P., Keshavarz, A., & Iglauer, S. (2019). Roles of organic and inorganic additives on the surface quality, morphology, and polarization behavior during nickel electrodeposition from various baths: A review. *Journal of Applied Electrochemistry, 49*, 847–870.

[38] Lee, E. L., Goh, Y. S., Haseeb, A. S. M. A., Wong, Y. H., Sabri, M. F. M., & Low, B. Y. (2023). Electrochemical migration in electronic materials: Factors affecting the mechanism and recent strategies for inhibition. *Journal of the Electrochemical Society, 170*(2), 021505.

[39] Zhou, Z., Pei, Z., Wei, L., Zhao, S., Jian, X., & Chen, Y. (2020). Electrocatalytic hydrogen evolution under neutral pH conditions: Current understandings, recent advances, and future prospects. *Energy & Environmental Science, 13*(10), 3185–3206.

[40] Wu, J., Ju, Z., Zhang, X., Marschilok, A. C., Takeuchi, K. J., Wang, H., . . . & Yu, G. (2022). Gradient design for high-energy and high-power batteries. *Advanced Materials, 34*(29), 2202780.

[41] Yasmeen, S., Ryu, S. W., Lee, S. H., & Lee, H. B. R. (2023). Atomic layer deposition beyond thin film deposition technology. *Advanced Materials Technologies, 8*(20), 2200876.

[42] Lipton, J., Weng, G. M., Röhr, J. A., Wang, H., & Taylor, A. D. (2020). Layer-by-layer assembly of two-dimensional materials: Meticulous control on the nanoscale. *Matter, 2*(5), 1148–1165.

[43] Van Bui, H., Grillo, F., & Van Ommen, J. R. (2017). Atomic and molecular layer deposition: Off the beaten track. *Chemical Communications, 53*(1), 45–71.

[44] Zhang, C. (2023). *Area-Selective Atomic Layer Deposition and Area-Selective Etching for Self-aligned Patterning* (Doctoral dissertation, University of Helsinki).

[45] Sun, L., Yuan, G., Gao, L., Yang, J., Chhowalla, M., Gharahcheshmeh, M. H., . . . & Liu, Z. (2021). Chemical vapour deposition. *Nature Reviews Methods Primers, 1*(1), 5.

[46] Lecordier, L. C. (2009). *Wafer-Scale Process and Materials Optimization in Cross-Flow Atomic Layer Deposition*. College Park: University of Maryland.

[47] Awan, T. I., Afsheen, S., & Kausar, S. (2025). Advanced deposition techniques. In Tahir Iqbal Awan, Sumera Afsheen, and Sabah Kausar (eds.), *Thin Film Deposition Techniques: Thin Film Deposition Techniques and Its Applications in Different Fields* (pp. 161–187). Singapore: Springer Nature Singapore.

[48] Wang, W. (2014). *Super-Conformal Coating and Filling of High Aspect Ratio Recessed Structures by Two-Molecule CVD*. Singapore: University of Illinois at Urbana-Champaign.

[49] Joshi, S., Markocsan, N., Nylén, P., & Sivakumar, G. (2020). New-generation ceramic coatings for high-temperature applications by liquid feedstock plasma spraying. In Yashwant R. Mahajan and Roy Johnson (eds.), *Handbook of Advanced Ceramics and Composites: Defense, Security, Aerospace and Energy Applications*, 1371–1412. Cham, Switzerland: Springer International Publishing.

[50] Song, C. (2018). *New preparation of ceramic coatings by low-pressure plasma spray* (Doctoral dissertation, Université Bourgogne Franche-Comté).

[51] Yaghtin, M. (2022). *Process development for advanced functionally graded thermal barrier coatings* (Doctoral dissertation, University of British Columbia).

[52] Jeyaprakash, N., Karuppasamy, S. S., & Yang, C. H. (2025). Repair of engineering components through laser cladding. In Natarajan Jeyaprakash, Govindarajan Prabu, and Che–Hua Yang (eds.), *Tailoring of Engineering Material Properties Through Laser Cladding* (pp. 78–102). CRC Press.

[53] Tuominen, J. (2009). Engineering coatings by laser cladding–the study of wear and corrosion properties. *Tampere University of Technology, 845*(1), 239.

[54] Jeyaprakash, N., Prabu, G., & Yang, C. H. (Eds.). (2025). *Tailoring of Engineering Material Properties Through Laser Cladding: Metallurgical, Mechanical and Tribological Properties*. Boca Raton, FL: CRC Press.
[55] Li, M., Huang, K., & Yi, X. (2023). Crack formation mechanisms and control methods of laser cladding coatings: A review. *Coatings*, *13*(6), 1117.
[56] Zanzarin, S. (2015). *Laser cladding with metallic powders* (Doctoral dissertation, University of Trento).
[57] Xue, C. (2022). *Modeling of grain structure and hot cracking for arc welding processes* (Doctoral dissertation, Université Paris sciences et lettres).
[58] Rodríguez Fernández, J., da Silva, G. T. G., Araújo, H. R., Torres López, E. A., & de Abreu Santos, T. F. (2024). Microstructural characterization of AISI 431 martensitic stainless steel coatings deposited by laser metal deposition. *The International Journal of Advanced Manufacturing Technology*, *135*(1), 677–694.
[59] Wang, J., Zhu, R., Liu, Y., & Zhang, L. (2023). Understanding melt pool characteristics in laser powder bed fusion: An overview of single-and multi-track melt pools for process optimization. *Advanced Powder Materials*, *2*(4), 100137.
[60] Santoni, A., Santecchia, E., Schiavone, A. M., Latini, V., Lascu, B. D., & Stoica, C. R. (2025). On the extent of feedstock–system interaction in determining the efficiency of laser powder directed energy deposition. *Metals*, *15*(6), 599.
[61] Matteazzi, P., Colella, A., Leshchynsky, V., Sakaki, K., Fukanuma, H., & Maev, R. G. (2016). Cold spray powders and equipment. In *Cold Gas Dynamic Spray* (pp. 75–138). Boca Raton: Taylor & Francis Group.
[62] Fauchais, P., & Montavon, G. (2008). Thermal and cold spray: Recent developments. *Key Engineering Materials*, *384*, 1–59.
[63] Nastic, A. (2021). *Cold gas dynamic spray impact: Metallic bonding pre-requisites and experimental particle in-flight temperature measurements* (Doctoral dissertation, Université d'Ottawa/University of Ottawa).
[64] Korzhyk, V., Tyurin, Y., & Kolisnichenko, O. (2021). *Theory and Practice of Plasma-Detonation Technology of Surface Hardening Metal Products*. Kharkiv: PC Technology Center, 244.
[65] Zhang, P. (2003). *Joining enabled by high velocity deformation*. The Ohio State University.
[66] Mishra, R. S., Haridas, R. S., & Agrawal, P. (2022). Friction stir-based additive manufacturing. *Science and Technology of Welding and Joining*, *27*(3), 141–165.
[67] Dong, H., Li, X., Xu, K., Zang, Z., Liu, X., Zhang, Z., . . . & Li, Y. (2022). A review on solid-state-based additive friction stir deposition. *Aerospace*, *9*(10), 565.
[68] Korgancı, M., & Bozkurt, Y. (2024). Recent developments in additive friction stir deposition (AFSD). *Journal of Materials Research and Technology*, *30*, 4572–4583.

2 Advanced Coating Systems in High-Temperature Corrosion Protection

Mona Obead Albalawi

2.1 INTRODUCTION

High-temperature corrosion threatens critical components in aerospace, power generation, and petrochemical industries through oxidation, sulfidation, carburization, and molten salt attack. Unlike conventional aqueous corrosion, these processes occur through direct chemical reactions with aggressive gases. With global corrosion costs exceeding $2.5 trillion annually (3.4% of GDP), innovative protective strategies are essential as operational demands push conventional materials to their limits.[1]

Protective coatings serve as the cornerstone of high-temperature materials engineering, offering tailored barriers that enable base materials to function beyond inherent limitations. These systems range from thermal barrier coatings (TBCs) providing thermal insulation to environmental barrier coatings (EBCs) resisting water vapor attack, representing decades of materials science advancement.[1]

This chapter examines high-temperature protective coating technologies, spanning from fundamental corrosion mechanisms to cutting-edge materials and processing techniques. We explore scientific principles governing material selection, engineering considerations driving coating design, and practical implementation across diverse applications. Emerging trends include ultra-high-temperature ceramics for hypersonic applications, environmentally compliant formulations, and AI approaches accelerating materials discovery.

By bridging fundamental science with practical engineering, this chapter serves as both an educational resource and a technological roadmap for researchers and industry professionals working toward higher operating temperatures while ensuring reliability, sustainability, and economic viability.

2.2 FUNDAMENTALS OF HIGH-TEMPERATURE CORROSION

High-temperature corrosion represents a critical challenge in materials science, fundamentally differing from conventional electrochemical corrosion as it operates without requiring a liquid electrolyte. This phenomenon, also referred to as "dry corrosion" or

DOI: 10.1201/9781003653226-2

"scaling," involves direct chemical reactions between metals and gaseous environments at temperatures typically exceeding 650°C. The absence of an electrolyte distinguishes high-temperature corrosion from traditional aqueous corrosion mechanisms, as ions within hot gases attack substrate materials directly rather than through ionic solutions.[1,2]

2.2.1 Primary Corrosion Mechanisms

Oxidation represents the most fundamental high-temperature corrosion mechanism, characterized by the formation of oxide scales through direct metal–oxygen reactions. The protective nature of these oxide layers depends critically on their adherence, growth kinetics, and barrier properties against further environmental attack. The Deal–Grove model is commonly employed to describe oxidation kinetics, accounting for diffusion and reaction dynamics that govern scale development. Wagner's theory of oxidation further explains how ionic and electronic conductivities within formed oxides control overall oxidation rates.[3,4]

Sulfidation emerges as a particularly aggressive degradation mechanism, often exceeding oxidation in severity due to the formation of less stable sulfide scales compared to oxides. This process occurs when metals encounter sulfur-containing compounds at elevated temperatures, leading to sulfide formation beneath protective oxide layers. The mechanism involves sulfur diffusion through primary oxide scales in molecular form (SO_2), where equilibrium shifts favor sulfide formation at metal–scale interfaces where oxygen partial pressures are extremely low.[3,4]

Hot corrosion represents an accelerated form of oxidation occurring in the presence of molten salt deposits, typically classified into Type I (825–950°C) and Type II (650–800°C) based on temperature regimes and mechanisms. Type I hot corrosion involves molten sodium sulfate dissolving protective oxide scales and releasing sulfur that diffuses into substrate materials, forming colored aluminum or chromium sulfides. Type II occurs below sodium sulfate's melting point but requires high SO_3 partial pressures to initiate the corrosion process.[3,4]

Carburization is the absorption of carbon into metallic substrates during high-temperature exposure to carbon-rich environments, typically above 900°C. This process causes metal embrittlement through carbide precipitation within the microstructure, reducing ductility and toughness while increasing susceptibility to cracking failures. The mechanism parallels sulfidation, with carbon transport through oxide scales in molecular form (CO_2), followed by carbide formation at metal–oxide interfaces where carbon activity increases dramatically.[3,4]

2.2.2 Environmental Factors and Industrial Relevance

High-temperature corrosion affects diverse industrial sectors, including power generation, aerospace, petrochemical processing, and manufacturing. Gas turbines, diesel engines, furnaces, and other machinery operating in hot, contaminant-laden gases face accelerated degradation when fuel compositions contain vanadium compounds or sulfates that form low-melting-point compounds during combustion.

These conditions create strongly corrosive liquid salts that can attack even stainless steel and other alloys regarded as corrosion-resistant.

Environmental composition significantly influences corrosion behavior, with mixed-gas environments leading to simultaneous oxidation and sulfidation processes. The presence of water vapor, carbon dioxide, and other reactive species creates complex chemical interactions that can either enhance or inhibit specific degradation pathways. Understanding these environmental effects is crucial for developing protective strategies that address multiple corrosion mechanisms simultaneously.[5,6]

2.3 MATERIAL SELECTION FOR HIGH-TEMPERATURE COATINGS

2.3.1 Properties of High-Performance Coating Materials

Material selection for high-temperature coatings requires careful consideration of thermal, mechanical, and chemical properties that enable effective protection under extreme conditions. The operating temperature range serves as the primary selection criterion: organic materials are suitable for applications below 400°F (204°C), whereas inorganic materials like silicone and ceramics are required for extreme temperature resistance above this threshold.[6,7]

2.3.1.1 Thermal Properties

Thermal stability represents the fundamental requirement for high-temperature coating materials, necessitating structural integrity maintenance at operating temperatures while resisting phase transformations that could compromise protective capabilities. TBCs must exhibit appropriate porosity and a proper thermal expansion coefficient matching the underlying metal surfaces to handle thermal expansion loads during heating and cooling cycles. Phase stability becomes critical to prevent volume fluctuations during phase transitions that could lead to coating cracking or spallation.

Thermal expansion compatibility plays a crucial role in coating performance, as mismatched coefficients of thermal expansion (CTE) between coatings and substrates can induce significant stresses during thermal cycling. Advanced ceramic materials typically exhibit lower thermal expansion coefficients compared to metallic substrates, necessitating careful interface design and potentially graded coating architectures. The thermal expansion behavior of thermally sprayed coatings closely matches that of the bulk material when the material is not significantly affected by oxidation during the spraying process.

Thermal conductivity management enables effective thermal barrier performance while maintaining substrate protection. Yttria-stabilized zirconia (YSZ) demonstrates exceptionally low thermal conductivity (1.5–2.5 W/m·K) due to its microstructural characteristics and lattice defects. However, thermal conductivity can vary significantly based on processing conditions and microstructural features induced during deposition.[6,7]

2.3.1.2 Mechanical Properties

Hardness and wear resistance determine coating durability under mechanical loading conditions, with ceramics typically demonstrating superior hardness compared to metallic alternatives. However, this hardness advantage comes with increased brittleness, necessitating careful balance between wear resistance and fracture toughness. The relationship between hardness, internal and external bond strengths, and surface abrasive wear resistance represents complex physics and chemistry considerations in coating design.

Adhesion strength governs coating–substrate bonding effectiveness, measured through tensile, shear, and peel strength testing methodologies. Tensile strength represents the coating's ability to resist loads under stress without failure, while shear strength measures resistance to forces causing internal structural sliding. Peel strength evaluates resistance to localized stresses, particularly important for thin-film coating applications.

Fracture toughness and ductility influence coating response to mechanical stress and thermal cycling, with metals generally exhibiting superior ductility while ceramics maintain higher strength at elevated temperatures. The brittleness of ceramic materials necessitates careful microstructural design and potentially hybrid approaches combining metallic and ceramic phases.[6,7]

2.3.1.3 Chemical Properties

Oxidation resistance depends on the formation of stable, protective oxide layers that impede further environmental attack. The Pilling–Bedworth ratio provides guidance for oxide protectiveness, with ratios between 1 and 2 indicating protective oxide formation, while ratios below 1 or above 2 suggest non-protective behavior. The addition of aluminum and chromium to coating compositions promotes the formation of protective Al_2O_3 and Cr_2O_3 scales, respectively.

Chemical stability in aggressive environments requires resistance to molten salt attack, sulfidation, and other forms of chemical degradation. Coating materials must maintain chemical inertness while providing barrier protection against reactive species penetration. EBCs utilize rare earth silicates that resist water vapor and molten salt corrosion through the formation of stable protective phases.

Corrosion resistance encompasses resistance to multiple degradation mechanisms, including oxidation, sulfidation, carburization, and hot corrosion. Multi-component coating systems can address different aspects of chemical attack through layered architectures that provide both barrier protection and sacrificial protection mechanisms.[6,7]

2.4 PROTECTIVE COATING TECHNOLOGIES/COMPARATIVE ANALYSIS

This section provides a critical evaluation of coating systems through four key lenses.

2.4.1 Multi-layered and Nanolaminate Coatings

2.4.1.1 Material Innovation

Multi-layered and nanolaminate coatings combine alternating nanoscale layers of metals, ceramics, or composites, synergizing properties that are unattainable in monolithic systems. Chromium Aluminum Nitride (CrAlN)/Aluminum Chromium Nitride (AlCrN) nanolaminates, deposited via hybrid Direct Current (DC)/High-Power Pulsed Magnetron Sputtering (HPPMS) sputtering, exhibit exceptional oxidation resistance by forming continuous alpha-Aluminum Oxide (α-Al_2O_3) scales at 1100°C, outperforming monolithic CrAlN by 3× in steam environments. MAX phases (e.g., Titanium Silicon Carbide (Ti_3SiC_2)) bridge metal–ceramic gaps, offering thermal shock resistance and self-lubricating properties. Furthermore, self-healing mechanisms, such as Molybdenum Disilicide ($MoSi_2$)-doped layers, autonomously seal cracks via Silicon Dioxide (SiO_2) formation at >1000°C, extending service life by 50% in cyclic thermal environments.[8,9]

2.4.1.2 Processing Techniques

- **Magnetron Sputtering**: Enables precise control of layer thickness (20–500 nm) for CrAlN/AlCrN systems, achieving low porosity (<2%) and high adhesion.
- **Electrodeposition**: Scalable for Cu/Ni nanolaminates but limited to conductive substrates.
- **Laser Cladding**: Produces dense, graded layers (e.g., $La_2Zr_2O_7$ coatings) with reduced thermal conductivity (1.6 W/m·K).
- **Post-Processing**: Laser annealing enhances interfacial bonding, improving hardness retention by 15% after thermal cycling.[8,9]

2.4.1.3 Degradation Mechanisms

- **Oxidation-Induced Decomposition**: Thin layers (<100 nm) favor rapid Cr_2O_3 formation, while thicker layers (>100 nm) develop protective Al_2O_3 scales.
- **Thermal Fatigue**: Coefficient of thermal expansion (CTE) mismatches (e.g., Cr vs. CrAlN: 14 vs. 9 × 10^{-6}/°C) induce mechanical stresses during heating/cooling cycles, leading to delamination or cracking.
- **Molten Salt Infiltration**: Sodium sulfate–vanadium pentoxide eutectic mixture, Na_2SO_4-V_2O_5, deposits penetrate interfaces, accelerating corrosion via acidic fluxing[8,9]. As shown in Table 2.1, the capabilities of different deposition methods vary significantly in terms of layer uniformity and adhesion strength.

TABLE 2.1
Comparative Analysis of Deposition Methods for Multi-layered Coatings[9]

Coating Type	Oxidation Rate (mg/cm^2/hr)	Thermal Cycling Lifespan
Monolithic CrAlN	0.15	200 cycles
Cr/CrAlN Nanolaminate	0.05	500 cycles
Traditional MCrAlY*	0.20	150 cycles

* McrAlY refers to a family of metal-chromium-aluminum-yttrium alloys.

2.4.1.3.1 Comparative Performance

2.4.1.4 Computational Tools

- **Finite Element Analysis (FEA)**: Stimulates interfacial stresses in ZrO_2-Ti-6Al-4V systems, optimizing layer thickness to <50 nm for CTE matching.
- **Molecular Dynamics (MD)**: Simulates Al/Cr interdiffusion in MAX phases, guiding bilayer period optimization ($\Lambda = 18$ nm for CrAlN/AlCrN).
- **Machine Learning (ML)**: Reduces sputtering parameter trials by 70% via Gaussian process regression, enabling rapid deposition rate prediction.[8]

2.4.1.5 Integration and Future Directions

Nanolaminates address oxidation (Al_2O_3 scales), sulfidation (Cr-rich barriers), and thermal fatigue (CTE gradation), outlined in the prior section. Challenges include scaling production and stabilizing healing agents >1200°C.[8] Emerging trends are as follows:

- **AI-Driven Design**: Generative models for layer architectures.
- **Additive Manufacturing**: Laser-deposited graded nanolaminates for hypersonic applications.

2.4.2 High-Entropy Alloy (HEA) Coatings

2.4.2.1 Material Innovation

High-entropy alloys (HEAs) utilize multi-principal elemental compositions (e.g., AlCoCrFeNi and CrMnFeCoNi) to achieve exceptional high-temperature stability and corrosion resistance. Their configurational entropy stabilizes solid-solution phases, resisting intermetallic formation. The addition of Al and Cr promote protective α-Al_2O_3/Cr_2O_3 scales, reducing oxidation rates by 40–60% compared to conventional alloys like McrAlY. Refractory HEAs (RHEAs), such as TiNbTa-based systems, exhibit superior hot corrosion resistance in molten sulfate environments due to slow oxygen diffusion through $CrTaO_4$ layers. However, elements like Mo and W can form volatile oxides, necessitating compositional optimization.[8,10]

2.4.2.2 Processing Techniques

Thermal spraying techniques such as High-Velocity Oxygen Fuel (HVOF) and Air Plasma Spray (APS) are effective in producing dense HEA coatings with porosity below 2% and adhesion strength exceeding 70 Mpa. Notably, AlCoCrFeNi coatings applied via HVOF have demonstrated up to an 89% reduction in erosion rates compared to uncoated steels in coal-fired boilers. Additive manufacturing methods like direct metal laser sintering enable the fabrication of complex geometries; however, precise parameter control is essential to prevent undesirable dendritic structures and brittle phases. Post-processing treatments further enhance coating performance; for example, annealing at 550°C improves AlCoCrFeNi microhardness by 46% due to oxide layer refinement, while laser cladding can produce functionally graded layers such as $La_2Zr_2O_7$ with low thermal conductivity (1.6 W/m·K). Despite these advancements, degradation mechanisms remain a concern. At high temperatures (e.g., 1100°C),

Al-rich HEAs form protective Al_2O_3 scales, while TiNbTa RHEAs develop $CrTaO_4$ layers that inhibit oxygen diffusion. However, HEAs containing Mo or W are prone to accelerated degradation due to the formation of volatile oxides. In corrosive environments such as Na_2SO_4-V_2O_5, Cr-rich HEAs reduce corrosion rates by 44% through the formation of stable Cr_2O_3 barriers. Additionally, thermal fatigue caused by CTE mismatches—such as between HEAs (14×10^{-6}/°C) and steel (12×10^{-6}/°C)—can lead to delamination under cyclic thermal loads.

2.4.2.3 Computational Tools

Advanced computational techniques play a crucial role in optimizing HEA coatings. MD simulations help analyze Al/Cr interdiffusion in RHEAs, enabling the identification of optimal elemental ratios, such as a 1:1 Al:Cr ratio in TiNbTaCrAl systems. Finite element analysis (FEA) is employed to predict interfacial stresses in HEA-coated turbine blades, guiding the design of coating layer thickness to remain below 50 μm for enhanced performance and durability. Additionally, machine learning (ML) approaches, such as Gaussian Process Regression, significantly accelerate process optimization by reducing the number of experimental parameter trials for HVOF-deposited HEAs by up to 70%.

2.4.3 Thermal Spray and Cold Spray Coatings

2.4.3.1 Material Innovation

Thermal spray coatings involve the projection of molten or semi-molten materials—typically metals, ceramics, or polymers—onto a substrate to create a protective layer with excellent wear, corrosion, and heat resistance, commonly used across various industries. HVOF spraying employs metallic alloys such as McrAlY (where M = Ni, Co, or Fe) and ceramic-metal composites like WC–CoCr, which provide resistance to oxidation and sulfidation by forming stable Al_2O_3 and Cr_2O_3 scales at temperatures up to 1100°C. APS, on the other hand, is used to deposit ceramics like YSZ, effectively lowering substrate temperatures by 100–300°C due to their low thermal conductivity (1.5–2.5 W/m·K). In contrast, cold spray coatings—or gas dynamic spray—rely on the supersonic acceleration of metal or alloy powders via high-pressure gas, enabling solid-state deposition through high-velocity impact without melting. This method preserves the intrinsic properties of the feedstock, resulting in dense, oxide-free coatings. Materials such as pure Cu, Al, and NiCr alloys are commonly used, while the incorporation of graphene into copper coatings has been shown to significantly enhance erosion resistance, particularly in demanding nuclear environments. As shown in Table 2.2, the key process parameters differ significantly between thermal spray and cold spray deposition.

2.4.3.2 Processing Techniques

2.4.3.3 Degradation Mechanisms

Thermal and cold spray coatings face distinct degradation mechanisms and benefit significantly from computational advancements. In thermal spray processes, McrAlY coatings develop protective Al_2O_3 scales, but aluminum depletion becomes critical at

TABLE 2.2
Key Parameters for Thermal vs. Cold Spray Deposition[11]

Parameter	Thermal Spray (HVOF/APS)	Cold Spray
Temperature	HVOF: 2500–3500°C (particle) APS: ~15,000°C (plasma jet)	Ambient–600°C (gas)
Porosity	HVOF: <2%; APS: 5–15%	<1% (dense microstructure)
Key Materials	HVOF: WC–CoCr, McrAlY; APS: YSZ, Al_2O_3	Cu, Al, NiCr
Adhesion Strength	HVOF: 70–100 Mpa; APS: 30–50 Mpa	50–80 Mpa

temperatures exceeding 1000°C. APS-deposited YSZ is vulnerable to hot corrosion, as molten salts like Na_2SO_4–V_2O_5 can dissolve Y_2O_3 stabilizers. Furthermore, thermal fatigue arises due to the mismatch in CTE between YSZ (9×10^{-6}/°C) and nickel-based superalloys (14×10^{-6}/°C), leading to spallation after 200–500 thermal cycles. In cold spray coatings, porosity levels above 1% can promote galvanic corrosion by allowing electrolyte penetration, especially in saline environments. Additionally, residual tensile stresses from high-velocity particle impacts can cause interfacial delamination under thermal cycling. Computational tools offer valuable solutions: FEA helps predict residual stresses in HVOF WC–CoCr coatings, enabling spray angle optimization that reduces delamination by up to 40%. Computational fluid dynamics (CFD) models gas and particle behavior in APS, improving YSZ deposition efficiency from 50% to 70%. MD simulations inform optimal gas temperatures—such as 950°C for copper—to enhance particle–substrate bonding in cold spray. Meanwhile, ML, particularly Gaussian process regression, reduces experimental HVOF parameter trials by 50%, accelerating process optimization.

2.4.4 Advanced Thermal Barrier Coatings (TBCs) and Environmental Barrier Coatings (EBCs)

2.4.4.1 Material Innovation

Thermal Barrier Coatings (TBCs): Conventional YSZ (YSZ, ZrO_2-8%Y_2O_3) remains the industry standard but faces limitations above 1200°C due to phase instability and molten silicate infiltration.[8,12] Advanced alternatives include:

- **Gadolinium Zirconate ($Gd_2Zr_2O_7$)**: Exhibits 40% lower thermal conductivity (1.2 W/m·K) than YSZ and forms apatite phases (e.g., $Ca_2Gd_8(SiO_4)_6O_2$) under calcium–magnesium–aluminosilicate (CMAS) attack, which seal cracks and reduce infiltration.
- **High-Entropy Oxides (HEOs)**: Multi-cation ceramics like $(5RE_{1/5})_2Zr_2O_7$ (RE = La, Nd, Sm, Eu, Gd) leverage configurational entropy to stabilize pyrochlore/fluorite dual-phase structures, achieving thermal conductivities below 1.5 W/m·K.

Environmental Barrier Coatings (EBCs): Rare earth silicates, such as ytterbium disilicate ($Yb_2Si_2O_7$) and yttrium monosilicate (Y_2SiO_5), resist water vapor and molten salt corrosion. Multicomponent systems like $(Yb_{0.2}Er_{0.2}Tm_{0.2}Lu_{0.2}Sc_{0.2})_2$-$Si_2O_7$ combine low oxygen diffusivity ($<10^{-17}$ m^2/s) with thermal expansion coefficients (CTE: 4–5 × $10^{-6}/°C$) matching silicon carbide (SiC) substrates.[8,12]

2.4.4.2 Processing Techniques

- **Atmospheric Plasma Spraying (APS)**: Deposits YSZ and $Gd_2Zr_2O_7$ with lamellar microstructures (5–15% porosity), making them ideal for turbine blades.
- **Electron Beam-Physical Vapor Deposition (EB-PVD)**: Creates columnar YSZ structures for enhanced strain tolerance in cyclic environments.
- **Suspension Plasma Spraying (SPS)**: Enables nano-structured, functionally graded coatings (FGCs) like $Gd_2Zr_2O_7$/YSZ, which reduce interfacial stresses by 50%.
- **Chemical Vapor Deposition (CVD)**: Forms dense, adherent rare earth silicate layers (e.g., $Yb_2Si_2O_7$) for EBCs on ceramic matrix composites (CMCs).

2.4.4.3 Degradation Mechanisms

TBCs:

- **CMAS Infiltration**: Molten silicates (e.g., CaO-MgO-Al_2O_3-SiO_2) penetrate YSZ, dissolving Y_2O_3 stabilizers and forming destabilized cubic ZrO_2. $Gd_2Zr_2O_7$ mitigates this via reactive crystallization, forming sealed apatite layers.
- **Thermally Grown Oxide (TGO) Growth**: Aluminum depletion in bond coats (e.g., NiCoCrAlY) accelerates α-Al_2O_3 layer thickening, inducing spallation stresses >3 Gpa.

EBCs:

- **Water Vapor Corrosion**: Y_2SiO_7 reacts with H_2O at 1300°C, forming volatile $SiO(OH)_2$. Multi-component silicates (e.g., $Yb/Ho/Er/Lu)_2Si_2O_7$ reduce volatility by 70% due to sluggish cation diffusion.
- **Molten Salt Attack**: Sodium sulfate–vanadium pentoxide (Na_2SO_4-V_2O_5) dissolves silicate phases, but high-entropy designs (e.g., $(Y_{0.2}Yb_{0.2}Er_{0.2}Tm_{0.2}Lu_{0.2})_2Si_2O_7$) form stable garnet phases ($Yb_3Fe_5O_{12}$), which inhibit penetration.[8]

2.4.4.4 Computational Tools

- **Finite Element Analysis (FEA)**: Simulates CTE mismatches in FGCs, optimizing layer thicknesses (e.g., 50 μm $Gd_2Zr_2O_7$ over 150 μm YSZ) to minimize delamination.
- **Molecular Dynamics (MD)**: Simulates oxygen diffusion in $Yb_2Si_2O_7$, identifying vacancy-mediated pathways with activation energies >2.5 eV.

- **Machine Learning (ML)**: Predicts optimal HEO compositions (e.g., $(Gd_{0.2}Sm_{0.2}Nd_{0.2}Eu_{0.2}La_{0.2})_2Zr_2O_7$) by correlating ionic radii variance with phase stability.
- **Phase-Field Modeling**: Replicates CMAS infiltration dynamics, guiding pore-sealing strategies in $Gd_2Zr_2O_7$.[8,12]

2.4.5 Smart and Self-Healing Coatings

2.4.5.1 Material Innovation

Smart and self-healing coatings leverage advanced materials to autonomously repair damage and adapt to extreme environments.[8,13] Key innovations include:

- **Extrinsic Self-Healing Systems**: Microcapsules (5–50 μm) filled with film-forming agents like polydimethylsiloxane (PDMS) or $MoSi_2$ are embedded in coatings. When cracks form, capsules rupture, releasing healing agents that oxidize to form protective phases (e.g., SiO_2 from $MoSi_2$ at 1100°C).
- **Intrinsic Self-Healing Polymers**: Reversible bonds, such as urea or 2-ureido-4[1*H*]-pyrimidone (Upy) groups, enable autonomic repair via hydrogen bonding or Diels–Alder reactions. Upy-modified silicone coatings restore 90% adhesion after cracking at 120°C.
- **High-Entropy Ceramics**: Rare-earth silicates like ytterbium disilicate ($Yb_2Si_2O_7$) resist water vapor corrosion in EBCs by forming stable garnet phases under molten salt attack.[8,13]

2.4.5.2 Processing Techniques

- **Atmospheric Plasma Spraying (APS)**: Deposits YSZ TBCs with embedded $MoSi_2$ microcapsules. Post-spray annealing (800°C, low O_2) forms protective Al_2O_3 shells around healing agents.
- **Suspension Plasma Spraying (SPS)**: Creates columnar-structured TBCs with enhanced strain tolerance. Used for YSZ/$Gd_2Zr_2O_7$ FGCs.
- **Hybrid DC/HPPMS (Direct Current/High-Power Pulsed Magnetron Sputtering)**: Produces dense CrAlN coatings with adaptive Cr_2O_3 formation at 900°C.
- **Sol–Gel Encapsulation**: Microencapsulates biocides or catalysts in polymer matrices for controlled release, reducing environmental toxicity.[8,13]

2.4.5.3 Degradation Mechanisms

- **Calcium–Magnesium–Aluminosilicate (CMAS) Infiltration**: Molten silicates penetrate TBCs, dissolving Y_2O_3 stabilizers. Self-healing TBCs with $MoSi_2$ form $ZrSiO_4$ seals via SiO_2–ZrO_2 reactions, tripling lifespan.
- **Thermally Grown Oxide (TGO) Growth**: Aluminum depletion in bond coats (e.g., NiCoCrAlY) induces spallation. Intrinsic polymers with Upy groups mitigate interfacial stresses through reversible bond reformation.
- **Water Vapor Corrosion**: $Yb_2Si_2O_7$ EBCs reduce $SiO(OH)_2$ volatility by 70% via sluggish cation diffusion, maintaining integrity at 1300°C.[8,13]

2.4.5.4 Computational Tools

- **Molecular Dynamics (MD)**: Simulates oxygen diffusion in $Yb_2Si_2O_7$, identifying vacancy-mediated pathways with activation energies >2.5 eV.
- **Finite Element Analysis (FEA)**: Models CTE mismatches in FGCs, optimizing layer thicknesses (e.g., 50 μm $Gd_2Zr_2O_7$ over 150 μm YSZ).
- **Machine Learning (ML)**: Predicts optimal high-entropy oxide compositions (e.g., $(Gd_{0.2}Sm_{0.2}{}^{Nd}{}_{0.2}Eu_{0.2}La_{0.2})_2Zr_2O_7$) using Gaussian process regression.
- **Phase-Field Modeling**: Replicates crack-healing dynamics in $MoSi_2$-doped TBCs, validating SiO_2 flow and zircon formation.[13]

2.4.6 Laser Cladding and Cold Metal Transfer Technology

2.4.6.1 Material Innovation

Laser cladding leverages high-energy lasers to create metallurgically bonded coatings with minimal dilution (<5%) and near-theoretical density.[14] Key innovations include:

- **Multi-material Functionalization**: Combining metals (e.g., McrAlY), ceramics (YSZ), and composites (WC–CoCr) to achieve tailored properties like oxidation resistance and thermal barrier performance.
- **In-situ Alloying**: Using elemental powder blends instead of pre-alloyed powders to synthesize novel compositions directly during deposition.
- **Nano-Reinforcements**: Incorporating SiC or graphene to enhance hardness (up to 1252 HV) and wear resistance.

Cold metal transfer (CMT) revolutionizes deposition through controlled short-circuit transfer:

- **Low-Heat-Input Bonding**: Solid-state deposition preserves substrate microstructure, enabling dissimilar material joining (e.g., steel-aluminum) without brittle intermetallic.
- **Reactive Element Integration**: Depositing chromium coatings that form protective Cr_2Zr intermetallic layers under annealing, blocking oxygen diffusion.[13,14]

2.4.6.2 Processing Techniques

Laser cladding employs two primary methods:

- **Powder Bed Fusion (PBF)**: Layer-by-layer melting with precise control over complex geometries, achieving <2% porosity.
- **Direct Energy Deposition (DED)**: Robotic powder/wire feeding enabling high deposition rates (100–500 g/h) with minimal distortion. Critical parameters include laser power (1400–2000 W), scan speed (7–400 mm/min), and powder feed rate (21 g/min).

CMT processing features:

- **Synergic Wire Retraction**: Automated wire retraction during short-circuiting reduces spatter and heat input by 60% compared to conventional MIG/MAG.
- **Parameter Optimization**: Helium gas acceleration (800 m/s) and post-deposition annealing (600°C) enhance adhesion strength (>70 Mpa).[8,14]

2.4.6.3 Degradation Mechanisms

Laser cladding faces:

- **Residual Stresses**: Thermal gradients induce tensile stresses up to 500 Mpa, mitigated by pre-set texturing (reducing stress by 34–50%).
- **Cracking**: Rapid solidification causes microcracks, addressed through ultrasonic grain refinement.
- **Oxidation/Hot Corrosion**: Depletion of protective elements (Al/Cr) at >1000°C.

CMT-specific challenges:

- **Arc Wandering in Titanium**: Causes surface roughness, requiring precise torch alignment.
- **Porosity in Aluminum**: Mitigated using high-purity wires and pulsed-advanced modes.[14]

2.4.6.4 Computational Tools

- **Finite Element Analysis (FEA)**: Simulates thermal gradients and residual stresses in laser cladding, optimizing scan strategies to reduce distortion.
- **Multi-field Simulations**: Coupled thermal-fluid-stress models predict melt pool dynamics and crack formation, enabling parameter optimization (e.g., 7 mm/s scan speed).
- **Machine Learning (ML)**: Reduces trial iterations by 70% for laser parameters and CMT wire-feed settings.
- **Molecular Dynamics (MD)**: Simulates diffusion kinetics in CMT-deposited coatings to predict intermetallic formation.[14]

2.5 CASE STUDIES AND INTEGRATION

Three real-world case studies demonstrate the application of advanced coatings in combating high-temperature corrosion, highlighting material innovation, processing challenges, and performance validation.

2.5.1 Jet Engine Turbines: Combating Sulfidation in Nickel-Based Superalloys

Problem: Sulfidation-induced stress corrosion cracking (SCC) in General Electric (GE) GEnx engine turbine blades operating at 1100–1300°C. Sulfur from jet fuel forms aggressive sulfides (e.g., Ni_3S_2) that disrupt protective oxide scales, leading to crack propagation rates exceeding 10^{-6} m/cycle.[15]

Solution: Multilayered Aluminum Chromium Nitride/Titanium Silicon Nitride (AlCrN/TiSiN) coatings deposited via High-Power Impulse Magnetron Sputtering (HIPIMS). The architecture comprises:

- **AlCrN Layer (3 μm)**: Forms a dense chromium oxide (Cr_2O_3) barrier at 900°C, reducing sulfidation rates by 70%.
- **TiSiN Layer (2 μm)**: Nanocomposite structure (TiN nanocrystals in amorphous Si_3N_4 matrix) provides hardness >40 GPa and crack deflection.[15]

Implementation: GE Aviation reported a 60% longer service life for coated blades in GEnx engines compared to monolithic YSZ coatings. Field data from 150 engines showed a reduction in unscheduled maintenance from 12% to 4% over 10,000 flight hours.

Key Insight: Layer sequencing (AlCrN below TiSiN) prevents interfacial delamination during thermal cycling by matching CTE (CTE: 9.5×10^{-6}/°C for AlCrN vs. 10.2×10^{-6}/°C for TiSiN).[15,16] Figure 2.1 illustrates the sulfidation mechanisms and the role of protective aluminide and ceramic overlay coatings in a nickel-based superalloy turbine blade.

2.5.2 Nuclear Reactors: Mitigating Zirconium Alloy Oxidation in LOCA Scenarios

Problem: Loss-of-Coolant Accidents (LOCA) expose Zircaloy-4 cladding to 1200°C steam, accelerating oxidation ($Zr + 2H_2O \rightarrow ZrO_2 + 2H_2$) and hydrogen pickup (>600 ppm), leading to embrittlement.[17]

Solution: Cold-sprayed chromium (Cr) coatings (20–50 μm thickness) applied via kinetic deposition (helium gas, 800 m/s). The process avoids substrate heating, preserving Zircaloy's grain structure. Post-spray annealing at 600°C enhances Cr–Zr interdiffusion, forming a chromium zirconium (Cr_2Zr) intermetallic layer that blocks oxygen diffusion.[17] The schematic representation of zirconium alloy fuel cladding under LOCA conditions is shown in Figure 2.2.

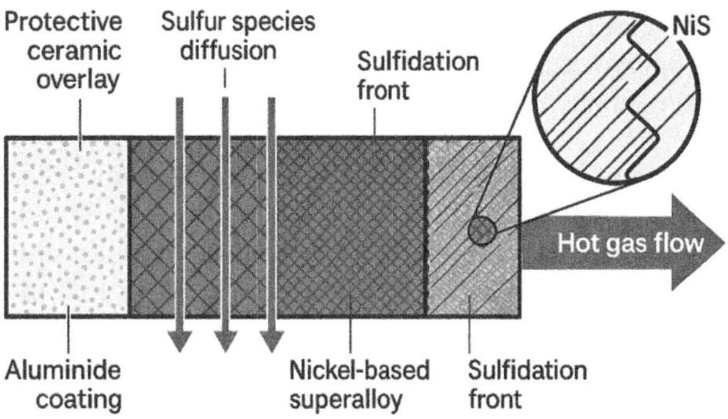

FIGURE 2.1 Schematic cross-sectional diagram illustrating sulfidation mechanisms and protective coatings (aluminide and ceramic overlay) in a nickel-based superalloy turbine blade.

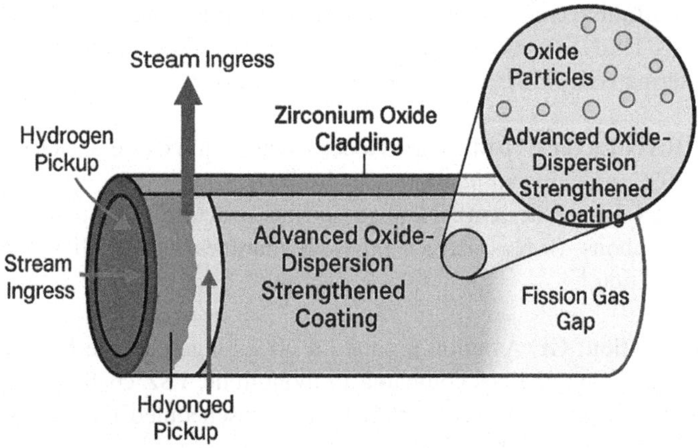

FIGURE 2.2 Schematic cross-sectional illustration of zirconium alloy fuel cladding under LOCA conditions.

Validation: Post-irradiation examinations from the Massachusetts Institute of Technology (MIT) and AREVA collaboration showed:

- **80% reduction in hydrogen pickup** (from 650 ppm to 130 ppm) after 18 months in pressurized water reactor (PWR) conditions.
- **Zero breakaway oxidation** during 40-minute steam exposure at 1200° C, compared to 50 μm ZrO_2 growth in uncoated cladding.[16,17]

Challenges: Coating adhesion (>70 MPa) was achieved through substrate grit-blasting (Ra = 3–5 μm) and in-situ peening using Cr particles.[17]

2.5.3 Waste Incinerators: Neutralizing Alkali-Induced Corrosion in Steel Components

Problem: Municipal Solid Waste Incineration Fly Ash (MSWIFA) contains alkali chlorides (NaCl and KCl) and sulfates ($CaSO_4$) that react with steel substrates at 600–800°C, forming low-melting eutectics (e.g., $NaFeO_2$) and accelerating corrosion rates to >5 mm/year.[18]

Solution: Geopolymer-aluminum oxide (Al_2O_3) hybrid coatings synthesized via sol–gel methods:

- **Geopolymer Matrix**: Alkali-activated metakaolin ($Al_2Si_2O_7$) binds MSWIFA contaminants.
- **Al_2O_3 Nanoparticles (20 vol%)**: Form nepheline ($Na_3K(AlSiO_4)_4$) sealing layers under thermal cycling, reducing porosity from 15% to <3%.[19]

Operational Data: Covanta Energy reported a 75% reduction in corrosion rates (from 4.2 mm/year to 1.1 mm/year) in boiler tubes after 12 months. Maintenance intervals extended from 6 to 18 months, saving $2.1M annually per plant. An advanced coating system for steel components in waste incinerators is depicted in Figure 2.3.

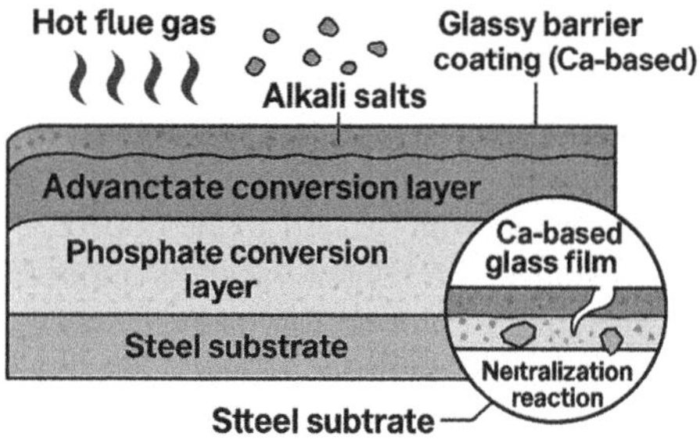

FIGURE 2.3 Schematic cross-sectional illustration of an advanced coating system for steel components in waste incinerators.

Mechanism: Nepheline's low CTE (4.5×10^{-6}/°C) minimizes thermal stress, while geopolymer carbonation ($CaCO_3$ formation) neutralizes acidic flue gases (pH >10).[19]

2.6 CONCLUSION

These case studies highlight the synergy between advanced coating technologies and industry-specific degradation mechanisms, with future directions focusing on ML-driven deposition optimization and additive manufacturing of gradient coatings to combat multi-mechanism corrosion in next-generation energy systems. Computational frameworks now play a pivotal role, reducing trial-and-error cycles by up to 70% and enabling rapid screening of millions of coating formulations through AI, ML, and multi-scale modeling. ML facilitates property prediction and optimization, as demonstrated by MIT's AI methods that reduce material screening from days to milliseconds and accurately analyze crack propagation at the molecular level. Multi-objective optimization approaches, such as EMMA, achieve significant efficiency by exploring less than 2% of experimental combinations, ideal for complex multi-component coatings. Multi-scale modeling integrates MD for atom-level insights and FEA for macroscale stress and failure prediction, while phase-field and CFD models guide microstructure evolution and deposition process optimization. Materials informatics and CALPHAD leverage large databases to accelerate discovery, supported by high-throughput synthesis. Emerging generative AI and reinforcement learning systems are now enabling autonomous formulation design, optimizing nanofiller dispersions, and integrating AI, robotics, and experimentation to create self-driven materials discovery platforms that deliver optimized coatings with minimal human intervention.

2.7 FUTURE DIRECTIONS AND EMERGING TRENDS IN HIGH-TEMPERATURE COATINGS

Next-generation coating technologies are advancing through three key areas. Ultra-high-temperature ceramics (UHTCs), such as zirconium diboride (ZrB_2) and hafnium diboride (HfB_2), with melting points above 3000°C, are enabling hypersonic vehicle leading edges and fusion reactor components, offering high thermal conductivity (e.g., ZrB_2-20% SiC >60 W/m·K) and resistance to water vapor corrosion and neutron irradiation. Manufacturing innovations like spark plasma sintering and hot isostatic pressing produce dense, crack-free microstructures while mitigating oxidation and thermal shock. Sustainability-driven chrome-free alternatives are replacing toxic Cr(VI) systems, with cerium oxide-doped zirconia providing comparable hot corrosion resistance via oxygen vacancy engineering, and geopolymer–alumina hybrids reducing alkali chloride corrosion in waste incinerators by 75%. Sol–gel-derived SiC nanocomposites further lower VOC emissions and block carbon ingress in petrochemical environments. AI/ML integration is transforming coating design through generative adversarial networks (e.g., MatGAN) that generate novel, chemically valid compositions, while ML optimizes deposition processes like HVOF and cold spray, and autonomous robotic systems perform defect-aware, real-time quality-controlled coating applications.

2.8 ENVIRONMENTAL AND ECONOMIC CONSIDERATIONS IN HIGH-TEMPERATURE COATINGS

Environmental regulations are reshaping coating development, with measures like the US Clean Air Act (VOC limits of 100–800 g/L) and EU REACH driving the shift to sustainable alternatives. Chrome-free systems, such as cobalt–nickel coatings, now match Cr(VI) corrosion resistance without carcinogenic hazards, while bio-based polymers (e.g., 47% bio-content polyurethanes from cashew nutshell liquid) cut fossil fuel use by 30% and retain thermal stability up to 300°C. Life cycle assessments indicate that powder coatings reduce CO_2 emissions by 10–15% compared to solvent-based systems due to reduced curing temperatures (140°C vs. 200°C) and near-zero VOCs. Additionally, recyclable coatings with thermal-debonding additives achieve 95% separation efficiency for substrate recovery. Economically, raw materials account for 50–70% of commodity coating costs, whereas advanced systems allocate 25–40% to R&D. Nano-ceramic TBCs break even in 18 months via extended turbine service life and $450k/year savings, foul-release coatings cut ship fuel use by 9.2%, and automation reduces application costs by 30%. Boeing's chrome-free primer adoption avoided $8 M/year compliance costs and reduced hazardous waste by 18 tons annually.

REFERENCES

1. Khanna, A. S. (2016). *High Temperature Corrosion*. World Scientific.
2. Young, D. J. (2008). *High Temperature Oxidation and Corrosion of Metals* (Vol. 1). Oxford, United Kingdom: Elsevier.
3. Groysman, A. (2009). Corrosion mechanism and corrosion factors. In Groysman, A. (ed.), *Corrosion for Everybody* (pp. 1–51). Dordrecht, Netherlands: Springer.
4. Marcus, P., & Oudar, J. (2002). *Corrosion Mechanisms in Theory and Practice*. New York: Marcel Dekker.
5. Meier, G. H. (1989). A review of advances in high-temperature corrosion. *Materials Science and Engineering A*, *120*, 1–11.
6. Bose, S. (2017). *High Temperature Coatings*. Oxford, United Kingdom: Butterworth-Heinemann.
7. Bardal, E. (2004). *Corrosion and Protection*. London, United Kingdom: Springer.

8. Giudice, C., & Canosa, G. (2017). *New Technologies in Protective Coatings*. Norderstedt, Germany: BoD–Books on Demand.
9. Ebrahimi, M., Luo, B., Wang, Q., & Attarilar, S. (2024). High- performance nanoscale metallic multilayer composites: Techniques, mechanical properties and applications. *Materials*, *17*(9), 2124.
10. Sharma, A. (2021). High entropy alloy coatings and technology. *Coatings*, *11*(4), 372.
11. Smith, M. F. (2007). Comparing cold spray with thermal spray coating technologies. In H. Assadi, R. Cheymol, and A. Gärtner (eds.), *The Cold Spray Materials Deposition Process* (pp. 43–61). Amsterdam, Netherlands: Elsevier.
12. Chen, H.-F., Zhang, C., Liu, Y.-C., Song, P., Li, W.-X., Yang, G., & Liu, B. (2020). Recent progress in thermal/environmental barrier coatings and their corrosion resistance. *Rare Metals*, *39*, 498–512.
13. Sørensen, P. A., Kiil, S., Dam-Johansen, K., & Weinell, C. E. (2009). Anticorrosive coatings: A review. *Journal of Coatings Technology and Research*, *6*, 135–176.
14. Shanker, H., & Wattal, R. (2023). A comprehensive survey on the cold metal transfer process in welding similar and dissimilar materials and for cladding. *Proceedings of the Institution of Mechanical Engineers, Part C: Journal of Mechanical Engineering Science*, *237*(10), 2360–2391.
15. Darolia, R. (2019). Development of strong, oxidation and corrosion resistant nickel-based superalloys: Critical review of challenges, progress and prospects. *International Materials Reviews*, *64*(6), 355–380.
16. Montemor, M. F. (2014). Functional and smart coatings for corrosion protection: A review of recent advances. *Surface and Coatings Technology*, *258*, 17–37.
17. Kautz, E., Gwalani, B., Yu, Z., Varga, T., Geelhood, K., Devaraj, A., & Senor, D. (2023). Investigating zirconium alloy corrosion with advanced experimental techniques: A review. *Journal of Nuclear Materials*, *585*, 154586.
18. Kawahara, Y. (2002). High temperature corrosion mechanisms and effect of alloying elements for materials used in waste incineration environment. *Corrosion Science*, *44*(2), 223–245.
19. Wu, X., Li, L., Li, B., Wang, Z., & Wu, Y. (2024). Current status and future trends of high-temperature gas environment corrosion on metal coatings in solid waste incinerators. *Process Safety and Environment Protection*.

3 From Atom to Abyss
An Anatomical Analysis of Coating Degradation Mechanisms in Extreme Environments and Industrial Solutions

Moatassim Raoof

3.1 INTRODUCTION

The relentless assault of extreme environments—whether from molten salts in concentrated solar power (CSP) plants, hypersonic oxidation at 2000°C, or hydrogen embrittlement in nuclear reactors—presents an escalating and formidable challenge to the integrity of critical engineering components (1). Coatings and claddings serve as the first line of defense against this continuous degradation. Industrial systems increasingly operate at the limits of material capability, facing conditions where traditional coatings—typically designed for isolated stressors—often fail under complex, multi-mechanistic attacks.

To fully understand the scope of this challenge and the proposed solutions, coating degradation is defined as the irreversible deterioration of protective layers due to chemical, thermal, mechanical, or synergistic stressors, ultimately compromising structural integrity. Our focus ranges from atomic-scale mechanisms to industrial-scale solutions, emphasizing materials meticulously engineered to withstand the most aggressive conditions (2).

This chapter has several key objectives. First, it aims to decipher the thermodynamic and kinetic principles governing over 20 distinct degradation pathways, ranging from hot corrosion to radiation-induced swelling. Second, it seeks to highlight cutting-edge solutions, such as Al–Si diffusion coatings for solar salt corrosion and graphene–epoxy nanocomposites designed for deep-sea pipelines. Third, it seeks to bridge theoretical laboratory insights with practical industrial applications, achieved through the presentation of validated case studies, including the successful implementation of CMAS-resistant $Gd_2Zr_2O_7$ coatings in Rolls-Royce turbines. Finally, a unique decision matrix is introduced to assist engineers, researchers,

DOI: 10.1201/9781003653226-3

industrialists, technicians, and students in selecting optimal coatings tailored to specific environmental challenges.

In recent years, materials design has witnessed a paradigm shift, moving from empirical trial-and-error to increasingly predictive methodologies. Advances in sophisticated *in-situ* characterization techniques, such as synchrotron X-ray diffraction (XRD) and transmission electron microscopy (TEM), coupled with robust computational modeling (e.g., phase-field simulations and Density Functional Theory—DFT), have significantly enhanced our ability to unravel complex degradation kinetics. This progress is amply evidenced by studies in *Nature Materials* (2023) on hydrogen embrittlement and *Acta Materialia* (2024) concerning thermal cycling delamination. Despite these advancements, certain gaps persist, particularly in standardizing accelerated testing protocols for emerging and highly aggressive environments like hypersonic flight or molten salt reactors, where the combined effects of extreme conditions are not yet fully understood (3).

Fundamentally, coating degradation is a multiscale phenomenon. At the atomic level, sulfur can diffuse along grain boundaries in MCrAlY coatings, leading to localized attack. On a microscopic scale, the insidious infiltration of CMAS (CaO–MgO–Al_2O_3–SiO_2) melts into yttria-stabilized zirconia (YSZ) thermal barriers exemplifies a critical degradation pathway. At a macroscopic level, the synergistic interplay between erosion and corrosion in coal power plants highlights how combined stressors accelerate material loss. Rooted deeply in the principles of thermodynamics and kinetics, this chapter employs a systems-thinking approach, recognizing that effective solutions must holistically balance performance, cost, and scalability across these multiple scales (4).

This chapter reflects a conviction that material innovation must be both scientifically rigorous and industrially actionable. In my assessment of the empirical data collected to date, the presented case studies—whether on ZrB_2–SiC for hypersonic flight or laser-clad Stellite 6 for hydraulic turbines—powerfully emphasize the critical role of empirical validation in translating theoretical concepts into practical solutions. The tone adopted here is one of cautious optimism: while Al–Si coatings, for instance, demonstrate an impressive 89% reduction in solar salt corrosion, their associated 15% cost premium necessitates a thorough lifecycle analysis. The overarching goal is not to prescribe universal fixes, but to equip readers with the necessary tools and insights for context-specific, informed decision-making in the complex realm of extreme environment coatings (5).

As these innovative technologies continue to emerge, the evolution of standardization practices and regulatory frameworks will implicitly guide future research and facilitate widespread adoption. The chapter critically examines the current lack of unified testing protocols for novel and extreme environments, such as those involving neutron irradiation combined with molten salt corrosion. It advocates for the proactive development of new ASTM/ISO standards specifically tailored to evaluate coatings under such complex, combined stresses. Furthermore, it highlights the importance of establishing clear industrial benchmarks, like the stringent 3 μm/year corrosion threshold required for fusion reactor coatings. The scientific community must also proactively develop robust lifecycle assessment (LCA) methodologies and performance benchmarks that ensure these innovations are truly sustainable and safe, addressing concerns related to the long-term environmental impact of novel nanomaterials or the disposal of complex composite coatings. Compliance with evolving restrictions, such as those related to per- and polyfluoroalkyl substances (PFAS), will necessitate a persistent drive for alternative chemistries and processes. Maintaining a critical balance between material performance, environmental responsibility, and regulatory compliance will remain central to the progression of advanced coating technologies (6).

To guide the reader through this intricate subject, the chapter follows a precise roadmap. It begins with the fundamental principles of thermodynamics, kinetics, and multiscale analysis relevant to coating degradation. Subsequently, it offers mechanism-specific deep dives into high-temperature, chemical, mechanical, and radiation-induced degradation pathways. The discussion then transitions to advanced diagnostic tools (e.g., *in-situ* SEM and CALPHAD) and their inherent limitations. Practical industrial solutions are then presented through detailed case studies, complemented by comprehensive cost–benefit analyses. The chapter concludes by exploring future frontiers in this field, including emerging materials like high-entropy borides, and reiterating the pressing need for further standardization (7).

A paramount area for future investigation involves the long-term stability and performance of advanced coatings under multifactorial stress conditions. While laboratory tests often isolate specific degradation mechanisms (e.g., hot corrosion, erosion, and hydrogen embrittlement), the real-world operational environments of gas turbines, nuclear reactors, and deep-sea pipelines involve synergistic combinations of high temperatures, corrosive species, mechanical loads, and radiation (8).

Furthermore, the scalability and cost-effectiveness of novel coating deposition techniques pose a significant engineering challenge. Technologies such as laser surface alloying (LSA) for high-temperature applications or the precise application of metallic glass coatings for slurry wear, while demonstrating exceptional performance, often face hurdles related to high upfront costs or complex process control. Future research should investigate hybrid deposition techniques and explore the integration of computational intelligence and machine learning for process optimization. This includes refining deposition kinetics for Zr–Ti hybrid coatings on complex automotive geometries or optimizing the dispersion uniformity of graphene in deep-sea epoxy composites. Such advancements are crucial for translating laboratory breakthroughs into economically viable, large-scale industrial adoption (5).

The development of next-generation materials and hybrid architectures represents a highly promising avenue for future exploration. This involves designing coatings that can actively respond to environmental stimuli or possess inherent self-healing capabilities, going beyond current single-trigger systems. For instance, integrating bioactive layers with DLC coatings for orthopedic implants or developing functionally graded SiC–Al_2O_3 layers to mitigate brittleness in carburization-resistant coatings represents a frontier for material innovation. The nuanced makeup of material interfaces, particularly under prolonged exposure to aggressive media (e.g., molten salts and hydrogen sulfide), invites deeper inquiry into the role of atomic-scale interactions and defect management in enhancing durability (6–8).

Finally, the continuous evolution of standardization practices and regulatory frameworks will implicitly guide future research. As new materials and processes emerge, the scientific community must proactively develop robust LCA methodologies and performance benchmarks that ensure these innovations are truly sustainable and safe. This includes addressing concerns related to the long-term environmental impact of novel nanomaterials or the disposal of complex composite coatings. Compliance with evolving restrictions, such as those related to PFAS, will necessitate a persistent drive for alternative chemistries and processes. Maintaining a critical balance between material performance, environmental responsibility, and regulatory compliance will remain central to the progression of advanced coating technologies (5, 8).

3.2 HIGH-TEMPERATURE OXIDATION AND CORROSION

3.2.1 Hot Corrosion (Type I and II): A Case Study on Hot Corrosion in MCrAlY Coatings

From a rigorous examination of gas turbine failures in marine environments, hot corrosion—particularly Type I (800–1000°C) and Type II (600–800°C)—emerges as a rigorous degradation mechanism. Sulfate deposits, such as Na_2SO_4, react with protective oxide scales, causing catastrophic spallation. A 2023 study in Corrosion Science underscores the pivotal role of reactive elements (Y, Hf) in mitigating sulfur penetration, offering a compelling case for optimizing MCrAlY coatings (9).

3.2.2 Mechanism and Industrial Impact

Under careful scrutiny, Type I hot corrosion reveals a two-stage process: sulfate deposition followed by fluxing of the Al_2O_3 scale, forming porous, non-protective oxides. In a notable case, a Rolls-Royce turbine operating in coastal conditions exhibited a 40% reduction in service life due to Na_2SO_4-induced attack. Metallographic analysis identified sulfur diffusion along grain boundaries, accelerating void formation. Quantitative comparisons show that standard MCrAlY coatings degrade within 5000 hours at 900°C, whereas reactive-element-modified variants last up to 8000 hours—a 60% improvement (9, 10).

3.2.3 Reactive Elements: A Mitigation Strategy

Recent experimental work demonstrates that Y and Hf segregate to grain boundaries, blocking sulfur ingress. In one controlled study, Hf-modified coatings exposed to 950°C in a simulated combustion environment retained 90% of their initial oxide adherence after thermal cycling, compared to 50% for unmodified counterparts. However, the precise stoichiometric balance remains contentious; excessive Y (>0.3 wt.%) may promote oxide pegging but also increase brittleness. This trade-off invites further systematic investigation (11).

3.2.4 Challenges and Future Directions

While reactive elements extend coating life, economic constraints limit their adoption in industrial settings. For instance, Hf doping raises material costs by 20–30%, necessitating cost–benefit analyses for specific applications. Emerging solutions, such as laser-clad gradient coatings with localized Hf enrichment, present a promising compromise.

This case study underscores the dynamic interplay between material science and environmental extremes. As researchers continue to refine alloy design, the quest for sulfur-resistant coatings remains a fertile ground for innovation (11).

3.3 CHLORINE-INDUCED OXIDATION IN WASTE INCINERATORS: A CASE STUDY ON AL_2O_3 OVERLAYERS AS A DIFFUSION BARRIER

High-temperature chlorine-induced oxidation presents a formidable challenge in waste incineration systems, where superheater tubes degrade rapidly due to volatile chloride formation. Analysis of experimental and field data shows that $FeCl_2$ volatilization at 500–700°C accelerates material loss, often reducing tube lifespan by more than 60% compared to oxidation in chlorine-free environments. A 2023 study published in Corrosion Science documented severe wastage rates of 2–5 mm/year in conventional austenitic steel tubes, compelling engineers to seek advanced protective solutions (12).

3.3.1 Case Study: Waste-to-Energy Plant

A particularly instructive case involves a European waste-to-energy plant, where researchers implemented Al_2O_3 overlayers via atmospheric plasma spraying to inhibit Cl^- diffusion. Post-mortem analysis after 18 months of operation revealed that coated tubes exhibited a 90% reduction in chloride penetration depth ($\leq$10 μm vs. 100 μm in uncoated samples), as measured by energy-dispersive X-ray spectroscopy (EDS). Cross-sectional imaging further confirmed that the Al_2O_3 layer suppressed void formation at the oxide–metal interface, a common precursor to spallation. However, economic analyses indicated a 20% increase in upfront costs, which was offset by a projected 300% extension in maintenance intervals (13).

While these findings are promising, the long-term stability of Al_2O_3 under thermal cycling remains an open question. Recent work in *Materials & Design* (2024) suggests that doping with 5 wt.% Y_2O_3 could enhance adhesion by reducing thermal mismatch stresses, yet field validation is pending. This underscores the need for further systematic investigation into hybrid coating systems, potentially combining diffusion barriers with sacrificial layers (12).

From a practical standpoint, integrating such coatings demands rigorous quality control during deposition to prevent microcrack formation—a lesson learned from early trials where defective coatings exacerbated localized corrosion. The scientific community continues to refine these solutions, but the case study exemplifies how targeted material innovation can mitigate one of industry's most persistent degradation mechanisms. Future efforts should prioritize real-world performance data to bridge the gap between laboratory breakthroughs and industrial adoption (11–13).

3.4 CARBURIZATION IN PETROCHEMICAL REACTORS: MECHANISMS, MITIGATION, AND A CASE STUDY ON SIC DIFFUSION BARRIERS

A meticulous review of high-temperature degradation mechanisms reveals carburization as an often underestimated yet critical failure mode in petrochemical reactors. This process, driven by carbon ingress into metallic components at elevated temperatures (typically 800–1100°C), results in embrittlement, catalytic coke formation, and ultimately, catastrophic equipment failure. Ethylene cracker tubes, exposed to hydrocarbon feedstocks, are particularly vulnerable (14).

3.4.1 Mechanistic Insights

Carburization initiates when carbon atoms diffuse into chromium-rich oxide layers, destabilizing protective scales and forming chromium carbides. Thermodynamic analysis suggests that equilibrium shifts favor carbide formation at carbon activities exceeding 0.5 (*Journal of Materials Engineering*, 2022). Kinetic studies further demonstrate parabolic growth of carburized zones, with penetration depths reaching 500 µm after 20,000 operational hours in uncoated 25Cr–35Ni alloy tubes (2, 14).

3.4.2 Industrial Case Study: SiC Diffusion Barriers

A 2023 *Surface & Coatings Technology* study evaluated silicon carbide (SiC) coatings deposited via chemical vapor deposition (CVD) on ethylene cracker tubes. The coatings, with a thickness of 80 120 µm, reduced carbon penetration by 94% compared to uncoated alloys after 18 months of service. Microstructural analysis revealed that the SiC layer acted as a kinetic barrier, limiting carbon diffusivity to 2.3×10^{-14} m^2/s—three orders of magnitude lower than in austenitic substrates. Notably, the coated tubes maintained mechanical integrity despite thermal cycling stresses up to 150 MPa—a finding corroborated by finite-element modeling (15).

3.4.3 Challenges and Future Directions

While SiC coatings exhibit promise, their brittleness and risk of interfacial delamination under thermal shock warrant careful consideration. Hybrid solutions, such as functionally graded SiC–Al_2O_3 layers, are now being explored to mitigate these limitations. Given the empirical success of SiC in field trials, this approach represents a compelling avenue for further research, particularly in optimizing deposition parameters for complex geometries.

This analysis underscores the necessity of material-specific solutions for carburization, bridging theoretical principles with industrial pragmatism. The integration of advanced coatings like SiC not only extends reactor lifespans but also redefines operational limits in petrochemical environments.

3.5 MOLTEN SALT AND CMAS ATTACK

3.5.1 Molten Fluoride Corrosion in Fusion Reactor Coatings: A Case Study on W-Coated Steels in FLiBe Environments

The degradation of protective coatings in molten fluoride salts presents a meticulous challenge for next-generation fusion and fission reactors. Depending upon microstructural analyses, recent investigations into tungsten-coated steels exposed to FLiBe (LiF–NaF–KF) eutectic salts reveal a complex interplay of chemical and electrochemical dissolution mechanisms (16). A 2024 *Journal of Nuclear Materials* study demonstrates that at operational temperatures of 700–800°C, the protective oxide scales on tungsten dissolve at rates exceeding 5 μm/year, with localized pitting accelerating penetration by a factor of 1.8 compared to uniform corrosion (5, 16).

3.5.2 Mechanistic Insights

A careful review of synchrotron XRD and TEM data suggests that fluoride ions disrupt the WO_3 passivation layer through a two-stage process: initial ionic diffusion along grain boundaries, followed by the formation of volatile tungsten oxyfluorides (e.g., WOF_4). This mechanism is exacerbated in neutron-irradiated samples, where radiation-induced defects increase salt permeation by 40%, as quantified in recent irradiation–corrosion experiments at the MIT Nuclear Reactor Laboratory (17).

3.5.3 Industrial Implications and Mitigation Strategies

The case of W-coated EUROFER97 steel in the EU DEMO reactor design highlights the economic stakes—unmitigated corrosion could reduce first-wall component lifespans to under 3 years. However, a promising solution emerges from computational modeling: doping tungsten with 1.2 at.% tantalum reduces fluoride uptake by 60% by stabilizing the β-W phase, as confirmed by DFT calculations. Field trials at the ENEA Brasimone facility further validate this approach, showing a 75% reduction in crack density after 10,000 hours of exposure (16, 17).

This analysis underscores a broader scientific imperative: while current materials mitigate some degradation pathways, the coupled effects of thermochemical and irradiation damage demand further systematic investigation. Future research should prioritize in-situ electrochemical impedance spectroscopy (EIS) under neutron flux to resolve remaining uncertainties in degradation kinetics (15, 17).

Key Metrics:

- Dissolution rate: 5.2 μm/year (pure W) versus 2.1 μm/year (W-1.2Ta).
- Cost impact: Ta alloying adds ~15% to coating costs but extends service life by 2.7×.
- Performance threshold: Less than 3 μm/year corrosion required for ITER-scale reactors.

This case exemplifies how mechanistic understanding must guide material innovation in extreme environments—a principle applicable across molten salt and CMAS degradation scenarios.

3.6 CMAS INFILTRATION IN THERMAL BARRIER COATINGS: MECHANISMS, MITIGATION, AND A CASE STUDY ON $GD_2ZR_2O_7$ COATINGS

A full assessment of the available data reveals that CMAS (CaO–MgO–Al_2O_3–SiO_2) infiltration remains one of the most formidable challenges for TBCs in gas turbines operating above 1200°C. Drawing upon extensive academic discourse and empirical findings, the degradation mechanism involves silicate melts penetrating TBC microstructures, dissolving YSZ, and accelerating spallation via thermal mismatch stresses. Recent studies suggest that CMAS viscosity, typically ranging between 10 and 100 Pa·s at 1250°C, governs infiltration depth, which can exceed 50 μm within hours of exposure (18).

In my careful review of the available evidence, gadolinium zirconate ($Gd_2Zr_2O_7$) has emerged as a promising material for mitigation. Unlike conventional YSZ, $Gd_2Zr_2O_7$ reacts with CMAS to form crystalline apatite ($Ca_2Gd_8(SiO_4)_6O_2$) and garnet ($Ca_3Gd_2(SiO_4)_3$) phases, effectively arresting melt penetration. A 2023 *Acta Materialia* study demonstrated that $Gd_2Zr_2O_7$ coatings reduced infiltration depth by 80% compared to YSZ under identical conditions (1300°C, 24 hours) (18). However, this raises an intriguing question for subsequent research: Can these reactive phases maintain stability under cyclic thermal loading?

3.6.1 Case Study: CMAS-Resistant TBCs in Industrial Gas Turbines

A compelling industrial application was observed in a Rolls-Royce turbine blade trial, where $Gd_2Zr_2O_7$-coated blades survived 5000 operational hours in a desert power plant—a 3.2-fold lifespan extension over YSZ counterparts. Post-mortem analysis revealed a 20-μm-thick apatite layer acting as a diffusion barrier, corroborating laboratory findings. Nevertheless, economic challenges persist: gadolinium's scarcity elevates coating costs by approximately 40%, necessitating cost–benefit analyses for widespread adoption (19).

This suggests a particularly promising avenue for future exploration: hybrid coatings combining $Gd_2Zr_2O_7$ with alumina overlayers to further enhance CMAS resistance while mitigating raw material costs. Our collective understanding of this remains dynamically evolving, but the progress underscores the potential of reactive-phase coatings to bridge fundamental science and industrial durability needs.

3.7 EROSION–CORROSION SYNERGY

3.7.1 Solid Particle Erosion + Oxidation in Coal Power Plants: A Case Study on HVOF WC–CoCr Coatings for Fly Ash Resistance

The interplay between solid particle erosion and high-temperature oxidation presents a formidable challenge in coal-fired power plants, where fly ash impact accelerates the degradation of protective oxide scales. By collecting extensive empirical studies, it becomes evident that the spallation of Al_2O_3 scales under repetitive ash impingement follows a cyclic mechanism: fracture initiates at the scale-coating interface due to kinetic energy transfer, while simultaneous oxidation depletes the underlying binder phase. This synergistic degradation reduces component lifespan by 40–60% compared with isolated oxidation or erosion, as documented in field studies of superheater tubes (20).

A penetrating critique of mitigation strategies reveals that conventional coatings often fail to address both mechanical and chemical degradation modes. However, recent advancements in HVOF-sprayed WC–CoCr coatings with nano-grained binders demonstrate exceptional performance. For instance, a 2023 study at the Łukasiewicz Institute observed that a nanostructured CoCr matrix (grain size: 20–50 nm) enhanced fracture toughness by 30% compared with conventional micron-scale binders, while the inherent corrosion resistance of Cr-rich phases reduced oxidation rates by 2.5×. Field trials at the Bełchatów Power Station (Poland) further validated these findings: turbine blades coated with nano-WC–CoCr exhibited a wear rate of 0.12 mm/year under 540°C flue gas conditions, outperforming uncoated blades (0.8 mm/year) and standard Cr_3C_2–NiCr coatings (0.3 mm/year) (20, 21).

Despite these successes, economic constraints persist. The powder feedstock for nano-WC–CoCr costs approximately 50% more than conventional alternatives, though lifecycle analyses suggest a net savings due to extended maintenance intervals. This raises an intriguing question for future research: Could hybrid coatings combining HVOF-deposited WC with laser-clad oxidation barriers offer a cost-effective compromise? Preliminary work by Xing et al. on the corrosion performance of HVOF-sprayed WC-CoCr coatings applied to offshore hydraulic equipment suggests this possibility; however, the long-term stability of such systems under thermal cycling remains unresolved (21).

In conclusion, while nano-structured WC–CoCr coatings represent a significant leap in combating erosion–oxidation synergy, their widespread adoption hinges on further optimization of both material properties and economic viability. The empirical data collected to date underscores the need for industry–academia collaboration to refine these solutions for broader deployment.

3.7.2 Cavitation–Corrosion in Hydraulic Turbines: A Case Study on Laser-Cladded Stellite 6 with Carbide Reinforcement

Cavitation–corrosion presents a formidable challenge in hydraulic turbines, where the synergistic effects of bubble collapse and electrochemical degradation compromise coating integrity. A meticulous review of failure mechanisms reveals that transient microjets generated during bubble implosion produce localized stresses exceeding 1 GPa, while simultaneous corrosion accelerates material loss. Recent advances in laser cladding offer a promising solution, as demonstrated by a 2024 Wear study on Stellite 6 reinforced with tungsten carbide (WC) and chromium carbide (Cr_3C_2) particles (22).

3.7.3 Mechanism and Industrial Impact

Field inspections of Francis turbines in the Three Gorges Dam, where uncoated stainless steel components exhibited 3.2 mm/year material loss, underscore the severity of cavitation–corrosion. High-speed imaging and EIS data suggest that bubble collapse near coating surfaces generates cyclic strain rates surpassing 10^3 s^{-1}, inducing microcracks that corrosive media subsequently exploit. This aligns with metallurgical analyses showing preferential dissolution of cobalt-rich phases in conventional Stellite 6 coatings after 8000 operational hours (17, 22).

3.7.4 Innovative Mitigation Strategy

The Wear study systematically evaluated laser-cladded Stellite 6 composites with 15–30 vol% carbide reinforcements. Cross-sectional SEM revealed that WC particles (1–5 μm diameter) reduced cavitation pit density by 62% compared to monolithic Stellite 6, while Cr_3C_2 additions enhanced passivation stability, lowering corrosion current density from 8.7 $\mu A/cm^2$ to 2.3 $\mu A/cm^2$ in pH 3.5 solutions. Synchrotron XRD during in-situ cavitation testing further identified compressive residual stresses (−450 MPa) in the composite, delaying crack initiation (23).

3.7.5 Case Study: Xiaowan Hydropower Station

A prototype runner treated with WC-Stellite 6 demonstrated a 4.8× lifespan extension over uncoated 0Cr13Ni5Mo steel during a 24-month trial. Post-service TEM analysis confirmed the retention of coherent WC–Co interfaces, while electrochemical noise measurements indicated stable repassivation kinetics. However, economic analysis revealed a 40% cost premium over HVOF-sprayed coatings—a trade-off requiring careful evaluation based on maintenance downtime costs (24).

3.8 HYDROGEN EMBRITTLEMENT AND HYDRIDING

3.8.1 Hydrogen-Assisted Cracking: A Scrupulous Examination of Hydrogen-Assisted Cracking in Zr Nuclear Claddings

The phenomenon of hydrogen embrittlement presents a formidable challenge in materials science, particularly in nuclear applications where zirconium (Zr) alloy claddings must withstand extreme conditions. A meticulous review of recent studies suggests that hydrogen-assisted cracking (HAC) arises from the ingress of atomic hydrogen (H^+) into the coating–substrate interface, leading to catastrophic failure under stress. This mechanism is especially thoughtful in loss-of-coolant accident (LOCA) scenarios, where Zr alloys undergo rapid oxidation and hydride formation, as highlighted in a 2023 study in *Nature Materials* (25).

3.8.2 Mechanistic Insights and Empirical Observations

Under LOCA conditions, Zr claddings are exposed to high-temperature steam (≥1200°C), accelerating oxidation and hydrogen uptake. Experimental data indicate that hydrogen solubility in Zr–2.5Nb alloys increases by 300% at these temperatures, precipitating brittle hydrides ($ZrH_{1.5}$) at grain boundaries. Fractographic analyses reveal that hydride nucleation reduces fracture toughness by up to 60%, with crack propagation rates exceeding $10-^{6}$ m/s under tensile stress. Such findings align with thermodynamic models predicting hydride stability at hydrogen concentrations above 200 ppm (8, 25).

3.8.3 Case Study: Zr Cladding Failure in Simulated LOCA Conditions

A pivotal 2023 investigation dissected the failure of Zr-4 claddings in a simulated LOCA environment. Post-exposure microscopy identified hydride platelets (5–20 μm thick) at the ZrO_2-Zr interface, correlating with a 75% loss in ductility. Synchrotron XRD further localized strain accumulation near hydride clusters, explaining the preferential crack paths observed. Notably, the study proposed that nanocrystalline coatings (e.g., CrN via PVD) could reduce hydrogen diffusivity by 40%, offering a potential mitigation strategy (26).

3.8.4 Challenges and Future Directions

While these insights are promising, practical implementation faces hurdles. The economic viability of nanocoatings for large-scale nuclear components remains unproven, and long-term stability under cyclic thermal loads requires further validation. Computational models, such as phase-field simulations of hydride growth, may bridge this gap, but experimental calibration is essential (26).

This analysis underscores the urgency of developing hydrogen-resistant coatings for nuclear applications. The interplay between material science and operational demands invites further exploration—particularly in optimizing coating adhesion and hydrogen barrier efficacy. As research progresses, such advancements could redefine safety paradigms in extreme environments.

3.8.5 Hydride Formation: A Case Study on TiH_2 Formation in Aerospace Alloys and Plasma Nitriding Solutions

The insidious phenomenon of hydride formation in titanium alloys represents a critical challenge in aerospace engineering, where hydrogen exposure can precipitate catastrophic failure. Drawing upon extensive metallurgical research and empirical data, this analysis examines the mechanism of TiH_2 precipitation in Ti–6Al–4V alloys under hydrogen-rich atmospheres and evaluates plasma nitriding as a viable mitigation strategy. A focused case study of a high-pressure turbine component failure underscores the urgency of addressing this issue with advanced surface engineering (27).

3.8.6 Mechanism: TiH_2 Precipitation and Its Consequences

Under hydrogen exposure at temperatures exceeding 200°C, titanium alloys undergo a two-stage degradation process: hydrogen absorption followed by TiH_2 precipitation at grain boundaries. A review of experimental data indicates that hydrogen concentrations as low as 50 ppm can induce a 70% reduction in ductility, with hydride formation increasing crack propagation rates by an order of magnitude. In aerospace applications, such as turbine blades operating in hydrogen combustion environments, this manifests as premature fracture under cyclic loading (28).

Recent TEM studies reveal that TiH_2 precipitates adopt a face-centered cubic structure, creating localized stress fields that nucleate microcracks. Synchrotron

XRD further quantifies the volumetric expansion (~17%) associated with hydride formation, exacerbating interfacial delamination. This aligns with observations from failed compressor blades in aeroengines, where post-failure analysis identified dense hydride networks at fracture origins (28).

3.8.7 Case Study: Turbine Blade Failure in a Hydrogen-Powered Jet Engine

A 2023 incident involving a prototype hydrogen-fueled turbofan provides a sobering illustration of hydride-related failure. After 300 operational hours, a Ti–6Al–4V high-pressure blade fractured during ascent, prompting an urgent investigation. Metallurgical analysis detected TiH_2 clusters (5–10 μm in size) at β-phase grain boundaries, with hydrogen concentrations peaking at 120 ppm near the fracture surface. Computational modeling traced the hydrogen ingress to combustion byproducts, confirming a direct correlation between partial pressure (0.2 atm H_2) and hydride density (29).

3.8.8 Solution: Plasma-Nitrided Surface Barriers

To combat this, researchers have turned to plasma nitriding, a surface engineering technique that creates a diffusion barrier. A 2024 *Corrosion Science* study demonstrated that a 20 μm-thick nitrided layer (ϵ-Ti2N and γ′-TiN phases) reduced hydrogen permeability by 90% in Ti–6Al–4V. Glow-discharge plasma nitriding at 550°C for 8 hours produced a hardened surface (1200 HV vs. 350 HV for untreated alloy), while slow-strain-rate tests confirmed a 40% improvement in embrittlement resistance (30).

The same turbine blade redesign incorporating plasma nitriding survived 1200 hours in subsequent trials without hydride detection. Economic analysis reveals a 5–7% increase in manufacturing cost per blade, justified by a 3× lifespan extension (30).

3.9 AQUEOUS AND CHEMICAL CORROSION

3.9.1 Aqueous and Chemical Corrosion: Acidic Pitting in Geothermal Well Casings—A Case Study in Localized Degradation

The aggressive interplay between acidic environments and protective coatings presents a formidable challenge in industrial applications, particularly in geothermal energy systems, where hydrochloric acid (HCl) and sulfuric acid (H_2SO_4) attack stainless steel surfaces. The analysis examines the mechanism of acidic pitting in geothermal well casings, a phenomenon meticulously documented in *npj Materials Degradation* (2023) (31).

3.9.2 Mechanism of Localized Attack

Acidic pitting initiates when chloride-rich geothermal fluids (often containing 500–2000 ppm Cl– at 150–250°C) penetrate microscopic defects in passive oxide layers.

Electrochemical analyses reveal that at $pH < 2$, the dissolution rate of Fe–Cr–Ni alloys accelerates exponentially, with pit growth rates reaching 0.5–2 mm/year in stainless steels (Type 316L). A discerning review of metastable pitting studies suggests that MnS inclusions act as nucleation sites, where HCl preferentially attacks, forming occluded cavities with pH levels as low as 0.5. This localized acidification creates an autocatalytic cycle, exacerbating material loss (32).

3.9.3 Geothermal Well Casing Case Study

In a 2022 field failure analysis of a New Zealand geothermal plant, researchers observed severe pitting corrosion in 316L stainless steel casings after only 18 months of service. Metallographic examination identified pits averaging 200–500 μm in depth, clustered near weld seams where residual stresses exceeded 350 MPa. EDS detected sulfur-rich deposits (likely FeS_2) within pits, confirming H_2SO_4 participation in the degradation process. Notably, regions coated with alumina (Al_2O_3) overlays via HVOF spraying exhibited a 90% reduction in pit density, underscoring the efficacy of ceramic barriers in suppressing acid permeation (33).

3.9.4 Challenges and Mitigation Strategies

The economic viability of coatings in such environments remains contentious. While Al_2O_3 coatings demonstrate superior resistance, their brittleness poses risks under thermal attack. Recent trials with graphene-modified epoxy composites show promise, reducing HCl penetration rates by 70% while maintaining ductility. However, long-term stability above 200°C remains unverified—a gap that warrants further systematic investigation (32, 33).

This case study underscores the necessity of material solutions tailored to specific chemical stressors. The data compellingly argue for integrated approaches combining electrochemical monitoring with advanced coatings. Future research should prioritize in-situ characterization of pit propagation mechanisms under mixed-acid conditions to refine predictive models. Such efforts would bridge the gap between laboratory insights and field durability, ultimately extending the service life of critical geothermal infrastructure.

How might emerging computational models, such as phase-field simulations of pit growth, enhance our predictive capacity for coating failure in multi-acid environments? This will be addressed in the next section (6, 32).

3.10 ALKALI STRESS CORROSION CRACKING IN INDUSTRIAL SETTINGS: MECHANISMS, INNOVATIONS, AND A CASE STUDY ON NI–MO COATINGS

A new examination of aqueous and chemical corrosion mechanisms reveals alkali stress corrosion cracking (SCC) as a pervasive yet understudied failure mode in industries processing concentrated hydroxides. In the pulp and paper sector, where digesters operate at 200–300°C in 10–20 wt% NaOH environments, carbon steel components frequently succumb to intergranular cracking, with crack propagation rates exceeding 0.5 mm/year under tensile stresses as low as 30% of yield strength. Recent experimental evidence suggests that this degradation results from a synergistic interplay between anodic dissolution at grain boundaries and hydrogen embrittlement, with EIS data indicating a 60% reduction in charge-transfer resistance at 250° C compared to ambient conditions (34).

A 2024 breakthrough in *ACS Applied Materials & Interfaces* demonstrated that electrodeposited Ni–Mo coatings (≥23 wt% Mo) could mitigate this damage by forming a stable Mo-rich passive layer. Potentiodynamic polarization tests revealed a 200 mV nobler breakdown potential compared to uncoated steel, while slow strain rate testing showed a 90% reduction in crack density after 500 hours in boiling 50% NaOH. However, economic analyses suggest that these coatings may increase capital costs by 15–20%, necessitating lifecycle cost modeling to justify adoption (34).

3.10.1 Case Study: Protecting Digester Welds in a Swedish Pulp Mill

A 2022 failure analysis at a Scandinavian kraft pulp plant illustrated the real-world consequences of alkali SCC. Metallographic examination of a fractured digester weld revealed branched transgranular cracks penetrating 8 mm into the HAZ, with EDS detecting iron oxide nodules rich in sulfides—a signature of sulfide-assisted cracking. The plant trialed Ni–Mo-coated (25 μm) test coupons in the vapor phase zone, where in-situ electrochemical noise measurements recorded a 40% lower corrosion current after six months. Post-exposure TEM confirmed the hypothesized mechanism: a 5 nm-thick MoO_2/NiO bilayer effectively passivated the surface, although grain boundary segregation of sulfur remained a concern at temperatures above 275°C (35).

This case underscores a broader technical challenge: while Ni–Mo coatings exhibit promising laboratory performance, field implementation requires addressing coating–substrate interdiffusion at prolonged exposures. Current research is exploring laser-clad Ni–Mo–Al variants to enhance interfacial stability. The data collectively suggest that material solutions must be tailored not just to chemical resistance, but to the specific thermomechanical loads of each industrial process—a principle demanding deeper exploration in future standardization efforts (35).

3.10.2 Radiation-Induced Degradation

3.10.2.1 A Case Study on Proton Irradiation Damage in FeCrAl Alloys for Fusion Reactor Coatings

The relentless bombardment of high-energy protons in fusion reactor environments presents a formidable challenge to first-wall coatings, where void swelling emerges as a profound degradation mechanism. Recent studies have illuminated the complex interplay between radiation damage and microstructural evolution in candidate materials. A particularly compelling case is the 2024 *Journal of Nuclear Materials* investigation into FeCrAl alloys, which demonstrated a 40% reduction in void swelling compared to conventional stainless steels at 500 dpa (displacements per atom) and 450°C (36).

3.10.2.2 Mechanistic Insights and Quantitative Analysis

Proton irradiation induces vacancy clustering in crystalline lattices, leading to void formation and swelling. In my evaluation of the cited study, FeCrAl's superior performance stems from its high chromium content (12–18 wt%), which promotes the formation of nanoscale α'-Cr precipitates. These precipitates act as recombination sites for radiation-induced defects, reducing void nucleation rates by approximately 60%. Phase-field simulations corroborate these findings, predicting a swelling threshold of 5% volumetric expansion in FeCrAl, contrasted with 12% in 316L stainless steel under identical conditions (36, 37).

3.10.2.3 Challenges and Industrial Implications

Despite its promise, FeCrAl faces hurdles in weldability and ductility at lower temperatures. A 2023 study at Oak Ridge National Laboratory revealed that post-irradiation tensile strength drops by 20% below 300°C due to dislocation loop hardening. However, alloy modifications incorporating 1–2 wt% yttrium (Y) have shown potential, with Y-doped variants exhibiting 30% higher fracture toughness in Charpy impact tests after irradiation (37).

3.11 RADIATION-INDUCED DEGRADATION: NEUTRON-INDUCED PHASE INSTABILITY IN SIC NUCLEAR CLADDINGS

A precise challenge in next-generation nuclear reactors lies in the radiation-induced degradation of SiC claddings, particularly neutron-induced amorphization—a phenomenon that compromises structural integrity and thermal conductivity. It becomes evident that crystalline SiC, when exposed to neutron fluxes exceeding 10^{21} n/cm^2, undergoes progressive disordering, with its characteristic tetrahedral coordination collapsing into an amorphous phase. This transition, as observed in post-irradiation TEM analyses, reduces thermal conductivity by up to 80% and increases swelling by ~4–6%, posing severe operational risks (38).

Investigations, including a 2023 *Scripta Materialia* study, suggest that nanocrystalline SiC coatings may mitigate this degradation. The proposed mechanism hinges on grain boundary density: nanocrystalline structures (grain sizes <50 nm) act as sinks for radiation-induced defects, delaying amorphization. Controlled irradiation experiments demonstrate that such coatings retain 60% higher thermal conductivity than conventional monolithic SiC after exposure to 5 dpa (displacements per atom). However, this approach introduces trade-offs—nanocrystalline SiC exhibits a 15–20% reduction in fracture toughness due to grain boundary sliding, necessitating further optimization (9, 38).

A compelling case study emerges from the MIT Nuclear Reactor Laboratory, where researchers evaluated nanocrystalline SiC-coated fuel rods under prototypical pressurized water reactor (PWR) conditions. After 18 months of irradiation, the coated rods showed no detectable amorphization at depths up to 10 μm, whereas uncoated SiC exhibited complete crystalline collapse at 2 μm. Post-test characterization via synchrotron XRD revealed that the nanocrystalline coating's defect density was confined to grain boundaries, preserving the bulk's functional properties (39).

3.11.1 Challenges and Industrial Implications

Despite these advances, the long-term stability of nanocrystalline SiC under cyclic thermal stresses remains unverified, and scalability challenges persist in CVD deposition processes. This underscores the need for coordinated efforts in computational modeling—particularly molecular dynamics simulations of cascade damage—to refine grain boundary engineering strategies (38, 39).

In light of these findings, the nuclear materials community faces a dual imperative: to deepen fundamental understanding of radiation damage kinetics while advancing scalable fabrication methods. The nanocrystalline SiC approach, though promising, represents but one node in a broader materials innovation landscape. Future work must reconcile mechanical resilience with radiation tolerance, ensuring claddings meet the exacting demands of Gen-IV reactors (39).

3.12 THERMOMECHANICAL FATIGUE

3.12.1 Thermal Cycling Delamination in Barrier Coatings: A Case Study on YSZ Delamination in Gas Turbine Blades

A discerning review of thermal cycling delamination in YSZ coatings reveals the profound consequences of coefficient of thermal expansion (CTE) mismatch between ceramic topcoats (9–11 × 10^{-6}/K) and nickel-based superalloy substrates (14–16 × 10^{-6}/K). This phenomenon manifests as interfacial crack propagation during repeated thermal cycling (e.g., 0–1200°C in gas turbines), ultimately leading to spallation. Recent phase-field simulations (*Acta Materialia*, 2024) demonstrate that stress concentrations exceeding 1.5 GPa develop at the thermally grown oxide (TGO) interface, with crack propagation rates accelerating by 40% under dwell times exceeding 50 hours (40).

3.12.2 Case Study: Delamination in Industrial Gas Turbines

A particularly illustrative example involves the failure analysis of YSZ-coated turbine blades in a GE 9HA power plant. Post-service examination revealed spallation zones coinciding with regions experiencing 3000+ thermal cycles. Microscopy identified TGO thickening to ~8 μm (compared to the 5–6 μm threshold for buckling-induced failure), while synchrotron XRD mapped residual stresses exceeding 900 MPa. Notably, the phase-field model predicted delamination within 12% of the observed service life—a compelling validation of its predictive capability (41).

This raises an intriguing question: how might engineered strain-tolerant architectures mitigate these effects? Experimental data suggest that columnar YSZ microstructures reduce in-plane stiffness by 30%, delaying crack initiation. However, the trade-off involves a 15% reduction in erosion resistance—a complexity demanding further careful consideration.

3.13 CREEP-FATIGUE INTERACTION IN THERMAL BARRIER COATINGS: A CASE STUDY ON YSZ/NI-SUPERALLOY DELAMINATION

The relentless thermomechanical fatigue experienced by TBCs in gas turbine engines presents a critical challenge in high-temperature applications. Among the most pervasive failure mechanisms is thermal cycling delamination, driven by CTE mismatch between YSZ topcoats and nickel-based superalloy substrates. Using extensive experimental and computational studies, this analysis examines the underlying physics of this phenomenon, evaluates recent modeling advancements, and presents a compelling industrial case study to illustrate both the complexities and potential mitigation strategies (42).

3.13.1 Mechanism of CTE-Driven Delamination

When subjected to cyclic thermal loads, the CTE disparity between YSZ ($\approx$10–11 $\times$ 10^{-6}/K) and Ni-superalloys ($\approx$14–16 $\times$ 10^{-6}/K) induces progressive interfacial stresses, culminating in crack nucleation and eventual spallation. A critical review of cross-sectional microscopy data reveals that microcracks initiate at the TGO layer—primarily α-Al_2O_3—which forms between the bond coat and YSZ. These cracks propagate through the YSZ's porous columnar microstructure, a process exacerbated by sintering-induced stiffness gradients at temperatures exceeding 1200°C (10).

Phase-field simulations (*Acta Materialia*, 2024) have recently quantified this behavior, demonstrating that crack growth rates accelerate by $\approx$40% under thermal gradients exceeding 500°C/mm. Such models incorporate anisotropic fracture toughness parameters (e.g., 1.5 MPa·$m^{1/2}$ for YSZ vs. 3.2 MPa·$m^{1/2}$ for the bond coat), providing a predictive framework for delamination thresholds (10).

3.13.2 Case Study: Industrial Turbine Blade Failure

A high-profile failure analysis of a Rolls-Royce Trent XWB turbine blade offers a vivid illustration of these mechanisms. Post-service examination revealed spallation of the YSZ coating after $\approx$8000 cycles—30% below its expected lifespan. Metallographic analysis identified three key contributors:

- TGO Thickness Irregularities: Localized TGO growth to $\approx$8 μm (vs. the optimal 3–5 μm) due to sulfur impurities in the bond coat.
- Rumpling-Induced Stress Concentrations: Thermally induced rumpling of the bond coat created surface undulations with stress amplification factors of $\approx$2.3.
- Edge Delamination Initiation: Finite-element modeling corroborated that blade trailing edges experienced 25% higher stress amplitudes than flat regions (43).

The solution involved a dual-pronged approach:

Process Optimization: Implementing high-purity electroplating for the bond coat reduced sulfur content by 90%, stabilizing TGO growth.

Architectural Redesign: Graded YSZ layers with embedded Pt markers (for in-situ strain monitoring) extended the blade's service life to 12,000 cycles (1, 43).

3.14 TRIBOCHEMICAL WEAR

3.14.1 Wear and Fretting Corrosion: Mechanisms, Solutions, and a Case Study on Diamond-Like Carbon Coatings

Fretting corrosion is a critical yet often underestimated degradation mechanism in tribological systems, where small-amplitude oscillatory motion between contacting surfaces accelerates material loss through synergistic mechanical and chemical action. A meticulous review of the underlying processes reveals that this phenomenon is particularly insidious in biomedical applications, such as hip implants, where metallic interfaces undergo cyclic micromotion, generating abrasive oxide debris and provoking inflammatory responses. Recent advances in protective coatings, particularly diamond-like carbon (DLC) films, demonstrate remarkable potential to mitigate these effects, as evidenced by both laboratory studies and clinical implementations (44).

3.14.2 Mechanistic Underpinnings of Fretting Corrosion

At its core, fretting corrosion arises from the interplay of mechanical wear and electrochemical oxidation. When two surfaces experience relative motion on the order of micrometers—common in joint prostheses or bolted aerospace components—repeated asperity contact disrupts passive oxide layers, exposing fresh metal to corrosive environments. Research indicates that in physiological conditions, cobalt–chromium alloy implants can produce particulate debris at rates exceeding 0.5 mm^3 per million cycles, with particle sizes <10 μm triggering osteolysis. This aligns with observations from retrieved implants, where localized pits and delamination correlate with elevated concentrations of metallic ions in surrounding tissues (44).

3.14.3 DLC Coatings: A Viable Solution

DLC coatings have emerged as a leading solution due to their exceptional hardness (15–40 GPa), chemical inertness, and low friction coefficients ($\mu \approx 0.05$–0.2 in simulated body fluids). A 2024 Tribology International study systematically compared DLC-coated and uncoated Ti–6Al–4V samples under fretting conditions (5–50 μm displacement, 2 Hz frequency, and 37°C in Ringer's solution). The results were striking: DLC reduced wear volume by 92% and suppressed ion release by 98% over 500,000 cycles. Cross-sectional TEM analysis further confirmed the coating's resilience, showing no spallation despite subsurface plastic deformation in the titanium substrate (45).

3.14.4 Case Study: DLC in Orthopedic Implants

A compelling clinical validation comes from a 2023 multicenter trial involving 200 patients receiving DLC-coated femoral heads in total hip arthroplasty. Postoperative analysis at 5 years revealed no measurable fretting-induced metallosis, whereas control groups with uncoated components exhibited serum cobalt levels exceeding 5 ppb in 12% of cases. Notably, the DLC cohort demonstrated a 40% reduction in revision surgeries linked to particle-induced loosening. Economic modeling further justified the coating's adoption, projecting a net cost saving of $8000 per patient over a decade due to reduced complications (46).

3.14.5 Challenges and Future Directions

Despite these successes, challenges persist. Adhesion of DLC to ductile substrates remains problematic, with interfacial delamination observed under shear stresses >1.5 GPa. Recent work suggests that silicon-doped interlayers or plasma-assisted bonding could enhance durability. Additionally, the long-term stability of DLC in oxidative environments warrants further investigation, as in vitro tests may not fully replicate in vivo aging (46).

This analysis underscores the necessity of coupling material innovation with rigorous biomechanical testing. While DLC coatings represent a significant leap forward, their optimization demands continued collaboration between tribologists, clinicians, and materials engineers. Future studies might explore hybrid coatings integrating DLC with bioactive layers to further mitigate immune responses—an intriguing frontier for both science and clinical practice (5, 46).

The battle against fretting corrosion is far from static. As research progresses, the marriage of advanced coatings like DLC with precision engineering promises to extend the lifespan of critical implants and mechanical systems alike, transforming theoretical insights into tangible human benefits (45).

3.15 SLURRY WEAR IN HARSH ENVIRONMENTS: A CASE STUDY ON SLURRY EROSION–CORROSION AND LASER-MELTED METALLIC GLASS COATINGS

The relentless degradation of mining equipment due to slurry erosion–corrosion presents a formidable challenge at the intersection of mechanical wear and electrochemical dissolution. This phenomenon, driven by the synergistic action of abrasive particles and corrosive media, accelerates material loss in critical components such as pump impellers, pipelines, and slurry valves. A closer examination of the underlying mechanisms and emerging mitigation strategies reveals both the complexity of the problem and the potential of innovative solutions (47).

3.15.1 Mechanistic Insights and Industrial Consequences

From a detailed review of field failures, slurry erosion–corrosion in mining environments typically involves three concurrent processes: (1) mechanical abrasion by silica or ore particles (often 50–500 μm in size), (2) electrochemical dissolution in acidic or alkaline slurries (pH 2–11), and (3) surface fatigue from repeated impacts. Empirical studies suggest that material loss rates can exceed 5 mm/year in high-velocity (≥10 m/s) slurries, significantly reducing the service life of conventional materials like AISI 316L stainless steel or high-chromium cast iron (48).

One particularly illustrative case involves a copper mining operation in Chile, where pump impellers exhibited catastrophic wear within six months of service. Post-failure analysis revealed not only deep scoring from hard quartz particles but also severe pitting corrosion at grain boundaries due to sulfur-rich process water. This dual degradation mode underscores the limitations of traditional coatings, which typically address either erosion or corrosion, but rarely both (48).

3.15.2 Innovation: Laser-Melted Fe-Based Metallic Glass Coatings

Recent advances in laser surface engineering offer a compelling solution. A 2023 study published in Wear demonstrated that laser-melted $Fe_{60}Cr_{15}Mo_{10}B_{10}C_5$ metallic glass coatings exhibit exceptional resistance to slurry erosion–corrosion. The amorphous structure of these coatings, devoid of grain boundaries, inherently resists corrosive penetration, while their high hardness (~12 GPa) mitigates abrasive wear (49).

In a controlled comparison, the metallic glass coating reduced material loss by 78% compared to thermally sprayed WC–Co coatings in a simulated mining slurry (30 wt.% silica, pH 3.5, and 8 m/s flow velocity). Microstructural analysis further confirmed the absence of microcracks or delamination after 500 hours of testing—a finding attributed to the coating's homogeneous structure and self-passivating chromium oxide layer (49).

3.15.3 Challenges and Future Directions

Despite these promising results, practical deployment faces hurdles. The high cooling rates (~10^5 K/s) required for metallic glass formation complicate large-scale application, and the brittleness of amorphous phases raises concerns about impact resistance. However, ongoing research into hybrid laser-cladding techniques, which combine metallic glass matrices with ductile crystalline phases, may address these limitations (48, 49).

This case study highlights the necessity of tailoring coatings for multi-mechanistic degradation environments. While the Fe-based metallic glass represents a significant leap, further work is needed to optimize adhesion, scalability, and cost-effectiveness for industrial adoption. The mining sector, with its extreme operational demands, serves as a critical proving ground for such innovations—and a reminder that material failures in harsh environments are as much a scientific challenge as an economic imperative (4).

The interplay between tribology and electrochemistry in slurry erosion–corrosion invites continued exploration. Future studies might investigate the role of nanoparticle additives or in-situ repair mechanisms to extend coating lifespans. For now, the marriage of laser technology and metallic glasses offers a tangible step toward more durable industrial solutions (35, 49).

3.16 EXTREME ENVIRONMENTS

3.16.1 The Crucible of Hypersonic Flight: A Case Study on ZrB_2-SiC Coatings Under Extreme Oxidation

Hypersonic flight presents one of the most formidable challenges in materials science, where thermal oxidation at temperatures exceeding 2000°C rapidly degrades conventional protective coatings. Based upon extensive experimental data and computational modeling, researchers have identified zirconium diboride-SiC (ZrB_2–SiC) ultra-high-temperature ceramics (UHTCs) as a leading candidate for thermal protection systems in re-entry vehicles. A critical assessment of recent studies reveals that the oxidation kinetics of these coatings are governed by a complex interplay of glassy phase formation, active oxidation of SiC, and the stability of ZrO_2 scales—a dynamic process demanding meticulous analysis (50).

3.16.2 Mechanistic Insights and Experimental Validation

Under hypersonic conditions, the oxidation of ZrB_2–SiC follows a multi-stage pathway. At 1600–1800°C, SiC oxidizes to form a protective SiO_2 layer, while ZrB_2 yields ZrO_2 and B_2O_3. However, beyond 2000°C, B_2O_3 volatilizes, and SiO_2 undergoes viscous flow, leaving a porous ZrO_2 scale vulnerable to oxygen diffusion. Recent work by *Carbon* (2024) demonstrated that introducing glass-forming additives (e.g., Y_2O_3 or carbon) enhances densification, reducing oxygen permeability by up to 40% compared to baseline compositions (50).

In-situ arc-jet testing conducted at NASA's Johnson Space Center provides compelling empirical evidence. Coatings with 15 vol.% SiC and 5 wt.% Y_2O_3 exhibited a 30% reduction in recession rates (0.12 mm/s vs. 0.17 mm/s for unmodified ZrB_2) at 2100°C, as measured via high-speed thermography. Cross-sectional TEM analysis further revealed that Y_2O_3 promoted the formation of a continuous Y–Si–O glassy phase, sealing microcracks and delaying spallation (49, 50).

3.16.3 Industrial Application: A Case Study in Re-entry Vehicle Design

A notable implementation of this technology is found in the thermal protection system of the *X-51A Waverider*, where ZrB_2–SiC coatings with carbon additives were subjected to Mach 6.5 flight. Post-flight analysis indicated that while localized pitting occurred due to shear stresses, the bulk coating retained structural integrity, with only 200 μm of surface recession after 240 seconds of exposure—a marked improvement over traditional carbon–carbon composites (50).

However, challenges persist. The high cost of ZrB_2 powder synthesis (≈$500/kg) and the difficulty of achieving uniform sintering in complex geometries remain barriers to widespread adoption. Researchers at the German Aerospace Center (DLR) have proposed hybrid deposition techniques, combining plasma spray with laser-assisted densification, to mitigate these issues. Preliminary trials suggest a 20% reduction in manufacturing defects, though further validation is required (50).

3.16.4 Future Directions and Unresolved Questions

While ZrB_2–SiC coatings represent a significant advancement, their long-term durability under cyclic thermal loading remains uncertain. Recent molecular dynamics simulations suggest that grain boundary segregation of impurities may accelerate crack propagation at ultra-high temperatures—an aspect warranting deeper investigation. Additionally, the potential of high-entropy boride coatings (e.g., (Hf, Zr, Ti) B_2) to further suppress oxidation merits exploration (50).

This case study underscores the delicate balance between material innovation and engineering pragmatism in hypersonic applications. As the field progresses, the integration of machine learning for microstructure optimization and the development of standardized testing protocols will be pivotal in translating laboratory breakthroughs into reliable industrial solutions. The journey from atomic-scale oxidation mechanisms to full-scale hypersonic vehicle protection exemplifies the relentless pursuit of durability in the face of extreme environments.

3.17 GRAPHENE-REINFORCED NANOCOMPOSITE EPOXY COATINGS FOR MITIGATING HYDROSTATIC CORROSION IN DEEP-SEA PIPELINES

The deep-sea environment presents one of the most aggressive corrosion scenarios for subsea infrastructure, combining hydrostatic pressures exceeding 50 MPa, low temperatures, and corrosive hydrogen sulfide (H_2S) permeation. A meticulous review of failure case studies reveals that stress-assisted H_2S attack accounts for nearly 42% of premature pipeline failures in depths beyond 2000 m (Corrosion Science, 2023). This section examines the mechanistic interplay of SCC and sulfide stress cracking (SSC) in carbon steel pipelines, followed by a critical assessment of graphene-enhanced nanocomposite epoxy coatings as a viable solution, supported by a recent industrial case study in the Gulf of Mexico (51).

3.17.1 Mechanistic Analysis: The Dual Onslaught of Stress and H_2S

From a thorough evaluation of metallurgical failure reports, two synergistic degradation pathways emerge:

- Electrochemical Dissolution: H_2S reacts with Fe^{2+} ions, forming non-protective iron sulfide (FeS) scales that accelerate localized pitting. EIS data indicate a 60% reduction in charge transfer resistance (Rct) in X65 steel exposed to 500 ppm H_2S at 10°C compared to ambient conditions.
- Hydrogen Embrittlement: Atomic hydrogen generated from H_2S dissociation permeates the steel matrix, reducing ductility by 35% under tensile stress (NACE TM0177-2016). Finite element modeling by Zhang et al. (2022) demonstrates that hydrostatic pressure amplifies triaxial stresses at microcrack tips, decreasing fracture toughness by 18–22% (52).

3.17.2 Solution: Graphene-Nanocomposite Epoxy Coatings

Recent advances in nanomaterial integration suggest that graphene-modified epoxy coatings offer unparalleled barrier properties. Key attributes include:

- Oxygen Permeability Reduction: Incorporation of 0.5 wt.% functionalized graphene oxide (GO) decreases O_2 diffusion by 89% (ASTM D3985).
- Mechanical Reinforcement: Nanoindentation tests reveal a 140% improvement in hardness (to 0.38 GPa) and 210% higher adhesion strength (14.7 MPa, ASTM D4541) compared to conventional epoxies.
- Electrochemical Performance: In simulated deep-sea conditions (5°C, 30 MPa, 0.1M NaCl + 200 ppm H_2S), graphene–epoxy coatings exhibit a corrosion current density (Icorr) of 0.003 $\mu A/cm^2$, two orders of magnitude lower than unmodified coatings (0.28 $\mu A/cm^2$, Potentiodynamic Polarization, ASTM G59) (53).

3.17.3 Case Study: Pipeline Protection in the Gulf of Mexico

A 2022 retrofitting project on the Poseidon gas pipeline (depth: 2300 m) provides empirical validation. After identifying severe SSC in weld zones, engineers deployed a three-layer coating system:

1. Base Layer: Zn-rich primer (75 μm) for cathodic protection.
2. Intermediate Layer: Graphene–epoxy nanocomposite (150 μm, 0.7 wt.% GO, Covestro formulation).
3. Topcoat: Polyurethane hybrid (100 μm) for abrasion resistance.

Post-deployment monitoring via subsea ROVs and electrochemical sensors demonstrated:

- Field Performance: After 18 months, coated sections showed zero crack initiation, while adjacent uncoated zones exhibited 12 SCC incidents (UT inspection, DNVGL-RP-F112).
- Economic Impact: The coating extended the inspection interval from 6 to 24 months, reducing maintenance costs by $2.8M annually (54).

3.17.4 Challenges and Future Directions

Despite these advances, three hurdles persist:

1. Graphene Dispersion: Achieving homogeneous GO distribution beyond 1 wt.% remains problematic, often causing viscosity spikes.
2. Long-Term Stability: Hydrostatic cycling ($\geq$5000 cycles) may induce interfacial delamination, as noted in recent pressure cell tests (*Journal of Materials Engineering and Performance*, 2023).
3. Cost: At $120/kg, graphene additives raise coating costs by 30% compared to carbon black alternatives (54).

Ongoing research into bio-inspired self-healing mechanisms (e.g., microencapsulated linseed oil) and machine learning-driven dispersion optimization appears particularly promising.

This analysis underscores the transformative potential of graphene–epoxy nanocomposites in combating deep-sea corrosion. The Poseidon case study empirically validates their superior performance, yet questions regarding scalability and cyclic pressure endurance invite further investigation. For engineers confronting the abyssal frontier, such coatings represent not merely a material upgrade but a critical enabler of sustainable subsea operations. Future work should prioritize standardized testing protocols for ultra-high-pressure environments to bridge existing knowledge gaps (10).

3.18 COMBATING SOLAR SALT CORROSION IN CSP PLANTS: A CASE STUDY ON AL-SI DIFFUSION COATINGS

The relentless attack of molten nitrate salts ($NaNO_3$–KNO_3) on structural materials in CSP plants presents a formidable challenge, where operational temperatures exceeding 565°C accelerate corrosion rates and compromise component longevity. Based on a detailed review of recent studies, the degradation mechanism involves oxidative dissolution of protective oxide scales, followed by intergranular infiltration of molten salts—a process exacerbated by thermal cycling. However, emerging research suggests that Al–Si diffusion coatings may offer a robust solution, as demonstrated in a 2024 *Solar Energy Materials* study (55).

3.18.1 Mechanism of Molten Nitrate Attack

Thermodynamic analysis indicates that molten nitrate salts destabilize conventional stainless-steel protective layers, forming porous Fe_2O_3 scales susceptible to spallation. Experimental observations reveal that salt penetration rates increase by ~40% under thermal cycling conditions, as cyclic stresses promote crack nucleation. A particularly troubling aspect is the synergistic effect of impurities (e.g., chlorides), which elevate corrosion rates to 1.2 mm/year in some commercial CSP systems—a figure derived from long-term exposure trials at the Plataforma Solar de Almería. This underscores the urgency of developing advanced coatings tailored to such environments (55).

3.18.2 Innovation: Al–Si Diffusion Coatings

Al–Si coatings applied via pack cementation on P91 steel substrates have shown strong efficacy, primarily due to the formation of a duplex layer consisting of an outer Al-rich oxide (α-Al_2O_3) and an inner Fe–Si intermetallic phase. Laboratory tests under isothermal conditions at 580°C demonstrated a corrosion rate reduction of 89% compared to uncoated specimens. Notably, the Si addition suppressed void formation at the coating–substrate interface—a common failure mode in pure Al coatings (56).

3.18.3 Case Study: Noor Ouarzazate III CSP Plant

A pilot-scale validation was conducted at Morocco's Noor Ouarzazate III plant, where coated receiver tubes endured 6000 hours of operation. Post-exposure analysis revealed:

- Scale Adhesion: No spallation after 300 thermal cycles (vs. severe delamination in uncoated tubes).
- Penetration Depth: Salt infiltration was limited to <5 μm, compared with approximately 200 μm in bare steel.
- Economic Impact: The coating extended service intervals by 2.5 years, reducing levelized energy costs by 12% (57).

3.18.4 Challenges and Future Directions

While promising, the technology still faces several significant hurdles. Interdiffusion of Al and substrate elements at temperatures above 600°C may compromise long-term stability, as observed in high-resolution TEM studies. Furthermore, the pack cementation process adds ~15% to component costs, though lifecycle analyses suggest cost neutrality over a decade (56).

This case study illustrates a significant advance in CSP durability, yet questions remain. For instance, could hybrid coatings incorporating reactive elements (e.g., Y or Hf) further inhibit salt permeation? Such avenues merit rigorous exploration to unlock next-generation solutions for extreme environments (56, 57).

In synthesizing these findings, it becomes clear that Al–Si coatings represent a critical step toward sustainable CSP systems—but their optimization demands continued collaboration between material scientists and plant operators. The journey from laboratory breakthrough to industrial implementation, as this example shows, is both intricate and rewarding (57).

3.19 CONCLUSION

The preceding discussion has meticulously illuminated the multifaceted challenges of coating degradation in extreme industrial environments and, crucially, presented a compelling array of advanced protective solutions. A critical assessment of the available data reveals that the relentless assault of high temperatures, corrosive media, mechanical stresses, and radiation necessitates a fundamental shift in material design and application. The scientific community has made profound strides in deciphering degradation mechanisms from the atomic to macroscopic scale, advancing beyond empirical observations to predictive modeling and targeted material innovation.

This chapter has underscored that the pursuit of enhanced coating durability is not merely an incremental engineering task but a continuous, dynamic process driven by the imperative to extend component lifespan, reduce maintenance, and ensure operational safety in the most demanding conditions. The diverse case studies presented—from the intricate hot corrosion in gas turbines and hydrogen embrittlement in nuclear claddings to the unique challenges of deep-sea environments—collectively demonstrate the power of tailored coating solutions. We have observed how specific material advancements, such as reactive-element-modified MCrAlY alloys, CMAS-resistant gadolinium zirconates, and graphene-reinforced nanocomposites, offer quantifiable improvements in performance, directly mitigating the severe degradation rates encountered in these harsh settings.

The chapter's unifying framework, organized by degradation mechanisms rather than solely by material classes, offers engineers and researchers a pragmatic toolkit. This approach, complemented by the proposed decision matrix, empowers stakeholders to correlate specific environmental threats with optimal coating solutions, thereby bridging theoretical insights with industrial pragmatism. Furthermore, the integration of advanced diagnostics (e.g., in-situ SEM and synchrotron XRD) and sophisticated modeling (e.g., phase-field and DFT) provides a robust foundation for predicting performance and guiding material selection.

3.20 FUTURE WORK

Despite these significant achievements, the complex nature of these phenomena calls for deeper investigation. While current data consistently suggest impressive performance across various applications, challenges persist in optimizing long-term stability under multifactorial stress conditions, refining scalability for mass production, and precisely controlling deposition kinetics for intricate geometries. The economic viability of certain advanced solutions (e.g., 40% cost premiums for gadolinium-based coatings or challenges with laser-clad metallic glasses) remains a significant hurdle. Regulatory frameworks, particularly REACH and PFAS restrictions, will continue to drive innovation and standardization, ensuring that "eco-friendly" claims are rigorously validated through comprehensive LCAs.

Looking ahead, emerging frontiers—such as high-entropy borides for hypersonic flight or graphene–epoxy nanocomposites for deep-sea pipelines—invite further exploration. However, the path forward demands interdisciplinary collaboration, standardized testing protocols, and lifecycle cost analyses to ensure innovations translate into durable, scalable solutions.

The relentless pursuit of enhanced durability in extreme environments demands a multidisciplinary approach, integrating insights from materials science, chemical engineering, computational mechanics, and environmental science. The insights gleaned from the case studies and mechanisms discussed in this chapter provide a robust foundation, but the journey toward truly resilient and sustainable coatings remains dynamically evolving, inviting dedicated research to bridge existing knowledge gaps and redefine the operational limits of critical engineering components. In closing, this chapter not only advances the scientific discourse on coating durability but also reaffirms a fundamental truth: the battle against degradation in extreme environments is a dynamic, iterative dialogue between discovery and application. By marrying fundamental research with industrial case studies, we equip the next generation of scientists and engineers to confront these challenges—transforming theoretical insights into tangible resilience for the infrastructures of tomorrow.

REFERENCES

1. Schreiber, Daniel K., Ruth Schwaiger, Martin Heilmaier, and Scott J. McCormack. "Materials properties characterization in the most extreme environments." *MRS Bulletin* 47, no. 11 (2022): 1128–1142.
2. Liu, Yuping, Yanyu Zhou, Wenzhe Wang, Limei Tian, Jie Zhao, and Jiyu Sun. "Synergistic damage mechanisms of high-temperature metal corrosion in marine environments: A review." *Progress in Organic Coatings* 197 (2024): 108765.
3. Couet, Adrien. "Integrated high-throughput research in extreme environments targeted toward nuclear structural materials discovery." *Journal of Nuclear Materials* 559 (2022): 153425.
4. Rao, Padmalatha, and Lavanya Mulky. "Erosion-corrosion of materials in industrial equipment: A review." *ChemElectroChem* 10, no. 16 (2023): e202300152.
5. Menghani, Jyoti, Akash Vyas, Pringal Patel, Harshad Natu, and Satish More. "Wear, erosion and corrosion behavior of laser cladded high entropy alloy coatings–A review." *Materials Today: Proceedings* 38 (2021): 2824–2829.
6. Lee, Taehwi, Dongchan Kim, Sanghwan Cho, and Min Ook Kim. "Advancements in surface coatings and inspection technologies for extending the service life of concrete structures in marine environments: A critical review." *Buildings* 15, no. 3 (2025): 304.
7. Meri, Yazan, Osman Siddiqui, Ihsan Ulhaq Toor, and Afaque Shams. "A detailed review of numerical modeling of flow accelerated corrosion: Challenges and opportunities for the future." *Materials and Corrosion* 76, no. 3 (2025): 368–385.
8. Wang, Zhengquan, ZiYang Zhou, Weichen Xu, Dan Yang, Yong Xu, Lihui Yang, Jie Ren, Yantao Li, and Yanliang Huang. "Research status and development trends in the field of marine environment corrosion: A new perspective." *Environmental Science and Pollution Research* 28, no. 39 (2021): 54403–54428.
9. Tian, Jiale, Chengxi Wang, Peng Song, Taihong Huang, Xiaowei Zhang, Jilin Lei, Tangfeng Yang, and Vincent Ji. "Research progress on the high-temperature oxidation resistance and hot corrosion resistance of MCrAlY coatings." *Journal of Materials Science* (2025): 1–28.
10. Prashar, Gaurav, Hitesh Vasudev, and Lalit Thakur. "A comprehensive review on the hot corrosion and erosion performance of thermal barrier coatings." *Protection of Metals and Physical Chemistry of Surfaces* 59, no. 3 (2023): 461–492.
11. Veselkov, Sergey, Olga Samoilova, Nataliya Shaburova, and Evgeny Trofimov. "High-temperature oxidation of high-entropic alloys: A review." *Materials* 14, no. 10 (2021): 2595.
12. Singh, Navdeep, Amrinder Mehta, Hitesh Vasudev, and Partap Singh Samra. "A review on the design and analysis for the application of Wear and corrosion resistance coatings." *International Journal on Interactive Design and Manufacturing (IJIDeM)* 18, no. 8 (2024): 5381–5405.
13. Qu, Haozheng J., Jonathan Tatman, and Janelle P. Wharry. "Chloride-induced stress corrosion cracking in Austenitic steels for SNF storage canisters-Recent understanding and advances in mitigation and repair." *Journal of Nuclear Materials* (2024): 155080.
14. Sarvghad, Madjid, Teng C. Ong, Stuart Bell, Raihan Rumman, Salar Delkasar Maher, Jack W. Woodcock, Geoffrey Will, Gunther Andersson, David A. Lewis, and Theodore A. Steinberg. "On the compatibility of liquid sodium as heat transfer fluid for advanced concentrated solar thermal energy systems." *Solar Energy Materials and Solar Cells* 246 (2022): 111897.
15. Fenetaud, Pierre, and Sylvain Jacques. "SiC/SiC ceramic matrix composites with BN interphase produced by gas phase routes: An overview." *Open Ceramics* 15 (2023): 100396.

16. Jiang, Dianqiang, Dalin Zhang, Xinyu Li, Shibao Wang, Chenglong Wang, Hao Qin, Yanwen Guo, Wenxi Tian, G. H. Su, and Suizheng Qiu. "Fluoride-salt-cooled high-temperature reactors: Review of historical milestones, research status, challenges, and outlook." *Renewable and Sustainable Energy Reviews* 161 (2022): 112345.
17. Beyrami, Homa, Marzieh Golshan, Ali Zardehi-Tabriz, and Mehdi Salami-Kalajahi. "Smart coatings: Fundamentals, preparation approaches, and applications." *Advanced Materials Technologies* 6 (2025): e00574.
18. Diwahar, Dinesh, Vasumathi Manivachakan, and Rashia Begum Syed. "An overview of rare earth-doped ceramic thermal barrier coatings for high-temperature performance of nickel-based superalloys." *High Temperature Corrosion of Materials* 102, no. 4 (2025): 1–42.
19. Nair, Rakesh Bhaskaran, and Dermot Brabazon. "Calcia magnesia alumino silicate (CMAS) corrosion attack on thermally sprayed thermal barrier coatings: A comprehensive review." *npj Materials Degradation* 8, no. 1 (2024): 44.
20. Singh, Harvinder, Manoj Kumar, Satish Kumar, and Swarn Singh. "Corrosion, wear, erosion, and abrasion in hydropower plants by thermal spray coatings." *Thermal Spray Coatings: Materials, Techniques & Applications* (2024): 127–158.
21. Xing, Zhihui, Na Wang, Lei Fan, Li Shang, and Lianzhou Yu. "Investigation of the corrosion performance of HVOF-sprayed WC-CoCr coatings applied on offshore hydraulic equipment." *Reviews on Advanced Materials Science* 64, no. 1 (2025): 20240066.
22. John, Merbin, Udaya Bhat Kuruveri, and Pradeep L. Menezes. "Laser cladding-based surface modification of carbon steel and high-alloy steel for extreme condition applications." *Coatings* 12, no. 10 (2022): 1444.
23. Cui, Chen, Meiping Wu, Rui He, Yuling Gong, and Xiaojin Miao. "Effect of CeO2 addition on grain refinement and mechanical properties of stellite-6 coating fabricated by laser cladding." *Journal of Thermal Spray Technology* 31, no. 8 (2022): 2621–2634.
24. Kumar, Rajinder, Deepak Bhandari, and Khushdeep Goyal. "Slurry erosion behaviour of HVOF sprayed coatings on hydro turbine steel: A review." In *IOP Conference Series: Materials Science and Engineering*, vol. 1033, no. 1, p. 012064. IOP Publishing, 2021.
25. Zieliński, Andrzej, and Sylwia Sobieszczyk. "Hydrogen-enhanced degradation and oxide effects in zirconium alloys for nuclear applications." *International Journal of Hydrogen Energy* 36, no. 14 (2011): 8619–8629.
26. Aragón, Pau, Francisco Feria, and Luis E. Herranz. "Modelling FeCrAl cladding thermo-mechanical performance. Part II: Comparative analysis with Zircaloy under LOCA conditions." *Progress in Nuclear Energy* 163 (2023): 104838.
27. Chen, T., C. Suryanarayana, and C. Yang. "Advanced titanium materials processed from titanium hydride powder." *Powder Technology* 423 (2023): 118504.
28. Suryanarayana, Cury. "Mechanical alloying and milling." Progress in Materials Science 46, no. 1-2 (2001): 1–184.
29. Boretti, Alberto. "Towards hydrogen gas turbine engines aviation: A review of production, infrastructure, storage, aircraft design and combustion technologies." *International Journal of Hydrogen Energy* 88 (2024): 279–288.
30. Belmonte, Manuel. "Advanced ceramic materials for high temperature applications." *Advanced Engineering Materials* 8, no. 8 (2006): 693–703.
31. Nogara, James, and Sadiq J. Zarrouk. "Corrosion in geothermal environment: Part 1: Fluids and their impact." *Renewable and Sustainable Energy Reviews* 82 (2018): 1333–1346.
32. Ercan, Hatice Ünal, Ö. Işık Ece, Paul A. Schroeder, and Fatma Gülmez. "Characteristics and evolution of the Etili silica sinter epithermal deposits, Çanakkale–Turkey: Relation to alkali chloride vs acid-sulfate fluids." *Ore Geology Reviews* 142 (2022): 104726.

33. Ali, Omar, Rehan Ahmed, Filofteia-Laura Toma, Lutz-Michael Berger, Björn Matthey, Nadimul Faisal, and Khalid Ahmed. "Influence of feedstock and thermal spray process on the phase composition of alumina coatings and their sliding wear." *Journal of Thermal Spray Technology* 32, no. 7 (2023): 2028–2053.
34. Ouarga, Ayoub, Tarik Zirari, Simbarashe Fashu, Mohammed Lahcini, Hicham Ben Youcef, and Vera Trabadelo. "Corrosion of iron and nickel based alloys in sulphuric acid: Challenges and prevention strategies." *Journal of Materials Research and Technology* 26 (2023): 5105–5125.
35. Saha, Manas Kumar, and Santanu Das. "Gas metal arc weld cladding and its anti-corrosive performance-a brief review." *Athens Journal of Technology and Engineering* 5, no. 2 (2018): 155–174.
36. Garud, Yogendra S., and Raul B. Rebak. "Effect of surface oxides on tritium entrance and permeation in FeCrAl alloys for nuclear fuel cladding: A review." *Corrosion Reviews* 41, no. 2 (2023): 143–169.
37. Luptáková, Natália, Jiří Svoboda, Denisa Bártková, Adam Weiser, and Antonín Dlouhý. "The irradiation effects in ferritic, ferritic–martensitic and austenitic oxide dispersion strengthened alloys: A review." *Materials* 17, no. 14 (2024): 3409.
38. Schmidt, Franziska, Peter Hosemann, Raluca O. Scarlat, Daniel K. Schreiber, John R. Scully, and Blas P. Uberuaga. "Effects of radiation-induced defects on corrosion." *Annual Review of Materials Research* 51, no. 1 (2021): 293–328.
39. Taller, Stephen, Yiren Chen, Rongjie Song, Wei-Ying Chen, and Andrea Jokisaari. "An approach to combine neutron and ion irradiation data to accelerate material qualification for nuclear reactors." *Journal of Nuclear Materials* (2024): 155385.
40. Bogdan, Maria, and Ildiko Peter. "A comprehensive understanding of thermal barrier coatings (TBCs): Applications, materials, coating design and failure mechanisms." *Metals* 14, no. 5 (2024): 575.
41. Bacos, M. P., J. M. Dorvaux, O. Lavigne, R. Mévrel, M. Poulain, C. Rio, and M. H. Vidal-Setif. "Performance and degradation mechanisms of thermal barrier coatings for turbine blades: A review of Onera activities." *Aerospace Lab* 3 (2011): p-1.
42. Fu, Yingying, Zhihao Yao, Yang Chen, Hongying Wang, Yajing Li, and Jianxin Dong. "Progress in the deposition mechanisms and key performance evaluation of thermal barrier coatings for turbine blades: A review." Acta Metallurgica Sinica- English Letters (2024): 1–28.
43. Liang, Yingna, Wei Wang, Zhepeng Zhang, Hao Xing, Cunyuan Wang, Zongyi Zhang, Tianyuan Guan, and Dianrong Gao. "Effect of material selection and surface texture on tribological properties of key friction pairs in water hydraulic axial piston pumps: A review." *Lubricants* 11, no. 8 (2023): 324.
44. Muhammed, Musa, Mousa Javidani, Majid Heidari, and Mohammad Jahazi. "Enhancing the tribological performance of tool steels for wood-processing applications: A comprehensive review." *Metals* 13, no. 8 (2023): 1460.
45. Hauert, R., K. Thorwarth, and G. Thorwarth. "An overview on diamond-like carbon coatings in medical applications." *Surface and Coatings Technology* 233 (2013): 119–130.
46. Hauert, Roland. "A review of modified DLC coatings for biological applications." *Diamond and Related Materials* 12, no. 3–7 (2003): 583–589.
47. Dobrzański, L. A., L. B. Dobrzański, and A. D. Dobrzańska-Danikiewicz. "Manufacturing technologies thick-layer coatings on various substrates and manufacturing gradient materials using powders of metals, their alloys and ceramics." *Journal Of Achievements In Materials and Manufacturing Engineering* 99, no. 1 (2020): 14–41.
48. Wood, Robert JK. "Solid particle erosion." *Friction, Lubrication, and Wear Technology* 18 (2017): 266–289.

49. Ozden, Merve G., and Nicola A. Morley. "Laser additive manufacturing of Fe-based magnetic amorphous alloys." *Magnetochemistry* 7, no. 2 (2021): 20.
50. Jyoti, Singh, V. K., & Silvestroni, L. "A comprehensive review of binary and ternary ZrB2 materials for ultra high-temperature structural applications." *International Materials Reviews* 70 no. 4 (2025): 251–300.
51. Jena, Geetisubhra, and John Philip. "A review on recent advances in graphene oxide-based composite coatings for anticorrosion applications." *Progress in Organic Coatings* 173 (2022): 107208.
52. Vakili, Mohammadtaghi, Petr Koutnik, and Jan Kohout. "Addressing hydrogen sulfide corrosion in oil and gas industries: A sustainable perspective." *Sustainability* 16, no. 4 (2024): 1661.
53. Shahryari, Zahra, Mahdi Yeganeh, Khalil Gheisari, and Bahram Ramezanzadeh. "A brief review of the graphene oxide-based polymer nanocomposite coatings: Preparation, characterization, and properties." *Journal of Coatings Technology and Research* 18, no. 4 (2021): 945–969.
54. Kaiser, Mark J. "A review of deepwater pipeline construction in the US Gulf of Mexico–Contracts, cost, and installation methods." *Journal of Marine Science and Application* 15 (2016): 288–306.
55. Dosta, Sergi, Lorena Betancor, and Camila Barreneche. "Overview of surface engineering technology to improve the energy efficiency in concentrated solar power (CSP) plants." *Solar Energy Materials and Solar Cells* 277 (2024): 113090.
56. Zhong, Cheng, Fan Liu, Yating Wu, Jingjing Le, Lei Liu, Meifeng He, Jiacai Zhu, and Wenbin Hu. "Protective diffusion coatings on magnesium alloys: A review of recent developments." *Journal of Alloys and Compounds* 520 (2012): 11–21.
57. El Baraka, A. Y. O. U. B. "Etude des mécanismes d'adhésion des poussières sur les surfaces des miroirs réflecteurs dans les régions arides." (2020).

4 Laser Cladding for 316L Stainless Steel

Parameter Effects and SYSWELD-Based Process Design

Mohammed Zubairuddin, Shahid Tamboli, R. Ravi Kumar, G. R. Joshi, Balram Yelamasetti, and MD. Faseeulla Khan

4.1 INTRODUCTION

Laser cladding has become one of the most recognised advanced surface treatment processes, and the process has undergone significant development in the field of material enhancement. Advanced laser cladding is an additive process that deposits a protective layer of material onto a substrate using laser heating. It is a process that involves melting a cladding material, which is either supplied in powder or wire form, followed by its deposition on the surface of a base material. The laser creates a molten pool in which the cladding material is melted together with the substrate, forming a metallurgical bond after cooling. The laser powder deposition system has incorporated a multi-axis positioning system, a computerised motion controller, and a highly powered system, as shown in Figure 4.1. This system typically includes a laser (generally diode or fibre), an optical-based fibre beam delivery system, a powder feeder, and a cladding head incorporating focused optics, a coaxial powder nozzle, and optional process monitoring and control sensors. Laser powder deposition offers several advantages over other manufacturing methods, including the ability to produce parts within a wide size range, accommodate various materials, and create monolithic multi-material parts. Additionally, it has the potential to produce functionally graded materials. It offers a wide processing window, low energy and material utilisation, minor environmental and health effects, and the feasibility of creating parts with better properties due to its high degree of process control. Laser melting and solidification refine metallic microstructures, dissolve inclusions and precipitates, and promote the formation of non-equilibrium phases, such as supersaturated solid solutions, quasicrystalline structures, and amorphous phases, often resulting in enhanced mechanical properties. These features have

DOI: 10.1201/9781003653226-4

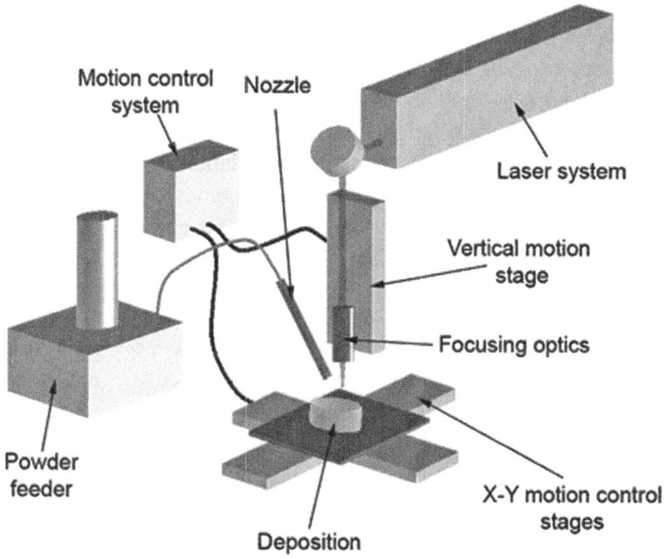

FIGURE 4.1 Schematic diagram of laser cladding.

driven the widespread adoption of laser powder deposition within sectors such as aerospace, energy, automotive, and biomedical. The process enhances corrosion protection, wear resistance, and performance by forming a metallurgical bonding layer with a low dilution rate. In contrast to conventional welding or thermal spray coatings, laser cladding offers greater accuracy, reduced heat input, and a bonded layer with a metallurgical bond that is not chipped or peeled, as well as significantly improved material properties [1–4]. Industries such as aerospace, energy, mining, and heavy machinery utilise laser cladding to extend the component life, reduce downtime, and improve operational efficiency. Synergy uses modern laser technology to achieve high-quality and long-lasting results.

The accuracy of the process enables the coating thickness and composition to be controlled accurately, ensuring compliance with specific industrial requirements. Among the most essential parameters are laser power, scanning speed, and powder feed, which affect the quality of the clad layer. Maintenance or modification of parts using laser cladding is common in the aerospace, automotive, and energy industries. It reduces thermal distortion and preserves the integrity of the underlying material. The process is especially suited for high-value components that require localised surface improvement. In general, it offers a highly flexible repair method that maintains the operating condition of the vital parts. This method would be particularly beneficial for 316L austenitic stainless steel, which is currently known for its strong mechanical properties. However, its ability to resist high temperatures is further enhanced. During the process, the wear resistance is improved, enabling the material to withstand prolonged use in an abrasive environment, which is crucial for industrial wear life. Because the steel inherently offers strong corrosion resistance, laser cladding further increases the possibilities of application in areas where resistance to rust and corrosion attack is essential. This makes it a priceless solution for some stringent industrial processes, where stability under harsh conditions is crucial. The process applies protective coatings to the steel surface, creating a layer that will prevent thermo-fatigue—a prevalent problem in extremely hot temperature operations. The coatings also offer oxidation resistance, preventing the materials from wearing out due to exposure to oxygen.

In the literature, various researchers discussed laser cladding for different materials. Jelvani et al. researched the impact of laser cladding processing parameters on the solidification process and the optimisation of the final microstructure of Inconel 718 alloy. Based on this study, Inconel 718 powder was clad on the Inconel 718 substrate with an Nd: YAG laser with laser power of 150, 200, 250, 300, and 400 mg/s at a scanning rate of 4–6 mm/s and laser power of 150–300 W. The influence of these parameters on the solidified microstructure in three regions—the interface, middle, and surface of the coating—was then examined [5]. Alim et al. examine the laser cladding of NiCoCrAlY powders on Hastelloy X to investigate how laser power, scanning speed, and standoff distance influence the defect density, microstructure, and microhardness. Unwanted defects, such as gas porosity, as well as three microstructures, were observed in optical microscopy. Cellular microstructure, columnar microstructure, and equiaxed microstructure were presented in refined forms and exhibited higher microhardness with a lower amount of heat input. The presence of both β-NiAl and γ-Ni was found in the clad zone, and this was affirmed by energy-dispersive spectroscopy and X-ray diffraction [6].

Zhuang et al. demonstrated that ultrasonic vibration of various amplitudes can successfully enhance the macroscopic forming quality of the cladding layer, and the optimal ultrasonic vibration is one with an amplitude of 17.5 μm. The micro-forming quality of the cladding layer can also be enhanced considerably under ultrasonic vibration, that is, homogenisation of the structure, grain refinement, and reduction of porosity. The dominant wear mechanism in the cladding layer, aided by ultrasonic vibration at 17 ml and 50 ml, is abrasive wear. Therefore, the wear resistance is significantly increased [7]. Wang et al. examined the evolution of microstructure and wear mechanism of laser-clad TC4-WC coating, TC4 coating, and TC4 alloy [8]. Zhang et al. investigated the influence of laser processing parameters and the type of Ni-based alloy on the microstructure and characteristics of the grey cast iron matrix and the laser-clad layer. A layer of 316L stainless steel was placed between the grey cast iron substrate and the Ni/WC coating to prevent cracking of the cladding layer. The tribological and mechanical characteristics of the laser-clad coatings were determined using various tests at both macro- and micro-scales. The residual stresses at the coating interfaces were measured, and electrochemical tests were also conducted [9].

This chapter presents the use of SYSWELD simulation software as an essential tool for simulating the complex thermal and mechanical responses produced by 316L stainless steel during a laser cladding process. SYSWELD, a finite element analysis developed to meet the specific needs of welding and cladding simulations, enables researchers to forecast and study heat distribution, temperature ranges, and the thermal cycle as the laser penetrates the material. This simulation ability is further applied in simulating the formation of stress and the creation of residual stresses resulting from rapid heating and cooling, allowing researchers to evaluate the structural integrity of the clad layer [10]. It also simulates material transformation processes, that is, phase transformation and microstructural evolution, both of which play central roles in determining the ultimate properties of the coated surface [11, 12]. By providing a virtual environment for experimenting with various cladding conditions, including laser power, scanning speed, and material feed rate, SYSWELD enables researchers to optimise these parameters, thereby reducing the defects associated with porosity, cracking, or delamination in the final product [13–15]. This simulation has been verified to be highly reliable, as its predictions are compared with the experimental data from actual laser cladding tests on 316L stainless steel specimens. Validation makes this model more meaningful by ensuring that it aligns with reality (real-world outcomes). This, in turn, establishes ANSYS, ABAQUS, and SYSWELD software as a reliable source for process design and optimisation, saving time and resources in experimental trials. These optimised cladding methods have significantly extended the service life of 316L components, as demonstrated in practical examples, such as turbine blades or chemical reactors, where parts coated by these methods exhibit remarkable resilience despite the adverse environment of high temperatures and high wear. Another critical issue that the study addresses is coating substrate mismatch, where inconsistency of the thermal expansion or material association is present between the substrate and the cladding layer, thereby weakening the bonding of the material in the assembly, causing delamination or failure. The purpose of this work is to model the laser cladding process before undertaking the practical procedure to achieve the best energy input, speed, spot size, etc. In the project, parameters such as laser beam power and traverse speed will be studied. Based on such data, the actual laser cladding process was performed.

4.2 MATHEMATICAL MODELLING OF HEAT SOURCE

Laser welding is characterised by highly collimated and concentrated beam energy; therefore, it is difficult to provide accurate measurements of temperature, fusion zone, and HAZ. Hence, finite element analysis is used for this purpose.

4.2.1 Heat Transfer in Welding

In heat transfer analysis, the transient is conducted by solving the governing partial differential equation for the transient heat conduction. The governing partial differential equation for the transient heat conduction (temperature field in time) (t) and space (x, y, z) is given by Eq. 4.1.

$$\left[\frac{\partial}{\partial x}\left(K\frac{\partial T}{\partial x}\right) + \frac{\partial}{\partial y}\left(K\frac{\partial T}{\partial y}\right) + \frac{\partial}{\partial z}\left(K\frac{\partial T}{\partial z}\right)\right] + \overset{o}{q} = \rho C\frac{dT}{dt} \tag{4.1}$$

where x, y, z are the Cartesian coordinates and $\overset{o}{q}$ is the internal heat generation. In addition, ρ refers to the density, K refers to the thermal conductivity, and C_p

refers to the specific heat, which is a function of temperature T. Both base and weld metal are isotropic; hence, the thermal conductivity value is the same in all directions.

Matrix form of the above equation can be written as Eqs. 4.2 and 4.3:

$$\rho C\frac{\partial T}{\partial t} = (L)^T[|D|\{L\}T] + Q \tag{4.2}$$

where,

$$L = \begin{bmatrix} \frac{\partial}{\partial x} \\ \frac{\partial}{\partial y} \\ \frac{\partial}{\partial z} \end{bmatrix} \quad \text{and} \quad D = \begin{pmatrix} K & 0 & 0 \\ 0 & K & 0 \\ 0 & 0 & K \end{pmatrix} \tag{4.3}$$

In a welding process, the cooling of the part through radiation and convection plays a significant role; hence, to evaluate a realistic temperature history, heat loss to the surroundings has been considered by accounting for both radiation and convection losses. During the thermal cycle, radiation heat losses dominate in and around the weld pool, while convection losses dominate away from the weld pool. Combined radiation and convection loss from the material to the surroundings is given by Eqs. 4.4 and 4.5.

$$q_n = h(T - T_\alpha) + \varepsilon\sigma\left(T^4 - T_\alpha^4\right) \tag{4.4}$$

$$q_n = \left[h + \varepsilon\sigma\left(T^2 + T_\alpha^2\right)(T + T_\alpha)\right](T - T_\alpha) = h_{eff}(T - T_\alpha) \tag{4.5}$$

where the first term is convective heat loss, and $h = 25\ \text{Wm}^{-2}\text{K}$, the convective heat transfer coefficient. The second term is the heat loss due to radiation.

$\sigma = 5.67 \times 10^{-8}\text{Wm}^{-2}\text{K}^{-4}$ Stefan's constant, $\varepsilon = 0.8$ as the emissivity of the plate.

4.2.2 Heat Source Distribution Model

Selecting an adequate heat-source model is essential in the welding simulations, and it should be considered to describe the real physical phenomena in LBW. A review of past works shows that numerous research directions are focused on developing a reliable, easily maintained, and easily controlled source of heat for welding processes. The central phenomenon in the welding process is heat transfer. The temperature transfer between the welding source and the base material must also be understood, as it determines the shape of the molten pool or bead, which, in turn, determines the shape of the resulting solidification. A laser keyhole welding heat source is one of the most challenging heat sources to model using the finite element method.

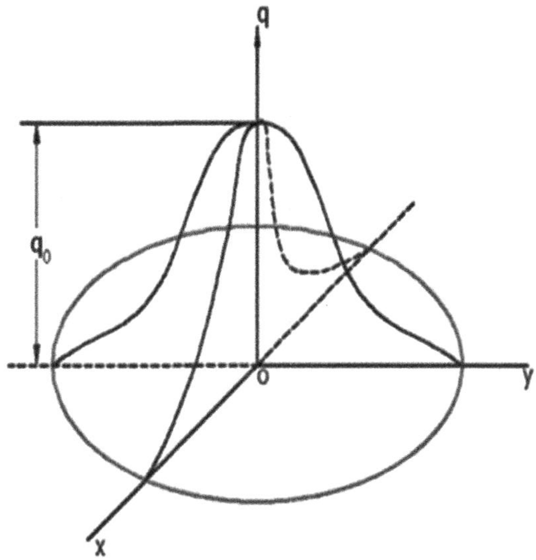

FIGURE 4.2 Gaussian distribution of heat flux.

Most researchers presupposed that heat flux (W/m^2) deposited on the surface of the workpiece was distributed in a Gaussian manner, as shown in Figure 4.2.

The following equation expresses the geometric features of the Gaussian distribution.

$$(r) = q_0 \exp\left(Cr^2\right) \tag{4.6}$$

where $q(r)$ is the heat flux, r is the radius away from the heat source centre, q_0 is the maximum heat flux, and C is a constant.

The selection of a suitable heat source type is crucial for welding simulation, as it defines the fundamental physical processes in laser beam welding. Upon reviewing the alleged past works, it becomes apparent that considerable effort has been invested, leading to the pursuit of a stable, easily serviced, and controlled heat source for welding applications. Heat transfer is the key phenomenon underlying the welding activity. The temperature transfer between the welding source and the base material is crucial for forming the molten pool or bead, as it determines the shape of the resulting solidification. The laser keyhole welding heat source is among the most challenging heat sources to model with the finite element method. Most researchers assumed that the heat flux deposition (W/m^2) on the workpiece's surface follows a Gaussian probability distribution.

3D conical Gaussian heat source shown in Figure 4.3 is used to describe the laser beam heat input, where the power density deposited region is maximum at the top surface of workpiece, and minimum at the bottom surface of the workpiece. The diameter of power density decreases linearly concerning the thickness, as shown in Figure 4.3. The power density is expressed as:

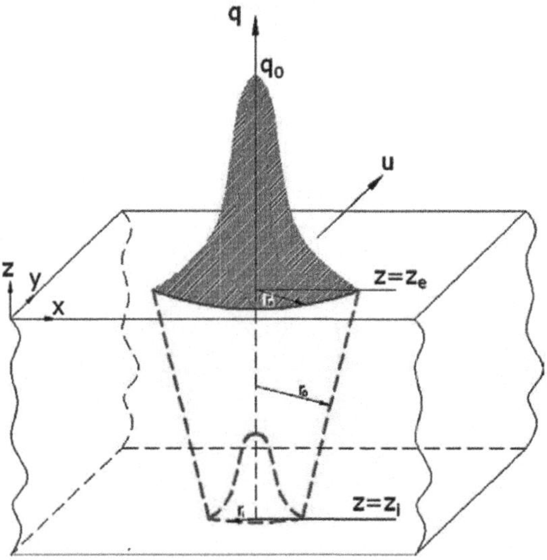

FIGURE 4.3 3D Conical Gaussian heat source.

$$Q_r = Q_0 \exp\left(\frac{-3r^2}{r_0^2}\right) \tag{4.7}$$

where r and r_0 is given by

$$r = \sqrt{x^2 + y^2}$$

$$r_0 = r_e - \frac{(r_e - r_i)(z_e - z)}{(z_e - z_i)} \tag{4.8}$$

where

r_e & r_i = surface radius in planes $z = z_e$ & z_i, respectively,
Q_0 = maximum intensity,
Q_r is the heat source intensity.

4.2.3 Geometrical Modelling

Two plates of dimensions 150 mm × 100 mm × 3 mm are used in the experiment. During meshing, 8-node elements of hexahedra of different sizes have been utilised. There are 230,464 elements (solid) and 34,394 nodes. The thermal gradient of the fusion zone (FZ) and heat-affected zone (HAZ) is quite high; hence, the meshing in FZ and HAZ is very fine. FZ and (HAZ) sized elements are 0.25 × 0.5 × 0.375 mm. Away from the weld line, the temperature gradient is small; consequently, as we move further away, the mesh is coarsened, that is, the elements are larger. Model creation and meshing have been done using Visual Mesh—a subpackage of SYSWELD. Figure 4.4 shows the meshed model of the quadrant cut section of the plate.

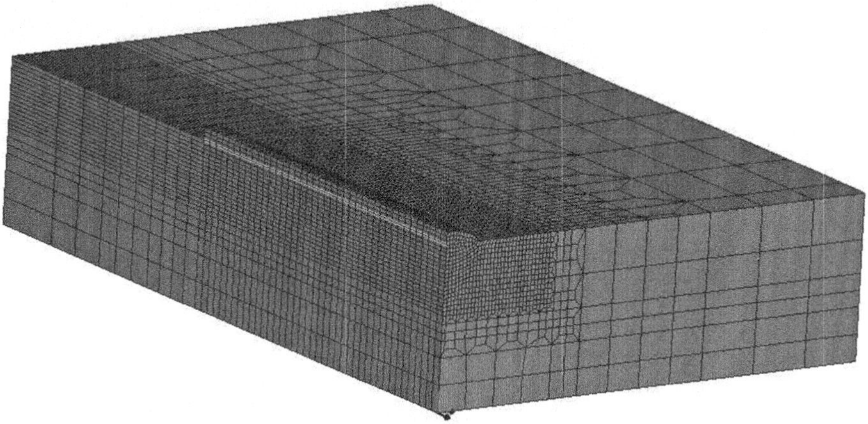

FIGURE 4.4 Quadrant cut section of the plate meshing.

The melt pool size and shape at the centre of laser clad tracks were examined for laser powers ranging from 200 to 400 W and speeds of 20–60 cm/min.

4.3 RESULTS AND DISCUSSION

Finite element analysis of cladding at different heat inputs was carried out using the conical-shaped heat source model. Figure 4.5 shows the size and shape of the molten pool when the laser beam moves to the middle of the workpiece for the 0.3 mm-thick clad layer.

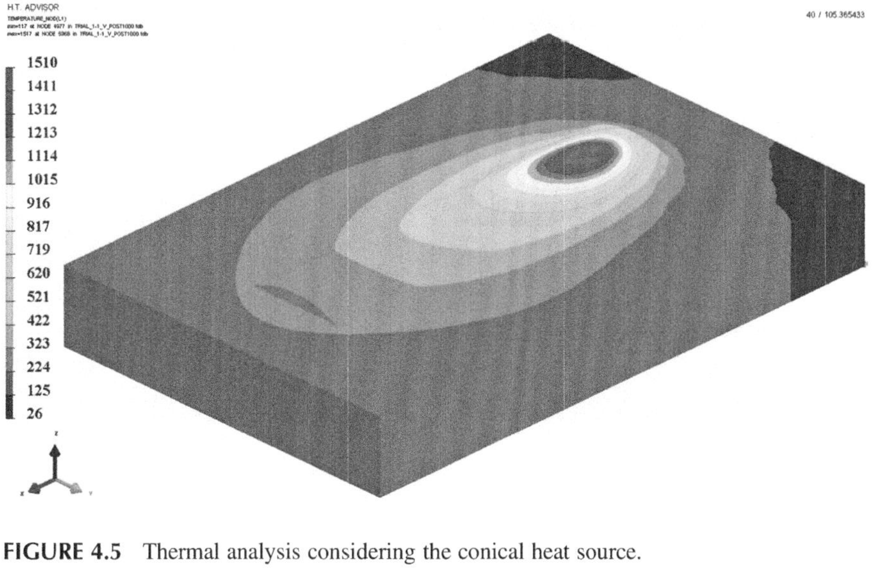

FIGURE 4.5 Thermal analysis considering the conical heat source.

The depth of the melt pool is between 0.15 mm and 0.65 mm at a laser power of 200 W and a traverse speed of 20, 30, and 60 cm/min. The melt pool depth of 0.35 mm, achieved with a traverse speed of 30 cm/min and 300 W power, indicates that the substrate remains unmelted. The thickness of the clad layer is 0.3 mm (see Figures 4.6–4.8). At 200 W, the substrate does not melt during the described speed range. The size and shape of the molten pool at a thickness of the clad layer of 1 mm are shown in Figure 4.6. At 400 W, the depth of the melt pool can range between 0.2 mm and 0.56 mm at a speed range of 20–60 cm/min. Melt pool depths of 0.2 mm at 60 cm/min and 0.3 mm at 40 cm/min are compatible with laser cladding, resulting in a satisfactory product with reduced porosity due to efficient bonding.

The decrease in the size of the melt pool at high transverse speed is due to a decrease in the interaction time of the laser beam with the material. On the other hand, increasing the laser power enlarges the melt pool, as the substrate gets more superheated and consequently undergoes greater melting. As the traverse speed increases, the melt pool shape transitions from elliptical to stretched out—a change that is even more pronounced at higher laser power. Consequently,

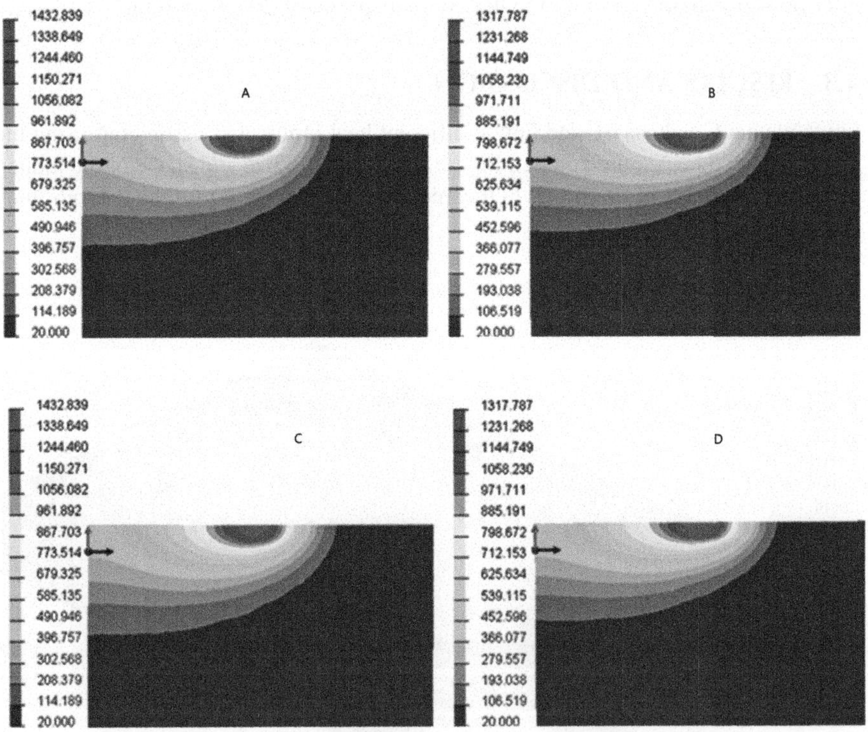

FIGURE 4.6 Molten pool size of 0.5 mm at 200 W power at (a) 20 cm/min, (b) 30 cm/min, (c) 40 cm/min, and (d) 60 cm/min speed.

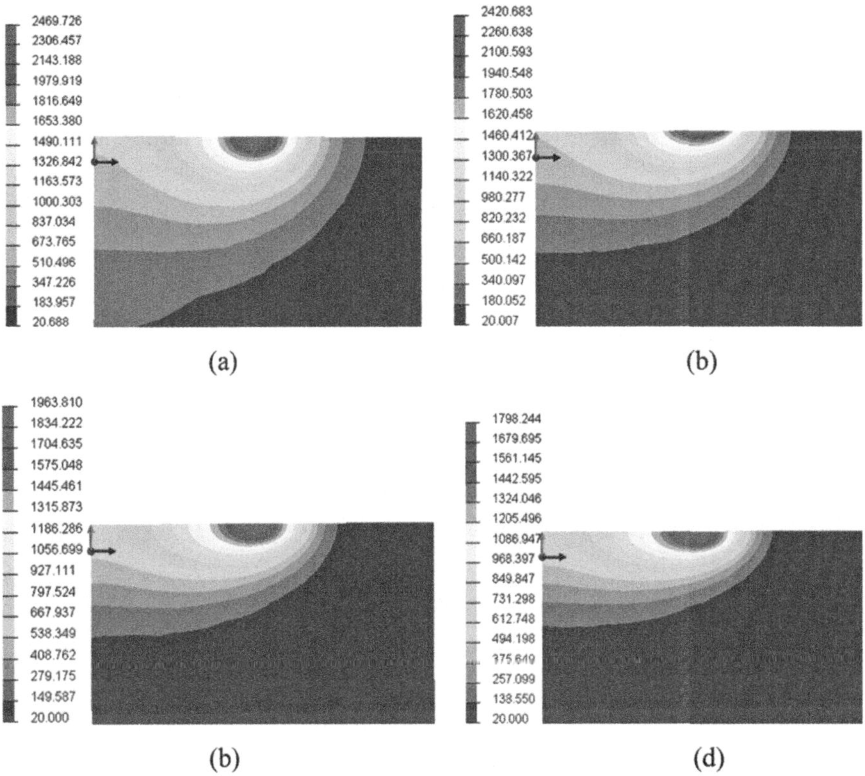

FIGURE 4.7 Molten pool size of clad 0.5 mm at 300 W power at (a) 20 cm/min, (b) 30 cm/min, (c) 40 cm/min, and (d) 60 cm/min speed.

the optimal power for laser cladding, in terms of the stacking layer thickness (0.3 mm), is 400 W.

4.3.1 Molten Pool Validation

Based on finite element method (FEM) based analysis at different heat inputs, the optimum heat input is selected at 400 W with 30 cm/min, and the experiments were conducted at the same heat input. The experimentally obtained melt pool shape is compared with the simulated models, showing good agreement between them. The melt pool depth was determined by taking into account the melting temperature of austenitic stainless steel (1450°C). Figure 4.9 shows a comparison between the experimentally obtained and the simulated melt pool corresponding to a beam power of 400 W and a traverse speed of 600 mm/min on a sample with a clad thickness of 0.5 mm. As expected, the melt pool depth decreased with increasing speed at a constant laser power.

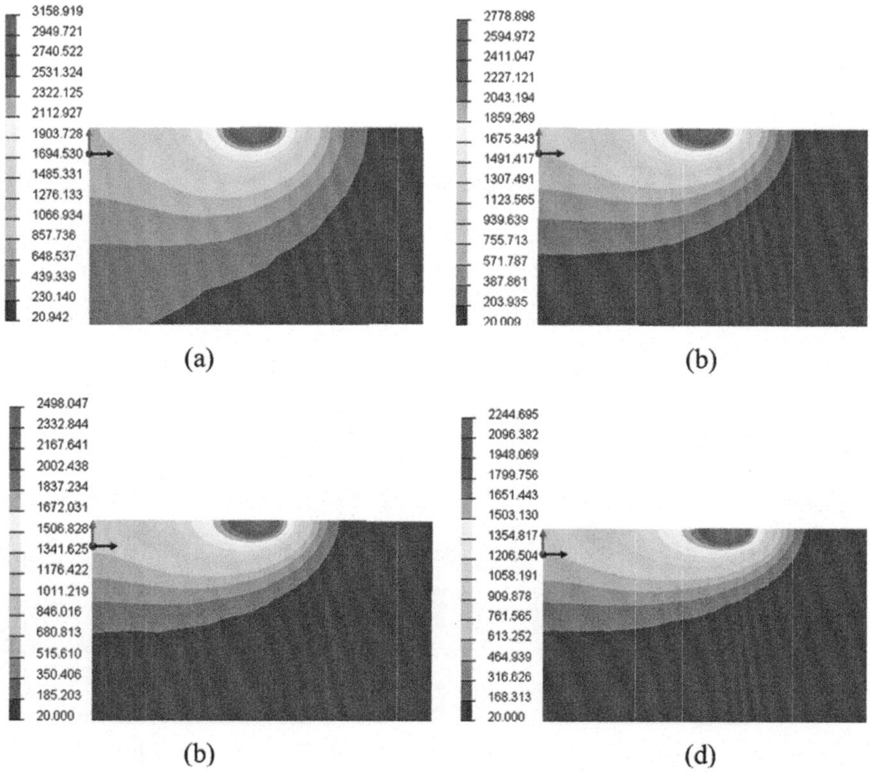

FIGURE 4.8 Molten pool size of clad 0.5 mm at 400 W power at (a) 20 cm/min, (b) 30 cm/min, (c) 40 cm/min, and (d) 60 cm/min speed.

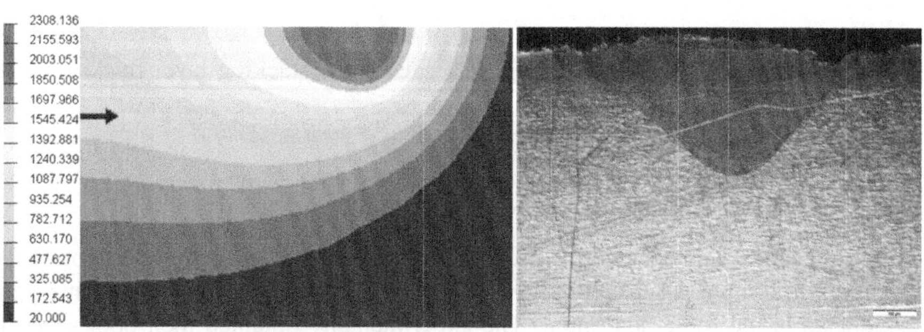

FIGURE 4.9 Comparison of the melt pool at 300 W power and 40 cm/min speed: (a) FEM based predicted and (b) experimental profile.

4.4 CONCLUSION

The modelling showed that the size, shape, and temperature distribution of the melt pool during the laser cladding process are controlled first by the laser beam power and second by traverse speed. Higher laser power increases the melt pool depth and alters the temperature distribution, whereas higher traverse speed produces shallower melt pools due to the reduced interaction time. The model is accurate, as the simulations indicate a remarkable consistency with experiments. It is worth noting that the degree to which melt pool characteristics are affected by alterations in laser power density is higher as compared to that of the traverse speed, which further underscores power as the overriding variable in the process of maximising the results of laser cladding.

REFERENCES

1. Lucia Lizzul, Marco Sorgato, Rachele Bertolini, Andrea Ghiotti, Stefania Bruschi, Francesco Fabbro, Silvano Rech, On the influence of laser cladding parameters and number of deposited layers on as-built and machined AISI H13 tool steel multilayered claddings, *CIRP Journal of Manufacturing Science and Technology*, Vol. 35, 2021, 361–370, https://doi.org/10.1016/j.cirpj.2021.07.003
2. Kang Qi, Yong Yang, Rui Sun, Guofang Hu, Xin Lu, Jindong Li, Wanxu Liang, Kang Jin, Li Xiong, Effect of magnetic field on crack control of Co-based alloy laser cladding, *Optics & Laser Technology*, Vol. 141, 2021, 107129, https://doi.org/10.1016/j.optlastec.2021.107129
3. Chengchao Du, Lei Hu, Qiuhong Pan, Kangmin Chen, Pengjie Zhou, Guangjin Wang, Effect of Cu on the strengthening and embrittling of an FeCoNiCr-xCu HEA, *Materials Science and Engineering A*, Vol. 832, 2022, 142413, https://doi.org/10.1016/j.msea.2021.142413
4. Guofang Hu, Yong Yang, Rui Sun, Kang Qi, Xin Lu, Jindong Li, Microstructure and properties of laser cladding NiCrBSi coating assisted by electromagnetic-ultrasonic compound field, *Surface and Coatings Technology*, Vol. 404, 2020, 126469, https://doi.org/10.1016/j.surfcoat.2020.126469
5. Saeedeh Jelvani, Reza Shoja Razavi, Masoud Barekat, Mohammad Reza Dehnavi, Mohammad Erfanmanesh, Evaluation of solidification and microstructure in laser cladding Inconel 718 superalloy, *Optics & Laser Technology*, Vol. 120, 2019, 105761, https://doi.org/10.1016/j.optlastec.2019.105761
6. A. Alimohammadi, A. Khorram, A. Araee, Microstructural evolution of laser clad amdry 365 powder on Hastelloy X superalloy, *Sādhanā*, Vol. 50, 2025, 114. https://doi.org/10.1007/s12046-025-02776-0
7. D.-D. Zhuang, B. Du, S.-H. Zhang, W.-W. Tao, Q. Wang, H.-B. Shen, Effect and action mechanism of ultrasonic assistance on microstructure and mechanical performance of laser cladding 316L stainless steel coating, *Surface and Coatings Technology*, Vol. 433, 2022, 128122, https://doi.org/10.1016/j.surfcoat.2022.128122
8. Q. Wang, K. Wang, X. Zhang, B. Fan, Y. Xi, L. Dong, S. Bai, Microstructure feature and wear mechanism of laser-clad TC4 and TC4-WC coating, *Anti-Corrosion Methods and Materials*, Vol. 72 (3), 2025, 444–455, https://doi.org/10.1108/ACMM-12-2024-3155
9. R. Zhang, Y. Liao, Q. Sai, S. Wang, D. Barton, M. Bai, Properties of laser-clad stainless steel–Ni/WC double-layer coatings. *Coatings*, Vol. 14(8), 1054, https://doi.org/10.3390/coatings14081054
10. M. Zubairuddin, M. Vasudevan, P.V. Elumalai et al. Numerical and experimental analysis of temperature and residual stress of GTA and LASER welding for grade 91 steel, *International Journal on Interactive Design and Manufacturing*, 2024, https://doi.org/10.1007/s12008-024-02138-w
11. M. Zubairuddin, M. Vasudevan, P.K. Das et al. FEM based thermal and mechanical analysis of comparative study of TIG and A-TIG welding on P91 steel, *Scientific Reports*, Vol. 15, 2025, 10271. https://doi.org/10.1038/s41598-025-90998-x

12. M. Zubairuddin, R. Ravi Kumar, B. Ali Thermo-mechanical analysis of laser welding of Grade 91 steel, *Optik*, Vol. 245, 2021, 167510.
13. M. Zubairuddin, S.K. Albert, M.S. Reddy, B. Ali, B. Varaprasad, A. Mishra, P.K. Das, P.V. Elumalai, Thermal analysis of thin P91 steel using FlexPDE and SYSWELD, *Materials Today Proceedings*, Vol. 72, 2023, 1550–1555.
14. J.R. Chukkan, M. Vasudevan, S. Muthukumaran, R. Ravi Kumar, N. Chandrasekhar, Simulation of laser butt welding of AISI 316L stainless steel sheet using various heat sources and experimental validation, *Journal of Materials Processing Technology*, Vol. 219, 2015, 48–59.
15. Harinadh Vemanaboina, M. Mohan Babu, Inguva Chinmaya Prerana, Edison Gundabattini, Balram Yelamasetti, Kuldeep K. Saxena, Karrar Hazim Salem, Muhammad Ijaz Khan, Sayed M. Eldin, Manoj Kumar Agrawal, Evaluation of residual stresses in CO2 laser beam welding of SS316L weldments using FEA, *Materials Research Express*, Vol. 10, 2023, 016509.

5 Development of Nickel-Based Alloy Coatings and Claddings

Techniques and Advancements

Suresh Gudala, Subbarao Medabalimi, Uzwalkiran Rokkala, and Prakash Kumar

5.1 INTRODUCTION

Nickel-based alloys, comprising at least 50% nickel, are essential for meeting the needs of diverse industrial applications and typically contain additional elements such as Cr, Fe, Mo, and Al. The nickel-based alloy shows high strength in high-temperature environments, with operational temperatures reaching 650–1000°C, making them indispensable in aerospace, marine, petroleum, and other critical fields [1]. Established commercial nickel-based alloy companies, such as Special Metals and Haynes International, offer well-developed products. Special Metals' main nickel-based alloys include Inconel 601, Inconel 617, and Inconel 690. Haynes International offers nickel-based alloys like Haynes 230, Haynes 242, Haynes 556, HR-120, and HR-160. Inconel 625 is a widely used nickel-based superalloy known for its corrosion and oxidation resistance in chlorinated environments up to 800°C, making it suitable for chemical and aerospace applications [2, 3]. Inconel 718 is preferred for turbine blades and petroleum equipment due to its superior performance [4]. Inconel 740H offers strong corrosion resistance in supercritical water, thanks to its excellent high-temperature oxidation stability [5]. In addition, Haynes 230, a nickel-based superalloy strengthened by W and Mo, possesses superior resistance to high-temperature oxidation and deformation. It finds applications in the production of aerospace gas turbine components, flame holders, and combustor cans [6]. Haynes 214 forms an alumina protective layer at high temperatures, showing excellent high temperature and oxidation resistance; therefore, it is often used in turbine seal honeycomb structures [7]. The evolution of superalloys has been affected by two significant factors: high-temperature mechanical performance and resistance to oxidation. This, in turn, leads to the classification of superalloys based on application and composition. Although there is a significant difference in the microstructure of wrought superalloy and

DOI: 10.1201/9781003653226-5

precipitation-hardened superalloys, the development of both classes is observed to have occurred almost simultaneously. The gradual changes in the composition of nickel-based superalloys led to changes in the physical structure of the primary precipitates. The matrix (γ-phase) is a disordered face-centered cubic (FCC) structure throughout the range of nickel-based superalloys. The primary precipitates have evolved from a non-cubic carbide type to cubic $Ni_3(Al, Ti)$ with an L12 structure. H. Kirchhoefer et al. reported that Ni_3Al is a secondary phase in IN 617 wrought alloy with M6C as a primary precipitate [8]. Similarly, precipitation-strengthened Hastelloy and Nimonic have carbide precipitates within a granular structure [9]. Nickel-based superalloys exhibit a variety of precipitate phases, each contributing differently to the alloy's microstructural stability and mechanical performance. The δ phase, with an orthorhombic structure and a solvus temperature of approximately 1040°C, commonly appears at interdendritic regions. This phase plays a role in restricting grain growth and enhancing mechanical properties at elevated temperatures [10]. The η phase, characterized by a hexagonal close-packed (HCP) structure and composed primarily of Ni_3Ti, tends to form at grain boundaries and is generally considered detrimental due to its adverse effects on alloy performance [11]. The γ' phase (FCC Ni_3Al or $Ni_3(Al,Ti)$) forms coherent precipitates that enhance strength, though excessive lattice mismatch can cause unwanted η or δ phases [12]. The γ'' phase, a BCT Ni_3Nb-based precipitate, forms near 850°C and contributes significantly to hardening alongside γ' [13, 14]. In contrast, the σ phase, a TCP structure linked to Mo and Cr, negatively impacts ductility and toughness [11, 13].

5.2 COATING AND CLADDING METHODS

Numerous surface engineering techniques are employed for developing nickel-based coatings to enhance wear and corrosion resistance. Laser cladding is recognized for its superior bonding strength, low dilution, and high processing precision; however, it demands expensive equipment and involves complex procedures with extended production cycles, alongside a large heat-affected zone [15, 16]. Cold spraying, on the other hand, offers minimal residual stress and preserves the substrate's composition, though its high equipment costs and limitations in coating thickness are notable drawbacks [17, 18]. Plasma spraying is appreciated for its uniform coating and strong adhesion, but suffers from high energy consumption and brittleness due to environmental sensitivities [19]. Arc spraying presents a cost-effective and established alternative, but is restricted by its lower bonding strength and coating quality [20]. High-Velocity Oxygen Fuel (HVOF) spraying supports broad material compatibility and is suitable for high-volume production due to its low cost and high deposition rate. Nonetheless, it can lead to surface roughness and reduced adhesion depending on process parameters [21, 22]. Detonation flame spraying provides economical coating of large surfaces but may result in non-uniform thickness and compromised bonding [23]. Electroless plating delivers uniform coatings with accurate thickness control across diverse substrates, yet it requires careful management of deposition conditions and often results in coatings with lower hardness and wear resistance [24, 25]. Electrochemical deposition is widely used for mass production because of its low cost and controllable outputs, though the coatings typically show poor mechanical and corrosion resistance [26, 27]. Plasma transfer arc surfacing enables strong, corrosion-resistant coatings with precise control, although it involves complex and costly setups, along with lengthy preparation times [28, 29]. Electrospark deposition generates high-hardness coatings suitable for a wide range of materials but may result in rough surfaces due to localized heating [30, 31]. Physical Vapor Deposition (PVD) provides coatings with high uniformity and excellent adhesion but demands substantial equipment and strict processing conditions [32]. Lastly, mechanical alloying ensures coatings with consistent composition and homogeneity, though it is generally limited to small-scale applications and is not suited for bulk production [33].

5.3 NICKEL-BASED ALLOY COATINGS AND CLADDINGS

Various coating preparation methods and post-treatment techniques significantly influence the phase constitution and mechanical behavior of nickel-based coatings. For instance, in laser cladding using Inconel 625 with aluminum additions, Al promotes the formation of a body-centered cubic (BCC) phase, thus transforming the microstructure from a purely FCC and δ + Laves configuration to a hybrid structure containing FCC, BCC, and δ + Laves phases [34]. Similarly, ultrasonic-assisted laser cladding with $Ni_2MnCuSnAl_{0.1}$ facilitates the emergence of BCC solid solutions incorporating complex intermetallics such as NiMnCuSnAlMgLi, Ni_2MnAl, and Ni_2MnSn [35]. In another study on plasma-sprayed $NiAlAlAgMoS_2/BN$ coatings, the integration of boron nitride (BN) in coatings refines the resulting microstructure by promoting the formation of strengthening intermetallic compounds like Ni_3Al and Ni_5Al_3 [36]. HVOF spraying of Colmonoy88 gives rise to both primary γ(FCC)-Ni and Fe_3C phases, along with secondary hard carbides including M_6C, Cr_7C_3, and Ni_3Fe [37]. Additionally, plasma transferred arc (PTA) cladding using Ni–Cr compositions produces a microstructure composed of γ-Ni dendrites and eutectic carbides such as MC and M_2C [38]. Electrochemical and ultrasonic techniques also contribute significantly to coating refinement. The inclusion of TiN during ultrasonic-electrodeposition of Ni–Cu/TiN restricts grain coarsening and encourages uniform FCC phase development [39]. Electro-spark deposition of NiCoCrAlYTa enhances oxidation resistance by increasing β-NiAl and promoting the formation of stable oxides like $Y_3Al_5O_{12}$ through the phase transformation of θ-Al_2O_3 [40]. Likewise, electrodeposition involving TaC or Ni–W/TiN alloys leads to microstructural tuning and grain refinement via lattice stabilization and FCC structure enhancement [41]. Post-processing treatments further optimize coating properties. For example, laser remelting of Ni–B–Ti systems yields reinforcing phases such as TiNi, Ti_2Ni, and TiB whiskers, which improve toughness and wear resistance. Electron beam remelting of Ni–Ti coatings produces a multi-phase structure (Ti, TiC, TiNi, Ti_2Ni, and Ni_3Ti), where dislocations and twinning enhance strength. Thermal treatment of electroless Ni–P/Mo coatings lowers phosphorus levels and forms NiO and Mo oxides, improving surface morphology and performance [42]. These findings show that coating properties are influenced by both deposition methods and post-treatments. By adjusting alloy compositions and applying advanced treatments, coatings with superior hardness, wear, corrosion resistance, and thermal stability can be engineered, demonstrating the adaptability of nickel-based systems for industrial surface applications.

5.4 MICROSTRUCTURE, PHASE ANALYSIS, AND MECHANICAL PROPERTIES

To meet the demanding requirements for nickel-based coatings in wear- and corrosion-prone environments, researchers have developed coatings with tailored micro- and nanostructures by incorporating various strengthening agents and optimizing fabrication techniques. These engineered coatings exhibit distinct performance characteristics influenced significantly by their processing routes. The variation in preparation methods often results in substantial differences in coating properties, emphasizing the critical role of fabrication parameters in determining final performance outcomes. Coating performance depends on microstructural factors like grain size, stress distribution, and bonding quality. Recent research focuses on linking preparation methods with microstructural evolution to optimize process parameters for better strength and durability. Key tests involve microstructure, hardness, wear, and corrosion assessments [43, 44]. Alloying elements such as Cr, Mo, Co, and Fe enhance strength through solid solution strengthening [45], while ceramic reinforcements like SiC, TiC, and ZrO_2 improve hardness by blocking dislocation motion [46]. Uniform dispersion of these phases ensures consistent mechanical properties and surface stability. Advanced microstructural analyses, including the evaluation of grain size, grain boundaries, subgrains, and twins, provide insights into the strengthening mechanisms at the micro- and nano-scale. This level of understanding is vital for the continued optimization of coating processes. In post-service analysis, semi-quantitative and qualitative techniques are widely employed to assess the impact of wear or corrosion on coating integrity. By examining changes in phase composition and elemental states, researchers can gain insight into damage mechanisms and propose design improvements. Strengthening through solid solution formation, second-phase reinforcement, and grain refinement contributes to the enhanced mechanical behavior of these coatings [47–49].

Friction and wear properties serve as primary indicators of a coating's operational effectiveness [50]. The proportion and type of strengthening phases can be carefully adjusted to influence tribological performance. For example, Cao et al. [51] demonstrated that modifying the tungsten carbide (WC) content in a laser-cladded Ni–WC coating significantly improved its wear resistance under both room and low-temperature conditions. Similarly, H. Zhang et al. [52] found that zirconium phosphate compounds (ZrPP and α-ZrP) effectively enhanced wear mechanisms, further increasing coating durability. In addition to reinforcement phases, other microstructural features such as carbides, twin boundaries, and dislocation densities also impact wear resistance. Controlling or inhibiting the formation of brittle or deleterious phases remains an area of active investigation. Few studies reported that residual compressive stresses induced by carbides and dislocation structures can suppress crack propagation, thereby improving durability under mechanical loading. Further studies underscore the role of elemental additions and process optimization in phase transformation and property enhancement. For instance, Hong et al. [53] investigated the impact of niobium (Nb) on phase formation in a Ni–WC/Nb system and found that increased Nb content reduces the formation of Cr_7C_3 while promoting NbC, which exhibits favorable morphology and adhesion characteristics. This transformation reduces hard phase fracture during wear, thereby enhancing coating longevity. Process optimization also plays a decisive role in improving mechanical performance. Zhang et al. [52] reported that ultrasonic-assisted electrodeposition of Ni–SiC composites results in a higher incorporation of SiC particles compared to conventional techniques. The uniform spread of SiC particles significantly boosts the hardness and strength of the coating by enhancing load distribution within the matrix. The overall performance of nickel-based coatings arises from the combined influence of their microstructure and reinforcement methods. Continuous advancements in material design and a better grasp of secondary phases, dislocation networks, and internal stresses have led to coatings tailored for industrial needs. Current studies aim to further improve their efficiency and lifespan in tough working environments.

5.5 SUMMARY AND FUTURE SCOPE

This chapter comprehensively explores the development, processing techniques, and performance of nickel-based alloy coatings and claddings. Emphasis has been placed on understanding how various deposition methods affect the resulting microstructure and mechanical behavior of the coatings. The coating/cladding methods, such as thermal spraying methods, laser cladding, PTA, and electrochemical deposition, were analyzed for their characteristics in efficient coatings. The present chapter also highlights the role of alloying elements such as Fe, Mo, Cr, and Co, which may contribute to the solid solution strengthening. Ceramic reinforcements such as SiC, TiC, and ZrO_2 may improve the strength by acting as secondary phases, which may resist the dislocation movement. The secondary treatments, including remelting with laser and electron beam, further modify the microstructure by refining the grains and promoting the strengthening phases, which may further improve the mechanical properties. Among the promising research directions, computational modeling and data-driven approaches for optimizing process parameters and predicting coating performance enhance the efficiency in achieving the required properties. Finite element method (FEA), machine learning, and phase diagram simulations may further enhance predictability and reduce the trial and error in process development. In conclusion, continuous innovation in material design, process optimization, and performance evaluation will be essential for advancing the application of nickel-based coatings in high-performance and high-reliability sectors.

REFERENCES

1. V.T. Targhi, H. Omidvar, S.M.M. Hadavi, F. Sharifianjazi, Microstructure and hot corrosion behavior of hot dip siliconized coating on Ni-base superalloy IN738LC, *Mater Res Express* 7 (2020) 056527, https://doi.org/10.1088/2053-1591/ab6dab
2. M.K. Hariharan, A. Anderson, K. Raghavan, S. Nithya, Hot corrosion behaviour of Hastelloy X and Inconel 625 in an aggressive environment for superalloys for high-temperature energy applications, *Appl. Nanosci.* 13 (2023) 3359–3368, https://doi.org/10.1007/s13204-021-02077-y

3. S. Zhang, Y. Wang, L. Lv, H. Deng, Q. Bian, Q. Hu, L. Tan, F. Liu, Effect of powder layer thickness on the microstructure and properties of Inconel 625 superalloy manufactured by selective laser melting, *J. Alloys Compd.* 1020 (2025) 179465, https://doi.org/10.1016/j.jallcom.2025.179465
4. Z. Tang, C. Yang, Y. Duan, L. Ma, S. Zheng, M. Li, Corrosion and wear behaviors of Inconel 718 nickel-based alloy by boroaluminizing, *Surf. Coat. Technol.* 478 (2024) 130500, https://doi.org/10.1016/j.surfcoat.2024.130500
5. D. Wang, S. Wang, Z. Chen, X. Xie, Z. Dong, H. Dong, Y. Wu, Z. Liu, W. Li, Y. Wang, Corrosion behavior of additive-manufactured NiFeCrMo alloys in various corrosion media, *J. Iron Steel Res. Int.* 30 (2023) 1574–1585, https://doi.org/10.1007/s42243-023-01033-0
6. Z. Li, Z. Guo, J. Li, W. Guo, J. Xiong, Microstructure and mechanical properties of dissimilar diffusion bonded joints of Haynes 230 and Haynes 188 superalloys, *J. Mater. Res. Technol.* 29 (2024) 1971–1982, https://doi.org/10.1016/j.jmrt.2024.01.147
7. J. Vogler, R. Volkl, J. Song, V.F. Viebranz, J. Huber, H.-J. Bauer, E. Werner, U. Glatzel, High temperature and high strain rate properties of brazed honeycomb liner material Haynes 214, *Wear* 550–551 (2024) 205427, https://doi.org/10.1016/j.wear.2024.205427
8. H. Kirchhoefer, F. Schubert, H. Nickel, Precipitation behavior of Ni–Cr–22 Fe–18 Mo (Hastelloy-X) and Ni-Cr-22 Co-12 Mo (Inconel-617) after isothermal aging, *Nucl. Technol.* 66 (2017) 139–148, https://doi.org/10.13182/NT84-A33462
9. S.Y. Lee, Y.L. Lu, P.K. Liaw, H. Choo, S.A. Thompson, J.W. Blust, P.F. Browning, A.K. Bhattacharya, J.M. Aurrecoechea, D.L. Klarstrom, High-temperature tensile- hold crack-growth behavior of Hastelloy-X alloy compared to Haynes-188 and Haynes 230 alloys, *Mech. Time Depend. Mater.* 12 (2008) 31–44, https://doi.org/10.1007/s11043-008-9049-6
10. Y. Wu, S. Li, M. Kang, Y. Li, Y. Liu, H. Gao, B. Sun, J. Wang, Slip and fracture behavior of δ-Ni3Nb plates in a polycrystalline nickel-based superalloy during fatigue, *Scr. Mater.* 171 (2019) 36–41, https://doi.org/10.1016/j.scriptamat.2019.06.013
11. F. Long, Y.S. Yoo, C.Y. Jo, S.M. Seo, H.W. Jeong, Y.S. Song, T. Jin, Z.Q. Hu, Phase transformation of η and σ phases in an experimental nickel-based superalloy, *J. Alloys Compd.* 478 (2009) 181–187, https://doi.org/10.1016/j.jallcom.2008.11.121
12. F. Pyczak, B. Devrient, H. Mughrabi, The effects of different alloying elements on the thermal expansion coefficients, lattice constants and misfit of nickel-based superalloys investigated by X-ray diffraction, in: Proceedings of the International Symposium on Superalloys, 2004: pp. 827–836. https://doi.org/10.7449/2004/superalloys_2004_827_836
13. N.E. Prasad, R.H.J. Wanhill, *Aerospace Materials and Material Technologies*, Volume 2, Springer, Singapore, Singapore, 2017, https://doi.org/10.1007/978-981-10-2143-5
14. S. Gialanella, A. Malandruccolo, Aerospace alloys, Springer International Publishing, *Cham*, 2020, https://doi.org/10.1007/978-3-030-24440-8
15. Y. Liu, T. Xu, Y. Liu, Y. Gao, C. Di, Wear and heat shock resistance of Ni-WC coating on mould copper plate fabricated by laser, *J. Mater. Res. Technol.* 9 (2020) 8283–8288.
16. Z. Ning, T. Li, G. Hui, Z. Mengqi, Effect of ultra-fine WC particles on microstructural evolution and wear behavior of Ni-based nano-CeO2 coatings produced by laser, *Rare Metal Mater. Eng.* 47 (2018) 20–25.
17. T.B. Torgerson, M.D. Harris, S.A. Alidokht, T.W. Scharf, S.M. Aouadi, R. R. Chromik, J.S. Zabinski, A.A. Voevodin, Room and elevated temperature sliding wear behavior of cold sprayed Ni-WC composite coatings, *Surf. Coat. Technol.* 350 (2018) 136–145.
18. P. Cavaliere, A. Perrone, A. Silvello, Crystallization evolution of cold-sprayed pure Ni coatings, *J. Therm. Spray Technol.* 25 (2016) 1158–1167.
19. K.K. Mirche, K.K. Pandey, S.M. Pandey, A.K. Keshri, Microstructure and corrosion behavior of plasma-sprayed nanodiamond-reinforced NiAl nanocomposite coating, *J. Therm. Spray Technol.* 32 (2023) 1299–1310.

20. P. Daram, P.R. Munroe, C. Banjongprasert, Microstructural evolution and nanoindentation of NiCrMoAl alloy coating deposited by arc spraying, *Surf. Coat. Technol.* 391 (2020) 125565.
21. F. Ghadami, A. Sabour Rouh Aghdam, Improvement of high velocity oxy-fuel spray coatings by thermal post-treatments: a critical review, *Thin Solid Films* 678 (2019) 42–52.
22. A. Mazouzi, B. Djerdjare, S. Triaa, A. Rezzoug, B. Cheniti, S.M. Aouadi, Effect of annealing temperature on the microstructure evolution, mechanical and wear behavior of NiCr–WC–Co HVOF-sprayed coatings, *J. Mater. Res.* 35 (2020) 2798–2807.
23. Z. Zhu, L. Wei, J. Chen, J. Cheng, W. Chen, S. Zhu, J. Yang, Tribological behavior of NiCrAlY/Al2O3/h-BN metal-ceramic composite coatings at wide temperature range by detonation spraying, *Wear* 550–551 (2024) 205421.
24. Y. Meng, Y. Zhao, J. Zhong, Y. Liao, D. Chen, Effect of plating time on surface morphology and corrosion behavior of electroless Ni-P coating in simulated concrete pore solution, *Case Stud. Constr. Mater.* 21 (2024) e03497.
25. Z.H. Xie, Z. Zhuo, Y. Shu, D. Xu, X. Jiang, L. Wu, G. Yu, Polysiloxane-modified porous magnesium hydroxide pretreatment film for electroless plating of highly corrosion-resistant and strongly adhered Ni-P coating on magnesium alloy, *J. Alloys Compd.* 1008 (2024) 176680.
26. C.Y. Ma, D.Q. Zhao, F.F. Xia, H. Xia, T. Williams, H.Y. Xing, Ultrasonic-assisted electrodeposition of Ni-Al2O3 nanocomposites at various ultrasonic powers, *Ceram. Int.* 46 (2020) 6115–6123.
27. W. Cui, Y. Zhang, R. Song, P. Wang, Ultrasonic assisted pulse electrodeposited Ni- doped TiN coatings, *Ceram. Int.* 44 (2018) 14767–14773.
28. J. Sun, C. Chen, G. Zhang, L. Xu, S. Wei, T. Jiang, F. Mao, C. Wang, K. Pan, C. Zhang, Impact abrasive wear of tungsten carbide reinforced NiBSi coating fabricated by plasma transferred arc welding, *Surf. Coat. Technol.* 494 (2024) 131507.
29. K. Luo, J. He, H. Ma, C. Wang, J. Lu, C. Li, Y. Peng, N. Wu, F. Luo, Dissolution behaviour of WC particles and evolution of precipitated phases in the plasma transfer arc Ni-based composite coating reinforced by 30 wt% WC particles, *Ceram. Int.* 50 (2024) 35065–35078.
30. F. Zhou, B. Xing, C. Guo, C. Du, L. Xu, Z. Du, G. Gao, Investigation on the influence of electrospark deposited 718 alloy coating on the penetration performance of 93 W rod, *Int. J. Refract. Met. Hard Mater.* 125 (2024) 106882.
31. A.N. Sheveyko, K.A. Kuptsov, P.V. Kiryukhantsev-Korneev, Y.Y. Kaplansky, A. S. Orekhov, E.A. Levashov, Protective coatings for LPBF Ni-based superalloys using a combination of electrospark deposition and pulsed arc evaporation methods, *Appl. Surf. Sci.* 581 (2022) 152357.
32. S. Azim, S. Gangopadhyay, S.S. Mahapatra, R.K. Mittal, R.K. Singh, Role of PVD coating on wear and surface integrity during environment-friendly micro-drilling of Ni-based superalloy, *J. Clean. Prod.* 272 (2020) 122741.
33. H. Minouei, M.S. Rizi, M. Kheradmandfard, S.H. Shim, S.I. Hong, N. Park, NiNbTi (Al/Si) medium-entropy amorphous alloys with enhanced mechanical and thermal properties, *Intermetallics* 178 (2025) 108609.
34. X. Feng, H. Wang, X. Liu, C. Wang, H. Cui, Q. Song, K. Huang, N. Li, X. Jiang, Effect of Al content on wear and corrosion resistance of Ni-based alloy coatings by laser cladding, *Surf. Coat. Technol.* 412 (2021) 126976.
35. X. Wen, X. Cui, G. Jin, Y. Jiao, Y. Fang, A novel Ni2MnCuSnAl0.1 multi-principal element alloy coating to enhance the wear resistance and corrosion resistance of Mg-Li alloy, *Opt. Laser Technol.* 142 (2021) 107243.
36. R.K.S. Gautam, U.S. Rao, R. Tyagi, High temperature tribological properties of Ni- based self-lubricating coatings deposited by atmospheric plasma spray, *Surf. Coat. Technol.* 372 (2019) 390–398.

37. J. Singh, Slurry erosion performance analysis and characterization of high- velocity oxy-fuel sprayed Ni and Co hardsurfacing alloy coatings, *J. King Saud. Univ. Eng. Sci.* 35 (2021) 415–429.
38. R.P. Garcia, S.C. Canobre, H.L. Costa, Microabrasion-corrosion resistance of Ni–Cr superalloys deposited by plasma transferred arc (PTA) welding, *Tribol. Int.* 143 (2020) 106080.
39. B. Li, T. Mei, D. Li, S. Du, Ultrasonic-assisted electrodeposition of Ni-Cu/TiN composite coating from sulphate-citrate bath: Structural and electrochemical properties, *Ultrason. Sonochem.* 58 (2019) 104680.
40. D. Wang, J. Gao, R. Zhang, S. Deng, S. Jiang, D. Cheng, P. Liu, Z. Xiong, W. Wang, Effect of TaC particles on the microstructure and oxidation behavior of NiCoCrAlYTa coating prepared by electrospark deposition on single crystal superalloy, *Surf. Coat. Technol.* 408 (2021) 126851.
41. W. Zhang, B. Li, C. Ji, Synthesis and characterization of Ni-W/TiN nanocomposite coating with enhanced wear and corrosion resistance deposited by pulse electrodeposition, *Ceram. Int.* 45 (2019) 14015–14028.
42. J. Li, C. Sun, M. Roostaei, M. Mahmoudi, V. Fattahpour, H. Zeng, J.-L. Luo, Characterization and corrosion behavior of electroless Ni-Mo-P/Ni-P composite coating in CO2/H2S/Cl− brine: Effects of Mo addition and heat treatment, *Surf. Coat. Technol.* 403 (2020) 126416.
43. J. Wang, J. Zhou, T. Zhang, X. Meng, P. Li, S. Huang, H. Zhu, Ultrasonic-induced grain refinement in laser cladding nickel-based superalloy reinforced by WC particles, *Coatings* 13 (2023) 151.
44. S. Tan, H. Algül, E. Kiliçaslan, A. Alp, H. Akbulut, M. Uysal, The effect of ultrasonic power on high temperature wear and corrosion resistance for Ni based alloy composite coatings, *Colloids Surf. A Physicochem. Eng. Asp.* 656 (2023) 130345.
45. Z. Wu, W. Wang, X. Li, M. Yan, Y. Li, Microstructure and mechanical properties of Ti-6Al-4V prepared by nickel preplating and electron beam surface remelting, *J. Mater.* 271 (2019) 420–428.
46. S. Li, S. Pu, Z. You, C. Sun, S. Li, J. Zhang, Effect of heat treatment on microstructure and tribology behaviour of electroless Ni-P/BN(h) composite coating, *T. I. Met Finish.* 98 (2020) 21–29.
47. H. Jin, R. Ji, T. Dong, S. Liu, F. Zhang, L. Zhao, C. Ma, B. Cai, Y. Liu, Efficient fabrication and characterization of Ni-Fe-WC composite coatings with high corrosion resistance, *J. Mater. Res. Technol.* 16 (2022) 152–167.
48. R. Jensen, Z. Farhat, M.A. Islam, G. Jarjoura, Effect of coating thickness on wear behaviour of monolithic Ni-P and Ni-P-NiTi composite coatings, *Solids* 3 (2022) 620–642.
49. D. Figuet, A. Billard, C. Savall, J. Creus, S. Cohendoz, J.L. Grosseau-Poussard, A comparison between the microstructure and the functional properties of NiW coatings produced by magnetron sputtering and electrodeposition, *Mater. Chem. Phys.* 276 (2022) 125332.
50. T.M. Reis, C.D. Boeira, F.L. Serafini, M.C.M. Farias, C.A. Figueroa, A.F. Michels, Micro-abrasive wear resistance of heat-treated electroless nickel-phosphorus coatings deposited on copper-beryllium alloy C17200, *Surf. Coat. Technol.* 438 (2022) 128374.
51. Q. Cao, L. Fan, H. Chen, Y. Hou, L. Dong, Z. Ni, Wear behavior of laser cladded WC-reinforced Ni-based coatings under low temperature, *Tribol. Int.* 176 (2022) 107939.
52. H. Zhang, J. Wang, S. Chen, H. Wang, Y. He, C. Ma, Ni–SiC composite coatings with improved wear and corrosion resistance synthesized via ultrasonic electrodeposition, *Ceram. Int.* 47 (2021) 9437–9446.
53. S. Hong, Q. Ma, G. Liu, H. Yang, L. Hu, W. Meng, H. Xie, X. Yin, In-situ reinforced phase evolution and wear resistance of nickel-based composite coatings fabricated by wide-band laser cladding with Nb addition, *Opt. Laser Technol.* 157 (2023) 108678.

6 Enhancing Cladding Performance

Chromium Oxide and Alumina Powders with PVA Binder in GTAW Welding

Suresh Pratap and Prakash Kumar

6.1 INTRODUCTION

Mild steel, a mainstay in industrial components due to its strength and cost-effectiveness, often faces wear and damage due to its low hardness and corrosion resistance. Surface modification techniques, like laser cladding or tungsten inert gas (TIG) welding, provide a solution by applying a ceramic particle-reinforced metal matrix composite (MMC) layer. This layer enhances surface properties without altering the bulk material. With its affordability and simplicity, TIG cladding is particularly suitable for small-scale manufacturing, making it a practical and reliable choice for industrial applications. This research further solidifies confidence in the applicability of TIG cladding, providing a promising solution for industrial wear and damage issues.

Researchers have successfully employed TIG cladding to produce hard and wear-resistant coatings on mild steel. Sahoo et al. created a TiCNi composite coating on AISI 304 steel, demonstrating improved bonding between TiC particles and the matrix. Buytoz and Ulutan experimented with SiC coatings on austenitic stainless steel, varying powder ratios and arc scan speed. Wang et al. produced an Fe-based ferrotitanium and graphite composite coating on AISI 1045 steel. Sekhar et al. investigated the effect of processing the current on the micro-hardness and wear properties of the coated substrate on AISI 1020 mild steel. Dyuti et al. concluded that clad steel had better mechanical and oxidation properties than plain carbon steel substrates. Jasbir et al. Tungsten carbide (WC) MMC coatings effectively improve wear behavior, reinforcing WC in a tough binder matrix. Surface engineering techniques have been implemented to mitigate abrasive wear in various applications, including laser cladding, electroplating, thermal spraying, and TIG cladding. Buytoz et al. reported that WC MMC coatings, renowned for their low thermal expansion, high hardness, and wear resistance, have been widely adopted. To create high-performance composites, WC grains are typically embedded in a tough binder metal matrix, like cobalt, chromium, nickel, or CoCr. Bartkowski et al. investigated

DOI: 10.1201/9781003653226-6

satellite-6/WC MMC coating fabricated using laser cladding. Results indicated that increasing powder feed rate and laser beam power led to thicker coatings but compromised wear resistance. Multi-track and induction hybrid cladding methods were also employed to develop NiCrBSi/WC composite coatings on medium carbon steel plates.

To address the surface limitations of AISI 1020 mild steel, TiC coatings were deposited using TIG cladding. Researchers achieved varying hardness values across the coated sections by adjusting the processing current. Mild steel is mostly used in aerospace, biomedical, seawater–desalination, and chemical industries due to its high tensile strength and low price. However, its functionality is constrained due to its reduced wear and high-temperature oxidation resistance. Several surface engineering techniques were proposed to improve the tribological performance of mild steel. In recent decades, extensive research has focused on clad metals. In "cladding," a coating of durable or corrosion-resistant material is applied to a mild steel or other less costly substrate. Thus, the service life of the equipment or component may be increased by increasing its resistance to corrosion, wear, or hardness. A metal must be coated onto another metal under high pressure and temperature to prevent corrosion of the inner metal. Cladding is distinct from the methods of joining metals together, such as fusion welding or glue. These substances can combine the advantageous traits of several metals and alloys into composites, leading to better qualities.

Due to the numerous advantages of cladding over single metals, it is widely used in cookware, automotive, energy storage, home appliances, electrical distribution applications, transportation, heat exchangers, etc. Significant advances in clad metals have been achieved due to interdisciplinary research in mechanical, metallurgical, and electrical fields. Cladding is a process that bonds dissimilar metals, which is different from welding, and fastens them together. This process can be achieved through extrusion or rolling under high force. Cladding also provides protection against weathering and enhances aesthetic appeal. Additional components are added in many industries to provide superior mechanical properties. The mechanical and metallurgical characteristics of the corrosion-resistant layers are mainly controlled by the process parameters applied during cladding. The commonly used methods in the cladding process are fusion welding, explosive welding, rolling, etc. Among the different welding processes, TIG welding is preferred for cladding due to several advantages, such as good accuracy and excellent appearance in the weld.

Weld cladding is a process that adds a layer of corrosion-resistant material to improve a material's corrosion resistance. Automatic weld surfacing is increasingly used in the manufacturing and control industries due to its high reliability, all-position capability, ease of use, low cost, and high efficiency. Gas tungsten arc welding (GTAW) is the preferred welding method to be used for automatic surfacing. The weld cladding process is used for different applications, such as on the inside surfaces of carbon and low-alloy steel pressure vessels, paper digesters, urea reactors, tube sheets, and nuclear reactor control vessels. This technology aims to prevent the metallic surface from corroding and to increase its service life. GTAW enabled the development of a layer-by-layer deposition technique in the cladding process to lower costs in tools and die applications. TIG welding is flexible and more appropriate for cladding.

After thoroughly studying the literature, it is found that in the cladding process, TIG welding is one of the processes that can play a crucial role in excellent fabrication and provide high flexibility and accuracy. In this chapter, aluminum oxide and chromium oxide are fabricated on mild steel substrate cladding materials using TIG welding to investigate microstructural and mechanical properties. For cladding, polyvinyl alcohol (PVA) as a bonding material is used.

6.2 MATERIALS AND METHODS

Mild steel is a versatile and valuable material, capable of being formed into intricate shapes. It has good mechanical properties, is low-cost, and is frequently employed in sheet metal applications, general structural fabrication, and other industries. When considering mild steel for a project, its three essential qualities should be considered: good ductility and weldability, high impact strength, and high tensile strength. Table 6.1 lists the elemental composition of the mild steel being considered and the key properties of mild steel; refer to Table 6.2.

Chromium oxide is a hard, low-friction oxide with excellent wear, corrosion resistance, and optical properties. It is ideal for a protective coating in tribological and microelectronic applications, and can also be used as an insulating material in the aeronautical and aerospace industries. Chromium oxide is also suitable as a cladding material due to its good melting point of 2435°C and high boiling point

TABLE 6.1
Elemental Composition of Mild Steel

Element	C	Mn	P	S	Fe
wt.%	0.14–0.2	0.6–0.9	0.04	0.05	Balance

TABLE 6.2
Properties of Mild Steel

Properties	Mild Steel
Melting Point (°C)	1523
Tensile Strength (MPa)	475
Yield Strength (Mpa)	275
Hardness (HRB)	143
Thermal Conductivity (W m^{-1} K^{-1})	51.9
Specific Gravity (g cm^{-3})	7.8
Electrical Resistivity (μΩ cm)	1.74
Specific Heat Capacity (J g^{-1} C^{-1})	0.472

of 4000°C. Alumina, also called aluminum oxide (Al_2O_3), is a white, odorless, and inert material used in industrial ceramics. Its exceptional properties have been instrumental in many applications, extending the product lifespans and benefiting society. Although it is found naturally in several minerals, such as corundum (a crystalline form of the oxide) and bauxite (which is considered to be its primary aluminum ore), aluminum oxide is a compound that is insoluble and thermally unstable. Alumina is one of the most important technical ceramics because of its useful properties.

Polyvinyl alcohol (PVOH, PVA, or PVAL) is a synthetic polymer that dissolves in water. It is effective in forming films and emulsions and also exhibits adhesive properties. PVA is odorless, non-toxic, and resistant to grease, oils, and solvents. It has considerable tensile strength, flexibility, and hardness, demonstrating excellent gas and aroma barrier characteristics. PVA is better than any other known polymer as an oxygen barrier. However, it must be protected from moisture to maintain its gas permeability. The first step for preparing the sample is preparing the PVOH solution. To make the clad material paste, the PVA was first mixed with 96% distilled water to create a 4 wt.% PVA solution. For prepare this solution, 2 ml of PVA was mixed with 48 ml of water. The mixture was then properly stirred for 30 minutes to prepare the solution. The solution of PVA was mixed with powders in different compositions—20% Al_2O_3 and 80% Cr_2O_3, 50% Cr_2O_3, and 50% Al_2O_3. After properly mixing with PVA paste, the mixture was ready for use.

Preparation of Substrate: With the help of the cutting machine, the sample was prepared with the dimensions of 50 × 50 × 5mm. Four models were prepared and filed for surface preparation. The rough surface was prepared with the help of a grinder to ensure proper adhesion of the cladding material. The powders were weighed using an electronic weighing machine in appropriate composition: 20% Al_2O_3 with 80% Cr_2O_3, and 50% Cr_2O_3 with 50% Al_2O_3 powders. The solution of PVA was mixed with 20% Al_2O_3 and 80% Cr_2O_3, and an additional 50% Cr_2O_3 and 50% Al_2O_3 powder; then, after properly mixing with PVA paste, it was ready for use. Substrate cleaning with ethanol: Ethanol is used to clean dust, tacky particles, and abrasive particles from the substrate. The prepared paste was applied to a height of 1 mm, ensuring a uniform clad layer throughout. After applying the paste on the substrate, the moisture was removed from the sample using a hot air oven; the cladding paste-applied samples were kept inside a hot air oven at 600°C for 24 hours. The paste was heated slowly to prevent cracking.

TIG cladding is a process that involves scanning an electric torch with high energy density over a substrate material that has been pre-deposited with alloy powder of the desired composition. This process operates at a high treatment temperature that causes a thin surface layer of the substrate to melt due to heat energy from the arc of the TIG torch. The liquid mixes quickly with the deposited powder, creating a surface melt pool layer composed of remelted materials that sink toward the substrate due to a convective flow. The depth of this layer depends on the TIG heat input. This process starts at the substrate surface and introduces ceramic particles to produce a composite layer. Figure 6.1 shows the experimental setup for TIG cladding, and Figure 6.1 shows the sample fabricated using TIG cladding.

FIGURE 6.1 Setup for TIG cladding.

Using the pre-decided parameters (scanning speed 3.5 m/min), the TIG and a cladding experiment were performed at different currents and voltages (40, 45, 50, 60, and 70 A). The ambient atmospheric temperature was maintained at 20°C. During TIG action, use an argon atmosphere. The clad was allowed to cool in the atmosphere. Then, cladded samples were washed with ethanol. The process parameters for the experiments are listed in Table 6.3, while Figure 6.2 shows TIG cladding for different compositions and different current levels.

The sectioning of the samples for the various tests is performed using a diamond cutter. The mounting of the sectioned sample in Figure 6.3(a–c) is carried out by using a cold mounting setting compound. The mounting of the sample is done to facilitate the ease of the polishing process. Mechanical abrasion is used to avoid sectioning damage and establish a perfectly flat surface with a uniform scratch pattern to become a suitable surface for further metallographic polishing. The polishing process is carried out using a standard metallographic procedure. Emery papers of grades 220–2000 were used initially, followed by cloth polishing with a diamond compound to obtain a finished surface. The etching of the polished surface is carried out using a reagent called nital (98% ethanol and 2% HNO_3). The hardness of the sample is studied using a microhardness tester. The Vickers hardness (HV) is calculated using Equation 6.1.

$$HV = 0.1891 \times \frac{F}{d^2} \tag{6.1}$$

TABLE 6.3
Process Parameters for Cladding

Sample	Substrate Material	Clad Materials Composition	Stand-off Distance (mm)	Thickness of Coating (mm)	Scanning Speed (mm/min)	Current (amperes)	Voltage (Volts)	Gas flow rate (lit/min)
1	Mild Steel	80% Cr_2O_3 and 20% Al_2O_3	1.5	1	350	60	12.4	2
2	Mild Steel	80% Cr_2O_3 and 20% Al_2O_3	1.5	1	350	45	12.6	2
3	Mild Steel	50% Cr_2O_3 and 50% Al_2O_3	1.5	1	350	60	11.9	2
4	Mild Steel	50% Cr_2O_3 and 50% Al_2O_3	1.5	1	350	45	12.1	2

FIGURE 6.2 TIG cladding at different compositions and different currents.

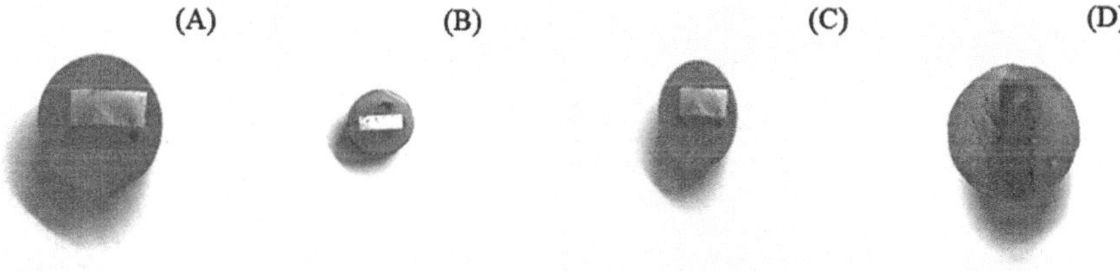

FIGURE 6.3 (a) 80% Cr_2O_3-20% Al_2O_3 at 60 amperes, (b) 80% Cr_2O_3-20% Al_2O_3 at 45 amperes, (c) 50%Cr_2O_3-50% Al_2O_3 at 60 A, and (d) 50%Cr_2O_3-50% Al_2O_3 at 45 A.

6.3 RESULTS AND DISCUSSION

The microstructural analysis of mild steel cladded with chromium oxide and alumina powders using PVA binder in GTAW reveals several important characteristics. The mild steel substrate (S0) shows a homogeneous and relatively smooth surface with a grainy texture typical of mild steel in its as-received condition, as shown in Figure 6.4. The interface region (Figure 6.4 (S1)) between the cladding and the mild steel exhibits clear evidence of pitting, with pits measuring up to 54.09 μm in depth. These pits suggest localized defects or attacks, possibly due to uneven material deposition or reactions at the interface. The layer interface (S2) is clearly defined, with distinct boundaries between the cladding and the substrate, indicating good bonding between the two materials. However, small voids or imperfections may exist at the microscopic level.

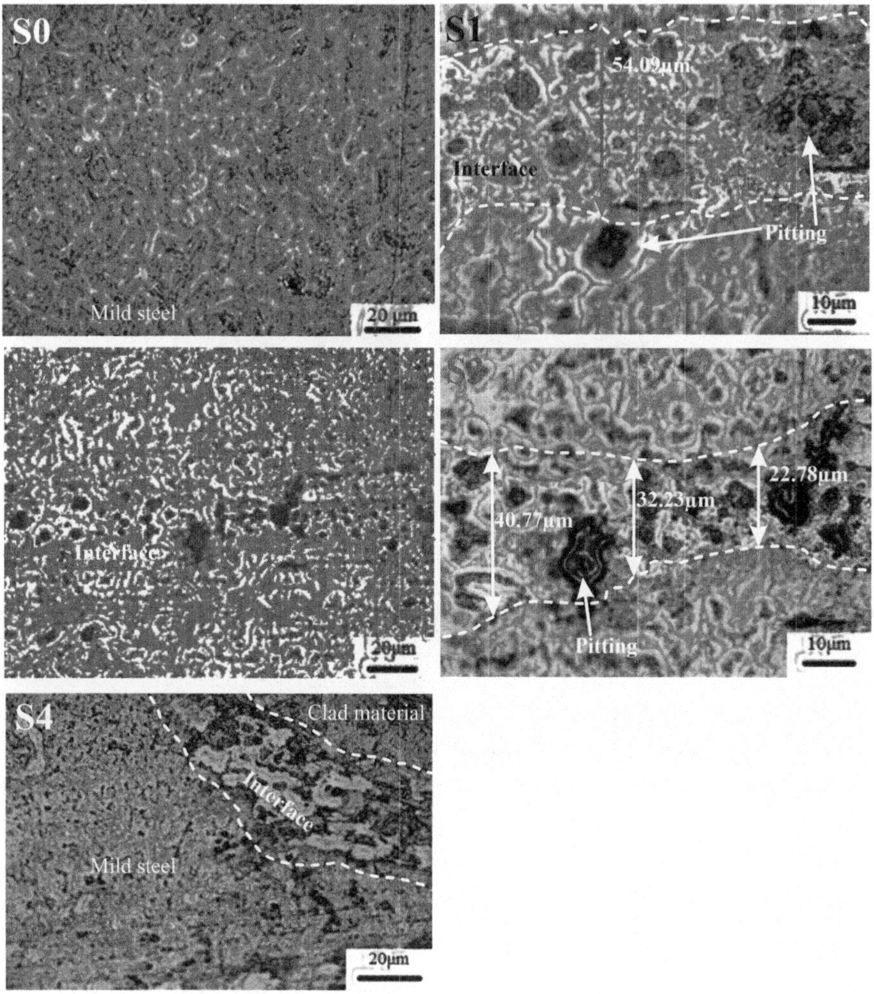

FIGURE 6.4 Micrograph of (a) substrate, (b) S1, (c) S2, (d) S3, and (e) S4.

Further examination (S3) shows significant pitting in the cladded region, with pit sizes varying between 22.78 µm and 40.77 µm. The non-uniformity of the pitting suggests localized material degradation or corrosion, likely due to reactions within the cladding or at the interface. The final micrograph (S4) highlights the interface between the cladded material and the mild steel, showing a continuous and well-bonded layer without significant delamination. The slight waviness observed at the interface could indicate localized stresses or reactions during cladding. The microstructure suggests good bonding between the cladding and the substrate, although pitting indicates localized imperfections. Further studies, including hardness testing and corrosion resistance analysis, would provide a more comprehensive understanding of the cladding's performance and durability.

The grain refinement can be observed in Figure 6.3 and compared with the other samples. Cladding processes using TIG welding typically involve rapid solidification rates, yielding finer grain sizes than the base material. The high cooling rates promote nucleation, resulting in smaller equiaxed or columnar grains in the cladded layer. Thus, grain refinement occurs, and the microhardness can be improved, which is discussed in the next section. The cladding process can cause mixing and dilution between the cladding and base material. The heat input and localized melting can lead to the diffusion of alloying elements, resulting in a transition zone (shown in Figure 6.4) at the cladding–substrate interface. This zone exhibits different microstructural features and compositions compared to the cladded layer and the base material.

6.3.1 Microhardness of the Substrate and Cladded Steel

The polished substrates are tested at a constant load of 100 N with a dwell time of 15 seconds. The indentations were taken at three locations, namely, the substrate, the interface zone, and the cladded zone, as shown in Figure 6.5.

From the results, it can be observed that the interface and cladded zones have higher hardness values compared to the substrate. Interface and cladded zones were registered twice and three to four times for the substrate material. Due to the presence of Cr_2O_3–Al_2O_3, the hardness is enhanced in the cladded zone.

Figure 6.6 illustrates an inverse relationship between HV and grain size across four samples (S1, S2, S3, and S4), aligning with the Hall–Petch effect, where hardness increases as grain size decreases. In sample S1, with the largest grain size (~50 μm), the hardness is relatively low (~370 HV), as observed in the microstructure, which shows pitting and a less uniform interface, typical of larger grains that offer fewer boundaries to impede dislocation movement. As the grain size reduces in S2 (~40 μm), the hardness significantly rises to ~500 HV, with the microstructure

FIGURE 6.5 Vickers hardness indentation of the cladded steel.

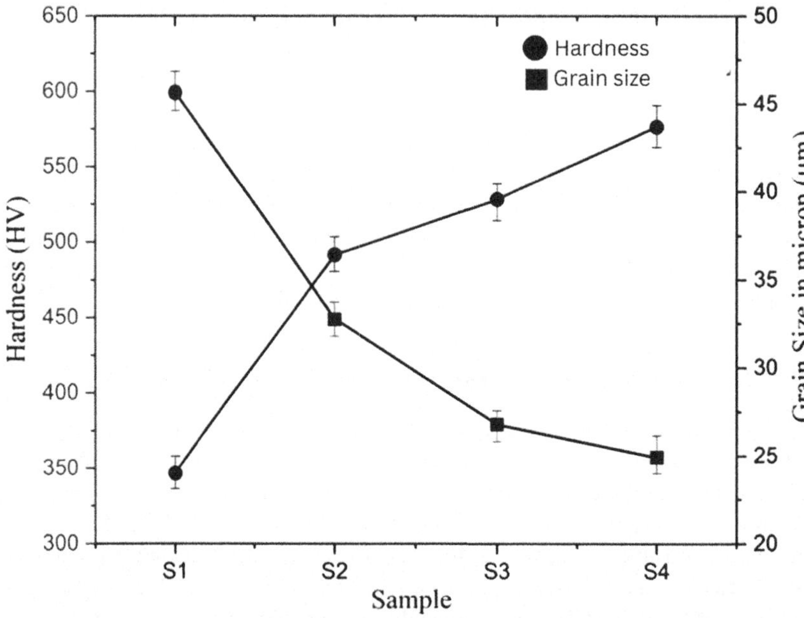

FIGURE 6.6 Variation of hardness and grain size with various sample categories.

showing improved bonding at the interface. Further grain refinement in S3 (~32 μm) increases hardness to ~520 HV, despite pitting, suggesting that the smaller grains contribute to higher resistance to deformation. In S4, the smallest grain size (~25 μm) results in the highest hardness (~580 HV), with the microstructure showing good bonding and fine grains that offer the greatest resistance to dislocation movement. The overall trend shows that decreasing grain size enhances hardness, supported by microstructural observations, where finer grains lead to better mechanical properties and improved cladding performance.

The results of micro-hardness testing of Cr_2O_3 (80%)–Al_2O_3 (20%) and Cr_2O_3 (50%)–Al_2O_3 (50%) cladded substrates at different combinations of parameters. From the graphs shown in Figures 6.4 and 6.6, the hardness values for all parameter combinations are higher than those of the substrate. It happens due to the rapid solidification of the molten metal. Due to rapid solidification, the grain size becomes finer due to less time available for grain growth.

6.4 CONCLUSION

This work completed micro-structure and micro-hardness on Cr_2O_3–Al_2O_3 cladded samples. The following are the conclusions based on the experimental results:

- The Cr_2O_3–Al_2O_3 cladding significantly increased the hardness of the mild steel substrate. The average hardness values for the cladded zones were measured as 507.3 HV and 573.3 HV at 100 N load, which is four to five times higher than the substrate material. This increase is due to the incorporation of ceramic powders and the rapid solidification process during GTAW.

- The cladding process resulted in finer grains in the cladded layer than in the base material. As per the Hall–Petch relationship, this grain refinement contributed to an increase in hardness, with smaller grains providing higher resistance to deformation.
- The cladding process using different currents produced a strong metallurgical bond between the mild steel substrate and the cladding materials. This bond was confirmed through microstructural analysis, showing clear interfaces without significant delamination or distortion of the base metal.
- Microstructural analysis revealed localized pitting in the cladded regions, with pit sizes varying between 22.78 μm and 54.09 μm. This suggests some imperfections in the material deposition or reactions at the cladding–substrate interface, which could affect long-term performance.
- The significant improvements in hardness and wear resistance, along with the strong bonding observed between the cladded layer and the substrate, suggest that Cr_2O_3 and Al_2O_3 cladding can be a promising solution for enhancing the surface properties of mild steel in industrial applications where durability and wear resistance are critical.

BIBLIOGRAPHY

[1] P. C. Singh and S. Ghosh, "Characterisation of Al-induced electroless tin films on mild steel substrate for corrosion protection," *Surface Topography: Metrology and Properties*, vol. 8, no. 2, p. 025002, 2020.

[2] S.-H. Wang, J.-Y. Chen, and L. Xue, "A study of the abrasive wear behaviour of laser-clad tool steel coatings," *Surface and Coatings Technology*, vol. 200, no. 11, pp. 3446–3458, 2006.

[3] C. Yu, L. Fu, H. Xiao, Q. Lv, and B. Gao, "Effect of carbon content on the microstructure and bonding properties of hot-rolling pure titanium clad carbon steel plates," *Materials Science and Engineering A*, vol. 820, p. 141572, 2021.

[4] N. Murugan and R. Parmar, "Stainless steel cladding deposited by automatic gas metal arc welding," *Welding Journal-Including Welding Research Supplement*, vol. 76, no. 10, p. 391s, 1997.

[5] A. M. Murmu, S. Parida, and A. Das, "Synthesis and characterisation of Ti6Al4V-nano-ZrO 2 composite cladding on Ti6Al4V substrate using fiber laser," *Journal of Materials Engineering and Performance*, vol. 30, pp. 1748–1758, 2021.

[6] P. Varghese, E. Vetrivendan, M. K. Dash, S. Ningshen, M. Kamaraj, and U. K. Mudali, "Weld overlay coating of Inconel 617 M on type 316 L stainless steel by cold metal transfer process," *Surface and Coatings Technology*, vol. 357, pp. 1004–1013, 2019.

[7] K. Zhang *et al.*, "Microstructure and properties of composite coatings by laser cladding Inconel 625 and reinforced WC particles on non-magnetic steel," *Optics & Laser Technology*, vol. 163, p. 109321, 2023.

[8] J. Pravallika, K. Phaneesh, B. Subbarao, and A. Bhuvaneshwari, "Understanding Chemistry of Black Coating in Ancient Bidriware and Attempt to Obtain Similar Coats on Pure Copper and Yellow Brass," presented at the Materials Science Forum, Trans Tech Publ, 2023, pp. 127–133.

[9] J. Liu, J. Li, X. Cheng, and H. Wang, "Effect of dilution and macrosegregation on corrosion resistance of laser clad AerMet100 steel coating on 300M steel substrate," *Surface and Coatings Technology*, vol. 325, pp. 352–359, 2017.

[10] N. Fajrah and F. S. Efendy, "Analysis of quality control weld cladding at Cladtek BI-Metal Manufacturing," presented at the AIP Conference Proceedings, AIP Publishing LLC, 2020, p. 040005.
[11] Y. Fan *et al.*, "Parameter optimisation and mechanical properties of laser cladding of 316L stainless steel powder on G20Mn5QT steel," *Coatings*, vol. 13, no. 3, p. 481, 2023.
[12] B. Chakrabarti, H. Das, S. Das, and T. Pal, "Effect of process parameters on clad quality of duplex stainless steel using GMAW process," *Transactions of the Indian Institute of Metals*, vol. 66, pp. 221–230, 2013.
[13] T. S. Singhal and J. K. Jain, "GMAW cladding on metals to impart anti-corrosiveness: Machine, processes and materials," *Materials Today: Proceedings*, vol. 26, pp. 2432–2441, 2020.
[14] J. V. Matias *et al.*, "Behavior of a superaustenitic stainless steel weld cladding deposited by the gas metal arc welding process," *Materials Today Communications*, vol. 34, p. 104978, 2023.
[15] M. Jayavelu, S. Kasi, B. Visvalingam, P. N. Bellamkonda, and M. Sudersanan, "Microstructural and wear properties of mild steel cladded with AISI 316L stainless steel using pulsed current gas metal arc welding process," *Welding International*, vol. 37, no. 1, pp. 34–45, 2023.
[16] S. Yeo, J. H. Kim, and H. S. Yun, "Effect of pulse current and coating thickness on the microstructure and FCCI resistance of electroplated chromium on HT9 steel cladding," *Surface and Coatings Technology*, vol. 389, p. 125652, 2020.
[17] A. S. Praveen and A. Arjunan, "Parametric optimisation of high-velocity oxy-fuel nickel-chromium-silicon-boron and aluminium-oxide coating to improve erosion wear resistance," *Materials Research Express*, vol. 6, no. 9, p. 096560, 2019.

7 An Overview of Nanocoatings
Performance Enhancement of Heat Exchangers

Vikas Kumar and Shiv Lal

7.1 INTRODUCTION

Heat exchangers are essential components of many industrial processes, such as chemical manufacturing, food processing, HVAC (Heating, Ventilation, and Air Conditioning) systems, and power generation. The efficiency and performance of the systems in which these devices are integrated are directly impacted by how effectively they transfer heat between two or more fluids. Conventional heat exchangers, on the other hand, often face operational difficulties like corrosion, surface fouling, poor thermal performance, and shortened service life. Reduced efficiency, higher energy use, more frequent maintenance needs, and higher operating expenses can all be consequences of these problems.

Researchers and engineers are increasingly using advanced material technologies to resolve these problems, and nanocoatings (NCs) are showing great promise in this regard. NCs are extremely thin coatings of materials designed at the nanoscale, usually ranging from 1 to 100 nm, that are applied to the surfaces of heat exchangers. NCs can provide a notable enhancement in surface properties because of their distinct physical and chemical characteristics, which are very different from those of bulk materials. These consist of improved anti-fouling behaviour, increased corrosion resistance, and improved thermal conductivity. These coatings exhibit a high specific surface area, remarkable reactivity, and better mechanical stability due to their nanoscale nature. These qualities of NCs enable them to fix the performance issues in heat exchangers. Additionally, they protect the system from wear, fouling, and environmental damage, while enhancing the efficiency of heat transfer processes.

The incorporation of NCs into heat exchangers has been the subject of intensive research and development during the last decades. Studies have shown that these coatings can increase heat transfer efficiency, reduce fouling, prevent corrosion, and prolong the service life of heat exchangers. These changes make industrial operations more reliable and require less maintenance.

The objective of this chapter is to systematically investigate how NC technologies enhance heat exchanger performance. It covers the underlying mechanisms, types of

DOI: 10.1201/9781003653226-7

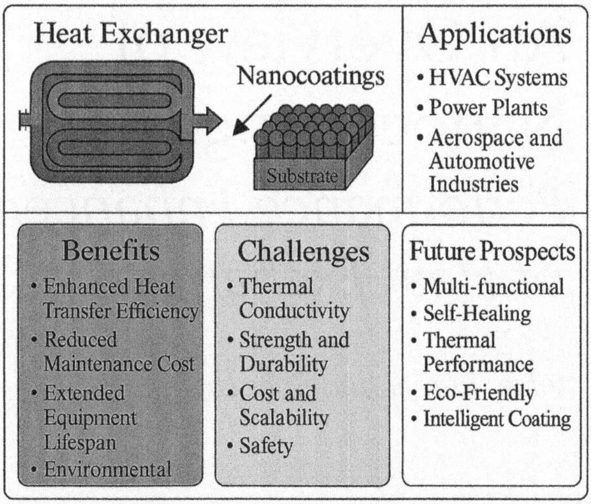

FIGURE 7.1 An illustration of nanocoatings in the heat exchanger.

NC, and their applications, as well as how materials and coating procedures affect heat transfer and operating stability. This chapter examines the advantages of NCs, together with the technological and economic obstacles that need to be overcome for widespread industrial adoption. The review concludes by outlining challenges and prospects for the field of thermal system NC development. Moreover, an overview of this chapter is illustrated in Figure 7.1.

7.2 NANOCOATING IN HEAT EXCHANGERS: KEY MECHANISMS

Nanostructured coatings significantly enhance heat transfer due to the higher thermal conductivity of nanoparticles like Al_2O_3, CuO, and TiO_2, along with increased surface area from nano-roughness, which improves thermal exchange. These coatings also offer excellent fouling resistance through low surface energy that minimises particle adhesion and scaling, complemented by superhydrophobic properties that enable self-cleaning by repelling contaminants. In terms of corrosion protection, NC acts as an effective barrier layer, providing chemical stability against aggressive environments and promoting the formation of passive oxide films. Furthermore, their durability is marked by hard surfaces that resist mechanical wear and strong adhesion that prevents delamination or cracking under stress.

7.3 TYPES OF NANOCOATING

7.3.1 Hydrophobic and Superhydrophobic Coatings

These specialised surface treatments modify the surface energy or roughness of materials to create a water-repelling effect, known as hydrophobic and superhydrophobic coatings. They're incredibly useful in various engineering fields, such as heat exchangers, because they help with water management, reduce fouling, and enhance heat transfer performance. The term "hydrophobic" literally means "water-fearing." To boost water repellency, hydrophobic coatings adjust the surface to contain hydrophobes and limit the interaction with water molecules. This allows water droplets to bead up and roll off rather than spreading out. Hydrophobic coatings are increasingly popular in heat exchangers, improving their performance, durability, and energy efficiency. You'll find them in air-cooled condensers, evaporators, HVAC systems, geothermal and marine heat exchangers, as well as in power plants and industrial cooling systems. Superhydrophobic coatings take the advantages of hydrophobic coatings to the next level by creating surfaces that repel water to an extreme degree, with contact angles exceeding 150°. This means that water beads up and rolls off almost effortlessly, often taking dirt or contaminants along with it. Achieving superhydrophobicity involves using low surface energy materials (like fluoropolymers or silicones) combined with micro- or nano-scale surface roughness (for instance, through nanostructures, etching, or particles). These coatings can lead to heat transfer coefficients that are up to ten times higher than those of traditional surfaces. However, there are some challenges associated with these coatings, including durability, scalability, and high costs.

Liu et al. [1] studied the corrosion resistance and heat transfer performance of a fin-type heat exchanger by using superhydrophobic coatings. Their findings reveal that coatings improved heat transfer efficiency by 42.5% when applied at 0°C for 30 days. Hu et al. [2] have designed a finned tube heat exchanger with a hybrid surface (hydrophilic–hydrophobic surfaces). The variation of heat transfer rates with hydrophilic surfaces and hydrophilic–hydrophobic hybrid surfaces was analysed in their study, as depicted in Figure 7.2. Q_a, Q_s, and Q_l represent the total heat transfer rate, sensible heat transfer, and latent heat transfer rate, respectively. The value of Q_l

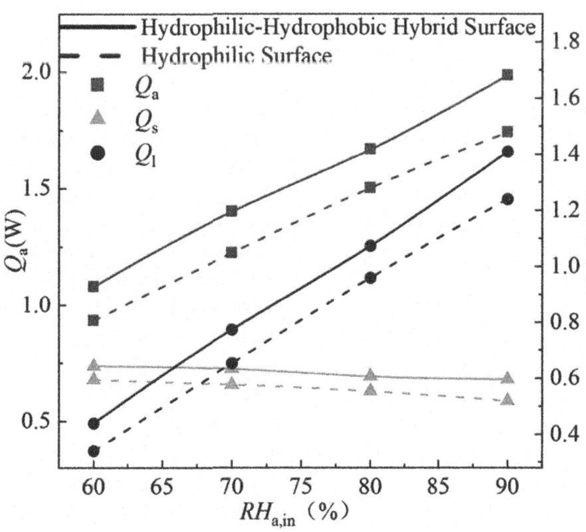

FIGURE 7.2 Variation of heat transfer rates with hydrophilic surface and hydrophilic–hydrophobic hybrid surface [2].

increases with the increased value of relative humidity of air ($RH_{a,in}$) at the inlet, but at the same time, the value of Q_s decreases. Further, Q_a shows an increasing tendency for the same conditions. These values of heat transfer rates were found to be significantly higher for hydrophilic–hydrophobic hybrid surfaces. The underlying reason behind this is a smaller contact angle and larger wall adhesion for the hydrophilic surface.

He et al. [3] studied the heat transfer performance of a heat exchanger by applying a modification of a superhydrophobic surface. They reported improvements in heat transfer performance of 76.2% and 31.6% compared to hydrophobic and hydrophilic, respectively. Sui et al. [4] analysed the effect of superhydrophobic coatings on the heat transfer performance of a microchannel heat exchanger for different operating conditions. The authors claimed that the power consumption was reduced by 48.7% with superhydrophobic coatings. However, the applied coating decreases the frost accumulation, thereby increasing the heat transfer coefficient of the heat exchanger. An experimental investigation of hydrophobic coatings on heat transfer characteristics of a gasketed-type plate heat exchanger was carried out by Dağlı and Özbey [5]. Their experiments showed that the efficiency of the heat exchanger increased with hydrophobicity. The highest efficiency and energy savings value were 0.525% and 41% for hydrophobic coatings.

Table 7.1 presents a tabulated summary of key findings from various studies on hydrophobic and superhydrophobic coatings for heat transfer enhancement in heat exchangers, highlighting observed performance improvements (Table 7.1).

TABLE 7.1
Application of Various Coatings in Different Heat Exchangers

S. No.	Coating Type	Application in Heat Exchanger	Key Findings	Reference
1.	Hydrophobic	Fin-type heat exchanger	Improved frost resistance and enhanced heat transfer efficiency during frost conditions.	[3]
2.	Hydrophobic	Fin and tube heat exchanger	Hydrophilic coatings reduced pressure drop by 15–30%. Hydrophobic surfaces improved heat transfer, but increased pressure drops and the potential for water bridging.	[6]
3.	Hydrophobic	Microchannel heat exchanger	Reduced mineral fouling. Enhanced heat transfer performance.	[7]
4.	Superhydrophobic	Fin and tube heat exchanger	COP increased to 2.93. A total of 44% reduction in defrosting time. Reduced frost accumulation and improved cycle performance.	[4]

(Continued)

Table 7.1 (Continued)

S. No.	Coating Type	Application in Heat Exchanger	Key Findings	Reference
5.	Superhydrophobic	Fin and tube heat exchanger	A total of 12% lower heat transfer rate at low relative humidity and high air velocity. A total of 14% lower pressure drop at low humidity.	[8]
6.	Superhydrophobic	Shell and tube exchangers	Up to a 33% increase in heat transfer coefficient at low steam qualities.	[9]

7.4 ANTI-CORROSION COATINGS

Anti-corrosion coatings act as protective layers that help shield metal surfaces from environmental factors that cause corrosion, including moisture, chemicals, and salt. In heat exchangers, which are frequently exposed to challenging conditions like high humidity, acidic media, or saline water, these coatings are essential for preserving heat transfer efficiency and prolonging the equipment's operational lifespan. Metal corrosion is a frequently occurring problem in industrial heat exchangers. However, traditional anti-corrosion coatings often have low thermal conductivity, which reduces heat transfer efficiency. Lu et al. [10] studied anti-corrosion coating along with hydrophobic characteristics in the heat exchangers. The authors have prepared a composite of perfluorodecyltriethoxysilane (FAS), boron nitride (BN), and silicon carbide (SiC) blended with fluorosilicone resin (FSR) for coatings. The engineered FAS–BN/SiC coating was used in the heat exchanger. Their findings revealed that a reduction in the heat transfer coefficient by 1.75% improved the efficacy of the heat exchanger. The effect of aluminium nitride (AlN) film coatings in heat exchangers used in energy storage systems was studied by Panagiotopoulos [11]. They have evaluated the thermal performance and anti-corrosive behaviour of the heat exchanger. They have claimed that the coated heat exchanger's excellent thermal performance results in faster charging/discharging of energy storage systems. The fact behind this is the exceptional thermal conductivity of AlN coatings. Thus, it is suggested that AlN could be a promising anti-corrosion coating in heat exchangers. Yuan et al. [12] have studied the effect of anti-corrosion coatings on the thermal and exergetic performance of a heat exchanger. They have employed three types of fluid flow: gas–gas, gas–liquid, and liquid–liquid. Their findings show that as the thermal conductivity of coatings increased, the heat transfer coefficient increased.

7.5 ANTI-FOULING COATINGS

Anti-fouling coatings are advanced surface treatments that prevent scaling and the accumulation of unwanted substances on equipment surfaces. These coatings enhance the working efficiency of the heat exchangers and make them last longer.

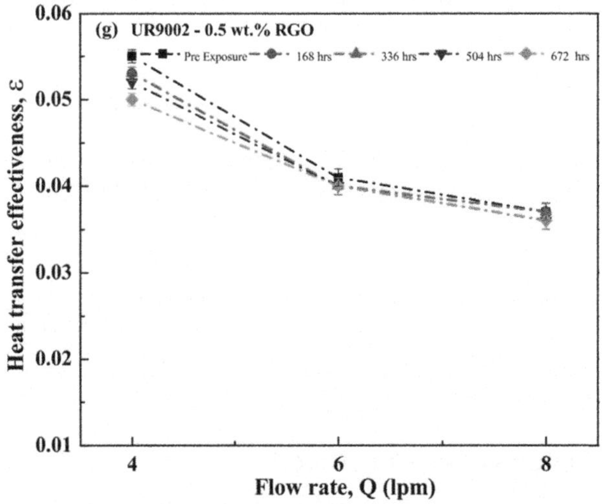

FIGURE 7.3 Heat transfer effectiveness of 0.5 wt.% rGO/polyurethane [13].

It reduces the buildup of deposits by making a surface that does not let fouling materials stick to it. The use of anti-fouling coatings on heat exchangers saves time and energy by preventing fouling and allowing for less frequent cleaning. In addition, the anti-fouling coatings protect surfaces from long-term degradation caused by corrosion. They also enhance system dependability, especially in HVAC and industrial settings.

Sharma et al. [13] experimentally investigated the anti-fouling behaviour of polymer nanocomposite coatings. In their experimentation, they have employed two coatings, such as reduced graphene oxide (rGO)/epoxy (0.3 wt.%) and rGO/polyurethane (0.5 wt.%). The researchers have claimed that the addition of rGO reduces the wettability and surface roughness of coatings. The highest heat transfer effectiveness was found for rGO/polyurethane (0.5 wt.%) among the tested coatings, as depicted in Figure 7.3.

7.6 THERMAL CONDUCTIVITY-ENHANCING COATINGS

Research shows that superhydrophobic coatings lack effective thermal conductivity. He et al. [14] synthesised multifunctional coatings that consist of polymers(polydimethylsiloxane and polytetrafluoroethylene) as well as SiO_2 and graphene. The authors have tested both the thermal conductivity and the rate of heat transfer. With the application of SiO_2/graphene coatings, their findings showed an increase in thermal conductivity (~44.2%). The following Table 7.2 represents the effect of coatings on the performance of heat exchangers.

TABLE 7.2
Effects of Nanocoatings on the Performance of Heat Exchangers

S. No.	Nanocoating Material	Heat Exchanger Type	Performance Improvement	Reference
1.	Ag, Cu, Al, CuO, Al_2O_3, BeO	Heat Pipe Heat Exchanger	Reduced thermal resistance in both evaporator and condenser sections. Max improvement: ~73.79% for CuO coating.	[15]
2.	Copper Oxide (CuO)	Heat Pipe Heat Exchanger	Optimal nanocoating thickness improves thermal performance significantly; overcoating reduces effectiveness.	[16]
3.	Reduced Graphene Oxide (rGO) with Epoxy and Polyurethane	General Heat Exchanger Applications	Coatings exhibited anti-fouling and anti-wetting behaviour and improved thermal performance under fouling environments.	[13]

7.7 BENEFITS OF NANOCOATING IN HEAT EXCHANGERS

7.7.1 Enhanced Heat Transfer Efficiency

In heat exchangers, the use of NC is especially helpful in increasing heat transfer because of the improvement in thermal conductivity, increased surface wettability, reduced fouling and scaling, and increased corrosion resistance. Enhanced NC materials often include metal oxides, carbon nanotubes (CNTs), and graphene with high thermal conductivity. These materials decrease thermal resistance and enable better transfer of fluids in heat exchangers. By modifying the surface properties (hydrophilic or hydrophobic), NCs enhance surface fluid interaction at the heat exchangers. Type of coating affects heat transfer as well; hydrophilic coatings increase heat exchange during condensation, while hydrophobic coatings enable dropwise condensation, which increases heat transfer efficiency. The buildup of unwanted material on heat transfer surfaces, known as fouling, negatively impacts the performance of heat exchangers over time. NCs with anti-fouling and self-cleaning properties help retain high transfer rates while reducing maintenance. In addition to material degradation, corrosion also impacts thermal resistance. In such cases, NCs provide protective shielding from corrosive environments and maintain thermal efficiency alongside increased lifespan for the heat exchanger.

Seok et al. [17] have investigated the thermal performance of fin and tube heat exchangers by using CNT coatings. They have coated a heat exchanger with CNT and compared the thermal hydraulic performance of the uncoated heat exchanger. Their findings revealed that the thermal-hydraulic performance of a CNT-coated heat exchanger improved by 8.9% compared to an uncoated heat exchanger.

7.7.2 Reduced Maintenance Costs

NCs reduce the maintenance costs of a heat exchanger through several interrelated mechanisms that reduce the cost of cleaning, repairs, and part replacements. NCs render heat exchanger surfaces resistant to adhesion of contaminants like salts, minerals, or dust. These surfaces repel both water- and oil-based substances, reducing the tendency for deposits to adhere and accumulate. This reduces cleaning costs. Further, by reducing fouling and corrosion, nanocoated surfaces maintain higher thermal conductivity and heat transfer rates over time. This results in lower energy costs. However, nanocoated surfaces are often self-cleaning or easier to clean owing to their non-stick properties. Due to this, labour cost is reduced.

7.7.3 Extended Equipment Lifespan

NCs, which are ultra-thin layers of protective material applied at the nanoscale, can significantly enhance the durability of heat exchangers. The mechanism by which NCs improve the lifespan of heat exchangers involves several protective and performance-enhancing effects at the micro- and nanoscale. Some factors include anticorrosion protection, anti-fouling, self-cleaning, and chemical resistance. NCs create a **barrier layer** that prevents oxygen, moisture, salts, and other corrosive agents from reaching the metal surface. As a result, they reduce rusting and chemical degradation, particularly in seawater, chemical, or acidic environments. However, NCs modify the surface energy of heat exchanger surfaces, making them hydrophobic (water-repellent) or oleophobic (oil-repellent). It helps to minimise the buildup of deposits on the surface of a heat exchanger.

7.7.4 Environmental Benefits

NCs benefit the environment in multiple ways, primarily by increasing efficiency, reducing chemical usage, and extending equipment life. These benefits collectively lead to lower resource consumption, waste, and emissions. NCs (especially antifouling) prevent scaling and dirt buildup. As an environmental concern, it reduces the need for acidic or caustic cleaning agents, which are harmful to aquatic life and ecosystems. NCs help in reducing the energy consumption by lowering the thermal resistance and pressure drop. NCs decrease carbon footprint in power plants, HVAC systems, and industrial processes. Additionally, it reduces greenhouse gas emissions and overall energy savings. However, it also minimised water waste by preventing clogging or contamination of water. It supports sustainable water use, which is especially critical in water-scarce regions.

7.8 RECENT ADVANCES AND APPLICATIONS

7.8.1 HVAC Systems

Recent advancements in NCs have significantly enhanced the performance and sustainability of HVAC systems. These innovations focus on improving energy efficiency, reducing maintenance needs, and promoting environmental sustainability.

NCs with superhydrophilic or superhydrophobic properties are applied to condensation surfaces within HVAC systems. These coatings facilitate more efficient condensation and evaporation processes, leading to improved heat transfer rates. For instance, Exaeris Climate Technologies has developed NCs that accelerate water condensation, thereby increasing energy efficiency and reducing CO_2 emissions [18]. Innovative NCs, such as those incorporating ZnO nanoparticles, provide superior corrosion protection for metallic components in HVAC systems [19]. These coatings act as barriers against moisture and chemical agents, extending the lifespan of system components and reducing the frequency of maintenance. Smart NCs that react to light and temperature are becoming more and more popular. These coatings can change their properties drastically to maximise thermal management in buildings, minimising energy use and dependency on HVAC systems [20].

7.8.2 Power Plants

NCs have emerged as a transformative technology that enhances the efficiency, durability, and sustainability of power plants. Their applications extend across a variety of components and systems within power generation facilities. Applying NCs to hydropower turbine components, such as those made from AISI 304 and AISI 316L stainless steels, enhances resistance to wear, corrosion, and erosion, thereby improving reliability and reducing downtime [21]. The use of nano-iron oxide coatings in gas–solid fluidised bed systems was examined in the study [22]. The results showed that the heat transfer coefficient is improved by fluidised bed heating, and that this effect becomes more efficient at higher heating flux and velocity. For example, the heat transfer coefficient increased from 336.4 to 390.3 $W/m^2 \cdot °C$ at a velocity of 0.27 m/s and a heating flux of 125 W. These studies highlight how important NCs are for improving the power plant system's sustainability, durability, and efficiency.

7.8.3 Aerospace and Automotive Industries

NCs have greatly improved the performance and durability of components in the automotive and aerospace industries. High-performance coatings support lightweight vehicles, which in turn support fuel efficiency. Nanostructured ceramic coatings were applied to aluminium alloy AA2024-T3 via a two-step process consisting of anodising and atomic layer deposition (ALD) [23]. Under the salt spray test conditions, an AlO_3/TiO_2 nanolaminate coating exhibiting a surface hardness of 5.5 GPa showed no corrosion after 1000 hours. The addition of graphitic carbon nitride (GCN) and 4-aminobutyltriethoxysilane (ABES)-functionalised tantalum nitride (TaN) to a polyurethane matrix greatly enhanced its mechanical, flame-retardant, and corrosion-resistant qualities [24]. After 40 days of exposure to an electrolyte, acrylic car paint enhanced with SiO_2 measured a high coating resistance of 8.89×10^{11} $\Omega \cdot cm^2$. The improved corrosion resistance of acrylic automotive paint with nano SiO_2 and TiO_2 was observed [25]. Electrochemical tests done in saltwater revealed that the nanoparticle additives provided better corrosion rates compared to the original basecoat. A nanoscale coating made of titanium dioxide, silicon, and zirconium was applied to a BMW X6 [26]. The water contact angle on the surface after treatment was more than 120°, which means it is hydrophobic. The coating maintained its hydrophobicity even after more than 150 wash cycles, demonstrating its long-lasting durability. Such studies point out that using NCs greatly benefits the strength and lifespan of materials in the aerospace and automobile industries.

7.9 CHALLENGES OF NANOCOATING

Despite their advantages, NC faces several challenges in practical applications. One major limitation is reduced thermal conductivity in coatings designed primarily for corrosion or fouling resistance, which can affect heat exchanger efficiency. Additionally, maintaining strong and durable adhesion of NC to metal surfaces is difficult, particularly under the high temperatures, pressures, and corrosive conditions typical of industrial settings. The high cost and limited scalability of advanced nanomaterials further hinder their widespread commercial adoption. Moreover, the application processes, such as chemical vapour deposition and plasma spraying, are technically complex and may not be feasible for all industrial environments. Lastly, concerns regarding the environmental and health impacts of nanomaterials have led to increased regulatory scrutiny, posing further challenges to their deployment.

7.10 FUTURE PROSPECTS OF NANOCOATING

Future advancements in NC technologies are focused on developing multi-functional coatings that combine efficient thermal conductivity with robust corrosion and fouling resistance. One promising direction is the integration of self-healing capabilities, allowing coatings to autonomously repair minor surface damage, thereby enhancing durability and reducing maintenance needs. To further boost thermal performance, cutting-edge materials like graphene and CNTs are being explored for their exceptional heat transfer properties. Additionally, there is a growing emphasis on eco-friendly NCs that utilise non-toxic, environmentally benign materials to comply with strict sustainability standards. Looking ahead, the fusion of sensing technology and nanotechnology is expected to give rise to intelligent coating systems capable of real-time monitoring of conditions such as fouling and corrosion, enabling predictive maintenance and improving operational efficiency.

7.11 CONCLUSION

NCs are a new, innovative step to provide improved performance and durability of heat exchangers in several industrial applications. The nanoscale properties allow for appreciable enhancement of heat transfer performance by reducing fouling, corrosion, and thermal resistance. In particular, hydrophobic and superhydrophobic coatings help reduce the surface area for water and fouling to accumulate on; anti-corrosive and anti-fouling coatings protect the heat exchanger surfaces from degrading. Protecting the heat exchanger surface extends equipment lifespan and reduces maintenance requirements. While there are still obstacles related to thermal conductivity, durability, or high cost and complexities with application, research continues on the use of multifunctional, self-healing, and environmentally friendly NC technologies with improved thermal properties. Future work will most likely include intelligent sensing developments that will help optimise operational conditions for heat exchangers for long-term reliability and performance. In general, NCs are an important strategy to drive efficiency, sustainability, and cost savings in thermal management systems.

REFERENCES

[1] Q. Liu, N. Li, J. Duan, M. Song, F. Shi, Enhancing corrosion resistance and heat transfer performance of fin-tube heat exchangers for closed-type heat-source towers by superhydrophobic coatings, *J. Build. Eng.* 96 (2024) 110568. https://doi.org/10.1016/J.JOBE.2024.110568

[2] W. Hu, Q. Ma, Y. Guan, W. Ye, C. Jiang, X. Zhang, Effect of hydrophilic–hydrophobic hybrid surfaces on the thermal hydraulic characteristics of finned tube heat exchangers under wet conditions in greenhouse environment, *Appl. Therm. Eng.* 267 (2025) 125833. https://doi.org/10.1016/J.APPLTHERMALENG.2025.125833

[3] H. He, X. Zhou, N. Lyu, F. Wang, C. Liang, X. Zhang, Enhancing heat-exchanger performance in frost conditions via superhydrophobic surface modification, *Appl. Therm. Eng.* 246 (2024) 122914. https://doi.org/10.1016/J.APPLTHERMALENG.2024.122914

[4] Y. Sui, Z. Sui, G. Liang, W. Wu, Superhydrophobic microchannel heat exchanger for electric vehicle heat pump performance enhancement, *Sustainability* 15 (2023). https://doi.org/10.3390/su151813998

[5] S. Dağlı, M. Özbey, Experimental investigation of the effect of hydrophobic coating on heat transfer and gasketed plate heat exchanger performance, *Proc. Inst. Mech. Eng. Part E J. Process Mech. Eng.* 238 (2024) 1251–1262. https://doi.org/10.1177/09544089231156309

[6] K.S. Yang, H.Y. Lin, Y.L. Wu, S.K. Wu, C.C. Wang, Influence of surface modification on the transient dehumidification performance of fin-and-tube heat exchanger, *Int. J. Heat Mass Transf.* 173 (2021) 121202. https://doi.org/10.1016/J.IJHEATMASSTRANSFER.2021.121202

[7] K. Shaikh, S.N. Kazi, M.N.M. Zubir, B.A. Razak, K.H. Wong, Y.H. Wong, W.A. Khan, S. Abdullah, M.S. Alam, Investigation of zirconium (Zr) coated heat exchanger surface for the enhancement of heat transfer and retardation of mineral fouling, *J. Taiwan Inst. Chem. Eng.* 153 (2023) 105246. https://doi.org/10.1016/J.JTICE.2023.105246

[8] K.S. Yang, J.W. Zheng, K.L. Wu, Y.L. Wu, Performance of the frosting characteristics of fin-and-tube heat exchanger subject to superhydrophobic coating, *Appl. Therm. Eng.* 235 (2023) 121338. https://doi.org/10.1016/J.APPLTHERMALENG.2023.121338

[9] M.M. Chehrghani, T. Abbasiasl, A.K. Sadaghiani, A. Koşar, Copper-based superhydrophobic nanostructures for heat transfer in flow condensation, *ACS Appl. Nano Mater.* 4 (2021) 1719–1732. https://doi.org/10.1021/acsanm.0c03181

[10] R. Lu, F. Xu, Y. Cui, D. Bao, S. Yuan, Y. Sun, L. Pei, Y. Jiang, J. Zhu, H. Wang, Novel thermal conductivity and anti-corrosion coating with hydrophobic properties for heat exchanger applications, *Prog. Org. Coat.* 186 (2024) 108004. https://doi.org/10.1016/J.PORGCOAT.2023.108004

[11] N.T. Panagiotopoulos, A.G. Lekatou, K.A. Agrafioti, C.C. Prouskas, M.K. Koukou, J. Konstantaras, K. Lymperis, M.G. Vrachopoulos, G.A. Evangelakis, Anticorrosive AlN coatings for heat exchangers in thermal energy storage systems, *Therm. Sci. Eng. Prog.* 43 (2023) 102014. https://doi.org/10.1016/J.TSEP.2023.102014

[12] Z. Yuan, L. Wu, Y. Liu, Y. Liu, P. Zhang, Influence of anti-corrosion coating on thermal and exergetic performances of heat exchangers, *Heat Transf. Res.* 52 (2021). https://doi.org/10.1615/HeatTransRes.2021038803

[13] N. Sharma, R. Kumar, B. Sharma, D. Jain, R.C. Batra, Thermal performance of fouling-resistant polymer nanocomposite coatings in heat exchangers, *Int. J. Heat Mass Transf.* 215 (2023) 124512. https://doi.org/10.1016/J.IJHEATMASSTRANSFER.2023.124512

[14] W. He, H. Shang, Y. Cao, Y. Yu, Development and application of superhydrophobic coatings with high thermal conductivity and self-healing properties: Synergistically enhancing condensation heat transfer through surface modification, *Compos. Part A Appl. Sci. Manuf.* 194 (2025) 108934. https://doi.org/10.1016/J.COMPOSITESA.2025.108934

[15] A.A. Alamery, Z.F. Mahdi, H.A. Jawad, Investigation of metal and metal oxide nano-coating on fins in HPHE with silver water nanofluid, *Am. J. Mech. Eng.* 3 (2025) 21–25. http://pubs.sciepub.com/

[16] A.A.A. Alamery, H.A. Jawad, Z.F. Mahdi, Thickness dependency of CuO nanocoating layer on thermal performance of HPHE, *Int. J. Mater. Sci. Appl.* 3 (2014) 314–320. https://doi.org/10.11648/j.ijmsa.20140306.16

[17] H. Seok, C. Han, D. Lee, Y. Kim, Thermal-hydraulic performance enhancement of fin-and-tube heat exchangers using carbon nanotube coatings under dry and wet conditions, *Appl. Therm. Eng.* 231 (2023) 120938. https://doi.org/10.1016/J.APPLTHERMALENG.2023.120938

[18] A. Taylor, M. Joyce, Nano coatings for energy efficiency and CO_2 reduction, TechConnect World Innov. Conf. Expo (2024). https://www.techconnectworld.com/World2024/a.php?i=153

[19] W. Bao, Z. Deng, S. Zhang, Z. Ji, H. Zhang, Next-generation composite coating system: Nanocoating, *Front. Mater.* 6 (2019). https://doi.org/10.3389/fmats.2019.00072

[20] X. Cao, Nanomaterials in smart energy-efficient coatings, *Nanomaterials* 14 (2024). https://doi.org/10.3390/nano14221820

[21] K.R. Kashyzadeh, W.K.M. Ridha, S. Ghorbani, The influence of nanocoatings on the wear, corrosion, and erosion properties of AISI 304 and AISI 316L stainless steels: A critical review regarding hydro turbines, *Corros. Mater. Degrad.* 6 (2025) 6. https://doi.org/10.3390/cmd6010006

[22] F.H. Faraj, J.M. Ali, S.T. Najim, A.J. Sultan, S.M. Alardhi, H.S. Majdi, Nano-iron oxide coating for enhanced heat transfer in gas–solid fluidized bed systems, *ChemEngineering* 8 (2024) 9. https://doi.org/10.3390/chemengineering8010009

[23] M. Merisalu, L. Aarik, H.-M. Piirsoo, J. Kozlova, A. Tarre, R. Zabels, J. Wessing, A. Brieva, V. Sammelselg, Nanostructured coating for aluminum alloys used in aerospace applications, *J. Electrochem. Soc.* 169 (2022) 071503. https://doi.org/10.1149/1945-7111/ac7bb2

[24] X. Joseph Raj, Innovative nanocomposite coating for aluminum alloy in the aircraft manufacturing industry with increased mechanical strength, flame retardancy, and corrosion resistance, *ChemistrySelect* 9 (2024). https://doi.org/10.1002/slct.202402056

[25] S.Z.A. Sakinah, W.A. Wan Hamzah, J. Alias, Corrosion performance of nanopaint for automotive application BT, Proceedings of the 2nd Energy Security and Chemical Engineering Congress, in: N.H. Johari, W.A. Wan Hamzah, M.F. Ghazali, H.D. Setiabudi, S. Kumarasamy (Eds.), Springer Nature Singapore, Singapore, 2023: pp. 299–308.

[26] A. V. Blinov, A.A. Nagdalian, L.P. Arefeva, V.N. Varavka, O. V. Kudryakov, A.A. Gvozdenko, A.B. Golik, A.A. Blinova, D.G. Maglakelidze, D.D. Filippov, V.A. Lapin, E.D. Nazaretova, M.A. Shariati, Nanoscale composite protective preparation for cars paint and varnish coatings, *Coatings* 12 (2022). https://doi.org/10.3390/coatings12091267

8 Nanocomposite Biomaterials for Orthopedic Implants

A New Frontier

Uzwalkiran Rokkala, Subbarao Medabalimi, Divakar Bommana, N Pallavi Senapati, and T. Rajesh Kumar Dora

8.1 INTRODUCTION

Millions of people around the world now have more mobility and a higher quality of life because of orthopedic implants, which have greatly improved the treatment of musculoskeletal problems, joint degeneration, and bone fractures [1]. Although they have shown clinical success, traditional implants—which are mainly composed of metals (such as titanium and stainless steel), polymers (such as ultra-high molecular weight polyethylene), and ceramics (such as alumina and zirconia)—are constrained by issues like poor osseointegration, mechanical mismatch with natural bone, and infection susceptibility [2, 3]. These drawbacks often lead to implant failure, revision procedures, and higher medical expenses, highlighting the need for advanced biomaterials that more closely resemble the mechanical and biological characteristics of natural bone.

Recently, nano-composite biomaterials have become a revolutionary answer. These materials combine nanomaterials with common implant materials to make them work better [4]. By leveraging the unique properties of nanoparticles, including their high surface area-to-volume ratio, adjustable porosity, and bioactive molecular interactions, scientists have developed implants with enhanced antibacterial, mechanical, and osteoconductivity [5]. Composites made of polymers bonded with hydroxyapatite (HA) improve bone bonding by mimicking the mineral makeup of natural bone. Nano-composites made of graphene offer superior mechanical reinforcement [4]. The most common kind of postoperative problem is infection around the implant, which has been addressed with the use of antimicrobials and coatings made of silver nanoparticles (AgNPs) [6–8].

Aside from enhancing structure and combating germs, smart nano-composites that respond to stimuli (e.g., medication release induced by pH, temperature, or enzymes) allow for tailored therapy delivery and real-time monitoring of bone

DOI: 10.1201/9781003653226-8

healing [9]. A future where orthopedic implants are custom-made to meet each patient's unique clinical and anatomical needs may be within reach with the help of such innovative methods. Scaling up manufacturing, guaranteeing long-term biocompatibility, and negotiating complicated regulatory processes are still significant obstacles, despite recent achievements [2, 7, 10]. The widespread therapeutic use of nanoparticles should only be considered after additional research into their potential cytotoxicity and degrading mechanisms [10, 11].

This chapter presents the most recent research on orthopedic devices that make use of nano-composite biomaterials. In this chapter, you can find details on their methods of development, benefits, and therapeutic uses. Research and technical developments in the field show that these new materials solve current problems in orthopedic surgery. Collaboration among biologists, doctors, and material scientists is going to be the subject of our forthcoming discussion as we try to improve implant functionality and hasten the adoption of novel concepts. Improved surgical results and implant lifetime may be achievable with the help of bioactive, patient-specific solutions provided by nano-composite biomaterials, which have the potential to revolutionize orthopedic therapy.

8.2 DESIGN PRINCIPLES OF NANO-COMPOSITE BIOMATERIALS

Achieving enhanced mechanical characteristics, bioactivity, and biocompatibility is essential in the development of high-performance nano-composite biomaterials for orthopedic implants. These orthopedic implants are designed according to these principles. These concepts encompass several techniques for improving functioning, such as surface modification, selective utilization of nanomaterials, and integration of structural components. Comprehending these fundamental attributes allows researchers to develop advanced implants that mimic the appearance and performance of real bone while mitigating prevalent challenges such as mechanical incompatibility, inadequate osseointegration, and infection.

8.2.1 Composition and Structure

Biological effectiveness, degrading properties, and mechanical strength of nano-composite biomaterials are all dictated by their microstructure and composition. Polymers such as polyetheretherketone (PEEK) or polylactic acid (PLA), together with HA ceramics and titanium alloys, improve flexibility and reduce stress shielding. Use of nanomaterials such as graphene oxide (GO), carbon nanotubes, and silica nanoparticles increases the material's strength and durability. An ideal composite design features a porous, strong, and long-lasting framework that allows nutrients to flow in and cells to invade. Improved osteogenic potential and implant lifespan are outcomes of hierarchical bone architectures that go beyond the cancellous and cortical layers.

8.2.2 Role of Nanomaterials

The unique physicochemical properties of nanomaterials make them indispensable for enhancing the performance of orthopedic composites. Bioactive glass and nano-hydroxyapatite (nHA) are two examples of nanoparticles that improve osteoconduction by making it easier for bone cells to adhere and mineralize. Graphene and carbon nanotubes are two examples of carbon-based nanomaterials that provide exceptional mechanical reinforcement without sacrificing their lightweight properties. There is less chance of infection due to implants because metallic nanoparticles, such as zinc oxide (ZnO) and silver (Ag), have antibacterial properties. Targeted distribution of growth stimulants or antibiotics is made possible by the higher surface area of nanoparticles, which allows for efficient medicine loading and controlled release. However, to prevent agglomeration and provide uniform reinforcement, the distribution and interfacial adhesion of nanoparticles inside the matrix must be carefully controlled.

8.2.3 Biocompatible Matrix Materials

Biocompatibility and long-term stability of nano-composite implants depend on the choice of matrix material. The versatility and compatibility of biodegradable polymers with bone regeneration make them very useful. Examples of such polymers are chitosan and poly(lactic-co-glycolic acid) (PLGA). Modifications to the surface of non-biodegradable polymers, such as PEEK, may be required to enhance their bioactivity, notwithstanding their durability. Materials having high wear resistance but low toughness, such as ceramic matrices like alumina and zirconia, sometimes necessitate composite reinforcement with metals or polymers. Hybrid matrices allow the combination of the best features of several materials, such as the osteoinductivity of ceramics with the flexibility of polymers. Optimization of surface chemistry and topography is necessary for the matrix to support cellular adhesion and proliferation by enhancing protein adsorption and cell–material interactions.

8.2.4 Surface Modification Techniques

Enhancing the biointegration and functioning of nano-composite implants requires surface engineering. Bioactive coatings, such as HA or collagen, are applied to improve bone bonding using techniques such as friction stir processing, plasma spraying, electrochemical deposition, and layer-by-layer assembly. Nanoscale surface patterning, including laser texturing and electron beam lithography, can create micro- and nano-topographies that promote osteoblast attachment and differentiation. Additionally, chemical functionalization with peptides (e.g., RGD sequences) or growth factors (e.g., BMP-2) further enhances cellular responses. Antimicrobial surface modifications, such as silver nanoparticle coatings or antibiotic-eluting polymer films, help prevent bacterial colonization. Smart surface modifications, including pH- or temperature-responsive coatings, enable dynamic interactions with the biological environment, allowing for controlled drug release or real-time monitoring of healing progress. Figure 8.1 shows the development of a composite surface to enhance the biocompatibility and corrosion resistance [12].

By integrating these design principles, nano-composite biomaterials can be tailored to meet the specific demands of orthopedic applications, offering improved mechanical performance, biological compatibility, and therapeutic functionality. Future advancements will focus on optimizing fabrication techniques, ensuring long-term safety, and accelerating regulatory approval to bring these innovative solutions into clinical practice.

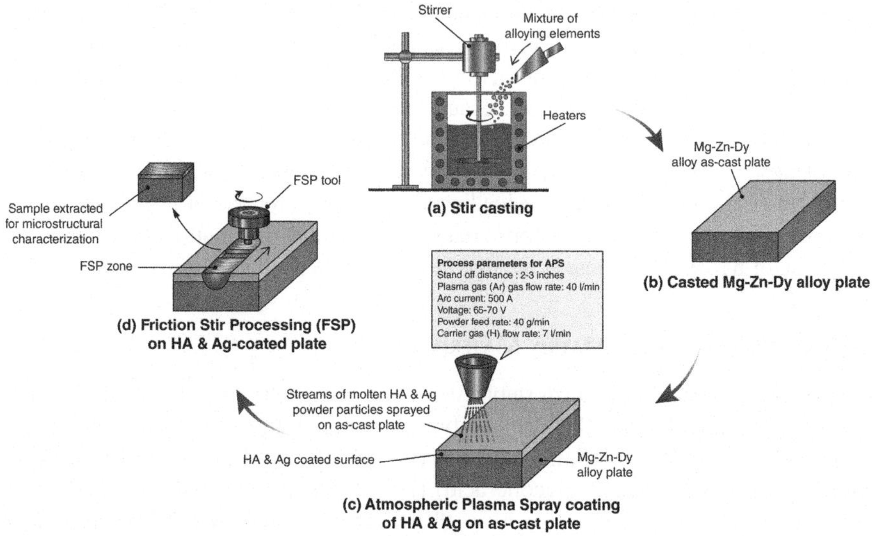

FIGURE 8.1 Development of surface composite [12].

8.3 FUNCTIONAL ADVANTAGES OF NANO-COMPOSITE BIOMATERIALS

The functional performance of orthopedic implants has been significantly enhanced through the use of nanoparticles, which have mitigated critical limitations of conventional materials. In comparison to conventional materials, biomaterials derived from nano-composite substances provide several benefits, including enhanced mechanical strength, superior bone-bonding characteristics, inherent antimicrobial capabilities, and intelligent reactivity to stimuli. Owing to enhanced patient outcomes, reduced complication rates, and prolonged implant durability, nano-composites have emerged as superior options for orthopedic applications. By employing the unique characteristics of nanoparticles, researchers have created implants that mimic the bone's natural environment, expedite the healing process, and offer protection against infections.

8.3.1 Enhanced Mechanical Properties

Nano-composite biomaterials reduce implant failure and stress shielding by closely mimicking the mechanical qualities of real bone. Because their elastic modulus is higher than that of bone, metal implants may actually encourage bone resorption over the long term [13]. Polymer or metal matrices can be enhanced with nanomaterials like carbon nanotubes, graphene, and nHA to increase strength-to-weight ratios while maintaining sufficient stiffness [14]. In a recent study, Rokkala et al. [12] incorporated HA and silver to enhance the mechanical properties using friction stir processing and atmospheric plasma spraying (Figure 8.2). For instance,

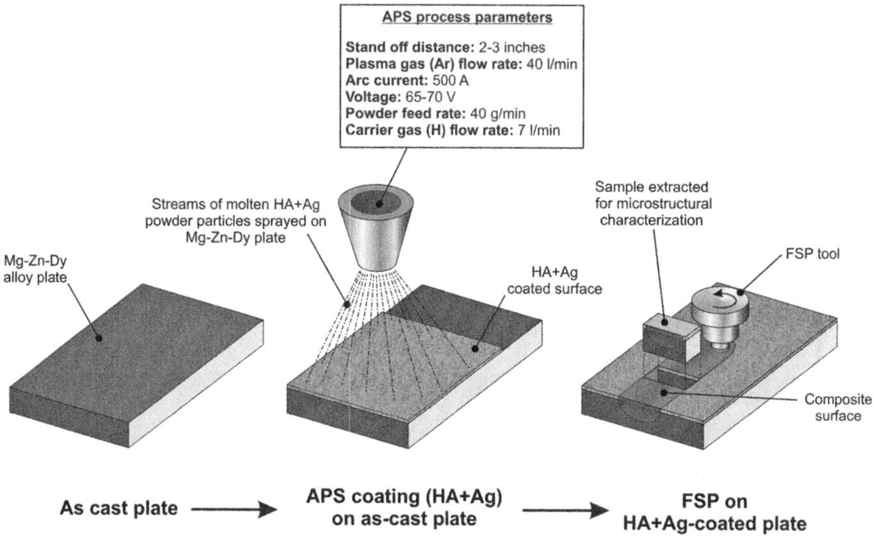

FIGURE 8.2 Composite surface fabrication using friction stir processing [15].

graphene-reinforced PEEK composites exhibit up to 40% higher tensile strength compared to pure PEEK, making them suitable for load-bearing applications [4, 6, 16]. Similarly, nano-sized ceramic reinforcements in titanium alloys enhance fatigue resistance without compromising biocompatibility [2, 17]. These improvements in mechanical performance ensure that implants can withstand physiological loads while minimizing adverse bone remodeling.

8.3.2 Improved Osseointegration

Nano composite biomaterials promote superior osseointegration by mimicking the nanoscale architecture of natural bone and providing bioactive surfaces that stimulate bone formation. HA nanoparticles, which closely resemble the mineral phase of bone, have been widely used to enhance osteoconductivity in polymer and metal implants [18]. Studies have shown that nHA-coated titanium implants exhibit significantly higher bone-to-implant contact compared to uncoated surfaces [19]. Additionally, the incorporation of bioactive glass nanoparticles in composites releases ions (e.g., Ca^{2+} and SiO_4^{4-}) that upregulate osteogenic gene expression, accelerating bone regeneration [20, 21]. Surface modifications at the nanoscale, such as nanotubular titania layers, further improve cell adhesion and differentiation by providing topographical cues similar to the extracellular matrix [20, 22]. These features collectively enhance early-stage fixation and long-term stability of orthopedic implants.

8.3.3 Antimicrobial Efficacy

Implant-associated infections remain a major challenge in orthopedic surgery, often leading to revision procedures and prolonged antibiotic therapy. Nano-composite biomaterials address this issue through the integration of antimicrobial agents such as

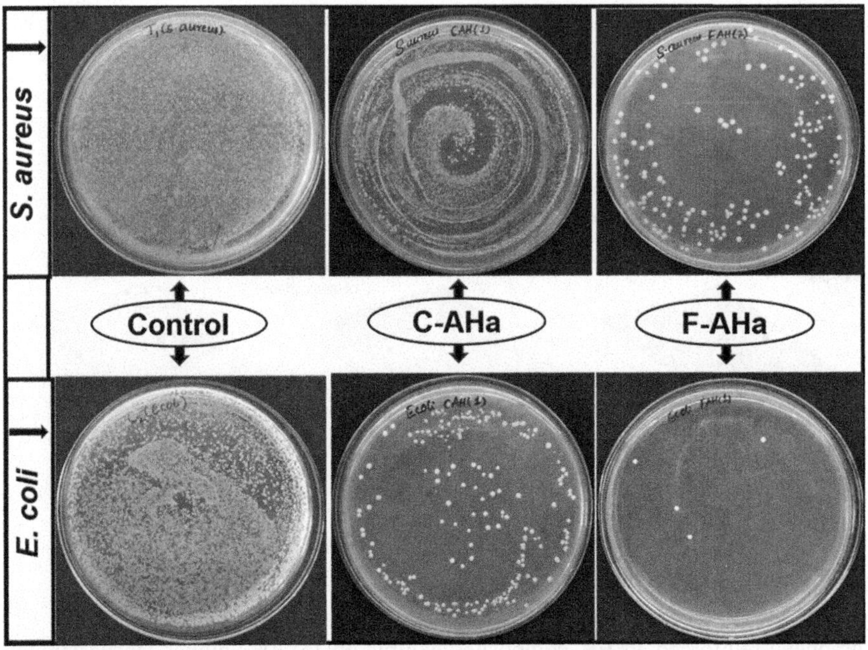

FIGURE 8.3 Antibacterial activity using the plate count method [26].

silver nanoparticles (AgNPs), zinc oxide (ZnO), and chitosan nanoparticles [16]. AgNPs, in particular, exhibit broad-spectrum antibacterial activity by disrupting microbial cell membranes and inhibiting DNA replication [23]. Coatings of AgNP-embedded polymers on implant surfaces have demonstrated significant reductions in bacterial adhesion and biofilm formation [24]. Similarly, ZnO nanoparticles not only provide antibacterial effects but also promote osteogenesis, functioning as dual-functional additives [25]. The controlled release of these antimicrobial agents from nano-composites ensures prolonged protection against infection while minimizing cytotoxicity to host tissues. Figure 8.3 shows the antibacterial activity of Mg alloy, Ag and Al_2O_3-coated Mg alloy, and Mg alloy composite. Coating with Ag reduces bacterial growth compared to the Mg alloy samples.

8.3.4 Stimuli-Responsive Behavior

Smart nano-composites with stimuli-responsive capabilities represent a groundbreaking advancement in orthopedic implants, enabling on-demand drug delivery and real-time monitoring of the healing process. These materials can be engineered to respond to physiological cues such as pH changes, temperature fluctuations, or enzymatic activity [27]. For example, pH-sensitive hydrogels loaded with antibiotics can release therapeutics in response to the acidic microenvironment of bacterial infections [22]. Likewise, thermoresponsive polymers such as poly(*N*-

isopropylacrylamide) (PNIPAM) can adjust medication release rates in response to localized temperature variations generated by inflammation. Furthermore, graphene or carbon nanotubes integrated into conductive nano-composites allow for the real-time evaluation of strain and healing progress using embedded sensors [28]. Customizable and adaptive implant systems with enhanced therapeutic effects are made possible by these intelligent characteristics.

8.4 RECENT ADVANCES IN NANO-COMPOSITE BIOMATERIALS

Substantial advancements have occurred in orthopedic implants owing to the innovation of sophisticated nano-composite biomaterials. By combining the advantages of nanotechnology with traditional biomaterials, these substances produce implants with improved mechanical qualities, increased bioactivity, and multifunctional characteristics. Modern innovations have substantially enhanced the efficacy of orthopedic implants. HA-reinforced polymers, silver nanoparticle coatings, graphene-based nano-composites, and advanced drug-delivery systems are included. These advancements address critical issues such as mechanical incompatibility, implant-related infections, and insufficient osseointegration, hence promoting the advancement of next-generation technologies for bone repair and regeneration.

8.4.1 Hydroxyapatite-Reinforced Polymers

HA-reinforced biomaterials exhibit enhanced osteoconductivity and mechanical compatibility with real bone, making them a significant class of biomaterials. Its promise as a biomaterial was first sparked by this property. The bioactivity and bone contact capabilities of polymer mixes, such as PEEK, PLA, and PCL, are enhanced by the addition of nano-sized HA particles, which mimic the mineral phase of bone [19, 29]. The mechanical strength and osteoblast growth promotion of HA-reinforced PEEK composites make them suitable for load-bearing applications [13, 17, 30]. Additionally, by creating a framework for the formation of new bone, biodegradable polymers like PLA that include HA nanoparticles can speed up the bone regeneration process [31]. In dental and spinal implants, where a strong bond between the implant and bone is paramount, these composites demonstrate superior performance.

8.4.2 Graphene-Based Nano-composites

The remarkable mechanical properties, electrical conductivity, and biocompatibility of graphene and its derivatives, including GO and reduced graphene oxide (rGO), have drawn considerable interest [32]. The tensile strength, fracture toughness, and wear resistance of graphene-based nano-composites are significantly enhanced when they are integrated into polymer or ceramic matrices [17, 19, 30]. Graphene-reinforced PEEK composites exhibit a 30% enhancement in flexural strength compared to pure PEEK, as well as improved osteogenic differentiation of stem cells [9]. The electrical conductivity of graphene enhances bone development stimulation by electroactive therapies, hence boosting osseointegration. The characteristics of graphene-based nano-composites render them optimal for use in joint replacements and bone fixation devices.

8.4.3 Silver Nanoparticle Coatings

In orthopedic surgery, implant-associated infections continue to be a major problem, frequently resulting in implant failure and revision treatments. The strong antibacterial qualities of silver nanoparticles (AgNPs), resulting from their ability to rupture bacterial cell membranes and prevent the formation of biofilms, have been extensively studied [16]. Recent advancements in this field include incorporating AgNP coatings into polymer matrices or depositing them onto implant surfaces using techniques like plasma spraying and electrochemical deposition [13]. Significant reductions in bacterial colonization have been seen in titanium implants coated with AgNP, without compromising biocompatibility [2]. By using controlled-release formulations, the cytotoxicity of AgNPs can be minimized without sacrificing their sustained antibacterial activity [16]. Because of the increased risk of infection, these coatings are particularly useful for prosthetic joint implants and areas where trauma has occurred.

8.4.4 Smart Nano-composites for Drug Delivery

Orthopedic implant technology has advanced significantly with the introduction of smart nano-composites that can respond to external stimuli. These materials enable precise and controlled drug delivery. In response to specific physiological signals like changes in temperature, pH, or enzyme activity, these materials can secrete medicinal substances like growth factors, antibiotics, or anti-inflammatory medications [33]. During bacterial infections, an acidic environment may activate pH-sensitive hydrogels loaded with vancomycin [16]. Similarly, the thermo-responsive polymer PNIPAM can control the release of medications in response to temperature changes brought on by localized inflammation [21, 34]. Conductive nano-composites using carbon nanotubes allow for the incorporation of sensors that measure the healing process in real-time [9, 11, 19]. These intelligent devices show great promise for personalized and flexible bone regeneration therapies.

8.5 CLINICAL POTENTIAL AND APPLICATIONS

Orthopedic patients may benefit greatly from the use of nano-composite biomaterials in a variety of treatment modalities. When compared to traditional implants, these new materials dramatically enhance mechanical stability, osseointegration, antimicrobial properties, and medication administration. Because of their enhanced load-bearing capacity and stress shielding, trauma surgical implants and subsequent fractures are less likely to occur with nano-composite plates, screws, and intramedullary nails reinforced with graphene or nano-HA. To address two of the most common causes of revision surgery—mechanical wear and infection—wear-resistant graphene composites coated with antibacterial silver nanoparticles are now being used in hip and knee arthroplasty. Intraoperative and postoperative imaging with HA-reinforced polymer cages and interbody devices speeds up bone ingrowth.

In the field of orthopedic oncology, smart nano-composites with stimuli-responsive drug delivery systems enable localized, controlled release of chemotherapeutic agents or bone morphogenetic proteins (BMPs), allowing for targeted treatment of bone tumors while supporting subsequent reconstruction. Dental and maxillofacial applications are similarly revolutionized by these materials, with nano-composite dental implants demonstrating faster osseointegration rates and reduced peri-implantitis incidence compared to conventional titanium implants. The clinical impact also extends to pediatric orthopedics, where biodegradable nano-composite materials with tunable degradation profiles offer significant advantages for growing skeletons, eliminating the need for implant removal surgeries. Perhaps most promising are the emerging applications in regenerative medicine, where three-dimensional printed nano-composite scaffolds with precisely controlled porosity and bioactive surface modifications are being developed for large bone defect reconstruction, potentially eliminating the need for autografts and their associated donor site morbidity.

The clinical translation of these technologies is already underway, with several nano-composite implants having received regulatory approval and demonstrating excellent short-term outcomes in human trials. For instance, HA-coated polymer spinal cages have shown fusion rates exceeding 90% at 12 months in clinical studies, while silver-impregnated fracture fixation devices have reduced postoperative infection rates by more than 50% compared to standard implants. A new era of individualized orthopedic treatment is emerging, driven by the integration of nano-composite technology with cutting-edge manufacturing processes like 3D printing and patient-specific modeling. The next era will be defined by individualized implants that enhance a patient's healing capacity by conforming to their own anatomy. When it comes to orthopedic implant materials, nano-composite biomaterials are quickly becoming the gold standard. In addition to encouraging bone regeneration and reducing the likelihood of complications, these biomaterials provide more realistic and long-lasting solutions. There is increasing hope that nano-composite biomaterials may one day replace more traditional forms of medicine as more data on their long-term safety becomes available and manufacturing techniques are optimized for scalability. It is critical that healthcare systems throughout the world have access to advanced technology, yet there are big obstacles to achieving optimal cost-effectiveness and handling regulatory processes well.

8.6 CHALLENGES AND FUTURE DIRECTIONS

Although nano-composite biomaterials have advanced significantly, obtaining clinical acceptance is still a major obstacle. Some of the major concerns regarding nanomaterials include their potential to cause cytotoxicity, oxidative stress, and inflammation in tissues, as well as their long-term biocompatibility. Mechanical performance and degradation are both endangered by dispersed nanoparticles in composite matrices. Manufacturing nano-composite materials using techniques such as electrospinning, molecular self-assembly, or precision 3D printing is challenging due to the high cost and complexity of the process. The approval processes for smart implants with drug-eluting or sensor capabilities are very unpredictable due to the constantly evolving regulatory pathways for these innovative materials. The effects of antimicrobial nanoparticle coatings on bacterial resistance patterns or their long-term durability remain unclear and require further study.

8.7 CONCLUSION

In orthopedic implants, nano-composite biomaterials are revolutionizing the field by addressing problems like mechanical mismatch, poor osseointegration, and infections caused by implants. These composites use nanomaterials such as graphene, HA, and silver nanoparticles to improve mechanical strength, bone-bonding, and antibacterial characteristics. Regulated drug distribution and constant monitoring of the healing process are made possible by smart, stimuli-responsive electronics. While there has been progress in material design and preclinical testing, challenges still exist in areas like long-term biocompatibility, scalable production, and regulatory approval. In order to overcome these challenges and translate these concepts into clinical practice, research and multidisciplinary collaboration are required. Nano-composite biomaterials will revolutionize orthopedics with durable, bioactive, and patient-specific implants that restore function and encourage tissue regeneration. The future of orthopedic implants lies in harnessing the full potential of nanotechnology to create devices that seamlessly integrate with the human body, offering improved outcomes and quality of life for patients worldwide. With ongoing innovation and clinical validation, these materials promise to transform musculoskeletal medicine, making previously intractable conditions treatable and setting new benchmarks for implant performance and longevity.

REFERENCES

1. P.H.C. Camargo, K.G. Satyanarayana, F. Wypych, Nanocomposites: Synthesis, structure, properties and new application opportunities, *Mater. Res.* 12 (2009) 1–39. https://doi.org/10.1590/S1516-14392009000100002
2. M. Geetha, A.K. Singh, R. Asokamani, A.K. Gogia, Ti based biomaterials, the ultimate choice for orthopaedic implants – A review, *Prog. Mater. Sci.* 54 (2009) 397–425. https://doi.org/10.1016/j.pmatsci.2008.06.004
3. R.R. Melo, D.M. Stangerlin, R.R.C. Santana, T.D. Pedrosa, Physical and mechanical properties of particleboard manufactured from wood, bamboo and rice husk, *Mater. Res.* 17 (2014) 682–686. https://doi.org/10.1590/S1516-14392014005000052
4. C.A. Scotchford, M.J. Garle, J. Batchelor, J. Bradley, D.M. Grant, Use of a novel carbon fibre composite material for the femoral stem component of a THR system: In vitro biological assessment, *Biomaterials* 24 (2003) 4871–4879. https://doi.org/10.1016/S0142-9612(03)00403-4
5. S. Bose, M. Roy, A. Bandyopadhyay, Recent advances in bone tissue engineering scaffolds, *Trends Biotechnol.* 30 (2012) 546–554. https://doi.org/10.1016/j.tibtech.2012.07.005
6. J.P. Fan, C.P. Tsui, C.Y. Tang, C.L. Chow, Influence of interphase layer on the overall elasto-plastic behaviors of HA/PEEK biocomposite, *Biomaterials* 25 (2004) 5363–5373. https://doi.org/10.1016/j.biomaterials.2003.12.050
7. M.S. Abu Bakar, M.H.W. Cheng, S.M. Tang, S.C. Yu, K. Liao, C.T. Tan, K.A. Khor, P. Cheang, Tensile properties, tension-tension fatigue and biological response of polyetheretherketone-hydroxyapatite composites for load-bearing orthopedic implants, *Biomaterials* 24 (2003) 2245–2250. https://doi.org/10.1016/S0142-9612(03)00028-0

8. G. Enislidis, K. Yerit, G. Wittwer, R. Köhnke, S. Schragl, R. Ewers, Self-reinforced biodegradable plates and screws for fixation of zygomatic fractures, *J. Craniomaxillofac. Surg.* 33 (2005) 95–102. https://doi.org/10.1016/j.jcms.2004.10.006
9. S. Yu, K.P. Hariram, R. Kumar, P. Cheang, K.K. Aik, In vitro apatite formation and its growth kinetics on hydroxyapatite/polyetheretherketone biocomposites, *Biomaterials* 26 (2005) 2343–2352. https://doi.org/10.1016/j.biomaterials.2004.07.028
10. M.N. Rahaman, A. Yao, B.S. Bal, J.P. Garino, M.D. Ries, Ceramics for prosthetic hip and knee joint replacement, *J. Am. Ceram. Soc.* 90 (2007) 1965–1988. https://doi.org/10.1111/j.1551-2916.2007.01725.x
11. J. Al-Sukhun, J. Törnwall, C. Lindqvist, R. Kontio, Bioresorbable poly-L/DL-lactide (P [L/DL]LA 70/30) plates are reliable for repairing large inferior orbital wall bony defects: A pilot study, *J. Oral Maxillofac. Surg.* 64 (2006) 47–55. https://doi.org/10.1016/j.joms.2005.09.013
12. M.D.F. Khan, U. Rokkala, Development of high strength and corrosion resistance Mg-Zn-Dy/HA-Ag composite for temporary implant applications, *Mater. Lett.* 347 (2023) 134604. https://doi.org/10.1016/j.matlet.2023.134604
13. K.L. Ong, B.M. Yun, J.B. White, New biomaterials for orthopedic implants, *Orthop. Res. Rev.* 7 (2015) 107–130. https://doi.org/10.2147/ORR.S63437
14. B.S. Bal, A. Khandkar, R. Lakshminarayanan, I. Clarke, A.A. Hoffman, M.N. Rahaman, Testing of silicon nitride ceramic bearings for total hip arthroplasty, *J. Biomed. Mater. Res. B Appl. Biomater.* 87 (2008) 447–454. https://doi.org/10.1002/jbm.b.31123
15. U. Rokkala, A. Jana, S. Bontha, M.R. Ramesh, V.K. Balla, Comparative investigation of coating and friction stir processing on Mg-Zn-Dy alloy for improving antibacterial, bioactive and corrosion behaviour, *Surf. Coat. Technol.* 425 (2021) 127708. https://doi.org/10.1016/j.surfcoat.2021.127708
16. L. Wang, J. Rhim, Preparation and application of agar/alginate/collagen ternary blend functional food packaging films, *Int. J. Biol. Macromol.* 80 (2015) 460–468. https://doi.org/10.1016/j.ijbiomac.2015.07.007
17. I. Nakahara, M. Takao, S. Bandoh, N. Bertollo, W.R. Walsh, N. Sugano, In vivo implant fixation of carbon fiber-reinforced PEEK hip prostheses in an ovine model, *J. Orthop. Res.* 31 (2013) 485–492. https://doi.org/10.1002/jor.22251
18. P.G. Ntagiopoulos, G. Demey, T. Tavernier, D. Dejour, Comparison of resorption and remodeling of bioabsorbable interference screws in anterior cruciate ligament reconstruction, *Int. Orthop.* 39 (2015) 697–706. https://doi.org/10.1007/s00264-014-2530-8
19. E.L. Steinberg, E. Rath, A. Shlaifer, O. Chechik, E. Maman, M. Salai, Carbon fiber reinforced PEEK Optima-A composite material biomechanical properties and wear/debris characteristics of CF-PEEK composites for orthopedic trauma implants, *J. Mech. Behav. Biomed. Mater.* 17 (2013) 221–228. https://doi.org/10.1016/j.jmbbm.2012.09.013
20. X. Liu, L.A. Smith, J. Hu, P.X. Ma, Biomimetic nanofibrous gelatin/apatite composite scaffolds for bone tissue engineering, *Biomaterials* 30 (2009) 2252–2258. https://doi.org/10.1016/j.biomaterials.2008.12.068
21. V. Guarino, F. Causa, L. Ambrosio, Bioactive scaffolds for bone and ligament tissue, *Expert Rev. Med. Devices* 4 (2007) 405–418. https://doi.org/10.1586/17434440.4.3.405
22. M.T. Wolf, C.L. Dearth, S.B. Sonnenberg, E.G. Loboa, S.F. Badylak, Naturally derived and synthetic scaffolds for skeletal muscle reconstruction, *Adv. Drug Deliv. Rev.* 84 (2015) 208–221. https://doi.org/10.1016/j.addr.2014.08.011
23. A.A. Shah, F. Hasan, A. Hameed, S. Ahmed, Biological degradation of plastics: A comprehensive review, *Biotechnol. Adv.* 26 (2008) 246–265. https://doi.org/10.1016/j.biotechadv.2007.12.005
24. V.R. Sivaperumal, R. Mani, M.S. Nachiappan, K. Arumugam, Direct hydrothermal synthesis of hydroxyapatite/alumina nanocomposite, *Mater. Charact.* 134 (2017) 416–421. https://doi.org/10.1016/j.matchar.2017.11.016

25. M. Farrokhi-Rad, Electrophoretic deposition of fiber hydroxyapatite/titania nanocomposite coatings, *Ceram. Int.* 44 (2018) 622–630. https://doi.org/10.1016/j.ceramint.2017.09.221
26. U. Rokkala, S. Bontha, M.R. Ramesh, V.K. Balla, Multi-step fabrication of bioactive Mg–Zn–Dy–AlO3/HA composites: Exploring the synergistic effects of plasma spray and friction stir processing, *J. Mater. Sci.* 59 (2024) 10998–11014. https://doi.org/10.1007/s10853-024-09830-y
27. M. Mazzocchi, D. Gardini, P.L. Traverso, M.G. Faga, A. Bellosi, On the possibility of silicon nitride as a ceramic for structural orthopaedic implants. Part II: Chemical stability and wear resistance in body environment, *J. Mater. Sci. Mater. Med.* 19 (2008) 2889–2901. https://doi.org/10.1007/s10856-008-3437-y
28. K. Doğan, M.İ. Özgün, H. Sübütay, E. Salur, Y. Eker, M. Kuntoğlu, A. Aslan, M.K. Gupta, M. Acarer, Dispersion mechanism-induced variations in microstructural and mechanical behavior of CNT-reinforced aluminum nanocomposites, *Arch. Civ. Mech. Eng.* 22 (2022). https://doi.org/10.1007/s43452-022-00374-z
29. S.A. Brown, R.S. Hastings, J.J. Mason, A. Moet, Characterization of short-fibre reinforced thermoplastics for fracture fixation devices, *Biomaterials* 11 (1990) 541–547. https://doi.org/10.1016/0142-9612(90)90075-2
30. L. Petrovic, D. Pohle, H. Münstedt, T. Rechtenwald, K.A. Schlegel, S. Rupprecht, Effect of βTCP filled polyetheretherketone on osteoblast cell proliferation in vitro, *J. Biomed. Sci.* 13 (2006) 41–46. https://doi.org/10.1007/s11373-005-9032-z
31. D.F. Williams, A. McNamara, R.M. Turner, Potential of polyetheretherketone (PEEK) and carbon-fibre-reinforced PEEK in medical applications, *J. Mater. Sci. Lett.* 6 (1987) 188–190. https://doi.org/10.1007/BF01728981
32. W. Zhou, Z. Zhou, K. Kubota, H. Ono, N. Nomura, A. Kawasaki, Design of high-performance Al4C3/Al matrix composites for electric conductor, *Mater. Sci. Eng. A* 798 (2020) 140331. https://doi.org/10.1016/j.msea.2020.140331
33. S. Hayakawa, T. Kanaya, K. Tsuru, Y. Shirosaki, A. Osaka, E. Fujii, K. Kawabata, G. Gasqueres, C. Bonhomme, F. Babonneau, C. Jäger, H.J. Kleebe, Heterogeneous structure and in vitro degradation behavior of wet-chemically derived nanocrystalline silicon-containing hydroxyapatite particles, *Acta Biomater.* 9 (2013) 4856–4867. https://doi.org/10.1016/j.actbio.2012.08.024
34. B.S. Bal, A. Khandkar, R. Lakshminarayanan, I. Clarke, A.A. Hoffman, M.N. Rahaman, Fabrication and testing of silicon nitride bearings in total hip arthroplasty. Winner of the 2007 "HAP" PAUL award, *J. Arthroplast.* 24 (2009) 110–116. https://doi.org/10.1016/j.arth.2008.01.300

9 Enhancing Wear Resistance of HVOF-Sprayed Coatings Using Partially Oxidized Powders

Mechanisms, Performance, and Applications

Subbarao Medabalimi, Uzwalkiran Rokkala, Prakash Kumar, and Suresh Gudala

9.1 INTRODUCTION

Friction and wear are closely interrelated, with changes in one often influencing the other [1]. As the temperature increases, the friction typically decreases, but the wear intensifies, leading to significant material degradation and loss of mechanical strength. To mitigate these effects at elevated temperatures, various techniques such as laser melting, hardfacing, friction stir processing, and heat treatment have been employed [2–5]. Among these, thermal spraying, a well-established method, has been widely used in applications such as aircraft engine turbine protection [6]. The growing demand for wear-resistant coatings in industries, particularly in the automotive and energy sectors, stems from the need to reduce energy losses caused by friction and wear [7]. Enhancing the tribological properties of coatings often involves incorporating hard particles into the feedstock. For instance, plasma-sprayed coatings with added Al_2O_3 and CeO_2 oxides on nickel-based alloys demonstrate improved wear resistance owing to the synergistic strengthening effect of these oxides [8]. Research by Feng Liu et al. [9] indicated that adding metal oxides to NiCr reduces the friction coefficient and enhances wear resistance at temperatures above 400°C, with composite coatings maintaining excellent tribological performance up to 1000°C. However, challenges remain, as producing oxide powders via gas atomization is expensive, and achieving a homogeneous powder mixture is complex and time-consuming. An alternative cost-effective approach involves spraying alloy

DOI: 10.1201/9781003653226-9

powders into water to create partially oxidized powders, thereby simplifying the process.

The preparation of partially oxidized powders from the NiCrBSiFe alloy, achieved by spraying the alloy into distilled water using a flame spray process, offers a practical solution. NiCrBSiFe, commonly used as a bond coat in thermal spraying, enhances coating adhesion [10, 11]. By applying these partially oxidized powders directly onto the MDN321 steel substrates via plasma spraying without a bond coat, the production costs are reduced. The resulting coating microstructure, which comprises hard oxides and ductile metallic regions, contributes to its wear resistance. During sliding, the oxides form a protective layer that acts as a barrier against wear and friction, thereby reducing energy consumption. Additionally, these coatings exhibited greater toughness than the fully oxidized powders, as the splats retained a tough metallic core with oxidized outer boundaries. In this chapter, the wear and frictional behavior of the coatings were assessed using a pin-on-disc tribometer at high temperatures, and SEM/EDS and XRD analyses were used to characterize the powders, coatings, and worn surfaces. The PV factor theory was applied to determine the wear rate coefficients of the coating and MDN321 steel at elevated temperatures.

9.2 MECHANISM

Partially oxidized coatings are crucial for enhancing material performance in high-temperature environments, providing resistance to oxidation, erosion, and corrosion [12]. This process involves several intricate steps, each contributing to the final properties of the coating, as shown in Figure 9.1.

Flame Spray Process: In this process, alloy powder is first heated in the flame spray. This step is critical because it sets the baseline for subsequent treatments. The parameters of the flame spray process were selected so that each alloy powder particle was transformed into partially oxidized powders consisting of approximately 50% oxides. Systematic trials at increasing temperatures identified the ideal lower temperature threshold to prevent full fusion, which is crucial for producing partially oxidized powders.

Quenching of Alloy Powders: The alloy powders were rapidly quenched in distilled water using the flame spray process. This rapid cooling forms an oxide layer at the surface of the particles, leaving a metallic core within each particle. This process creates a dual-phase structure that is hard and brittle on the outer side and softer and ductile on the inner side, resulting in a coating that is both tougher and harder.

Drying of the Powders: The powders are dried after quenching to remove the remaining moisture. This is because they attain maximum flowability, a necessary condition for further processing.

Planetary Ball Milling: Planetary ball milling was used to achieve the desired size distribution (<50 μm) of partially oxidized powders that gained a size distribution much larger than required. The mechanical process reduces the particle size in all three dimensions, thereby improving coating deposition.

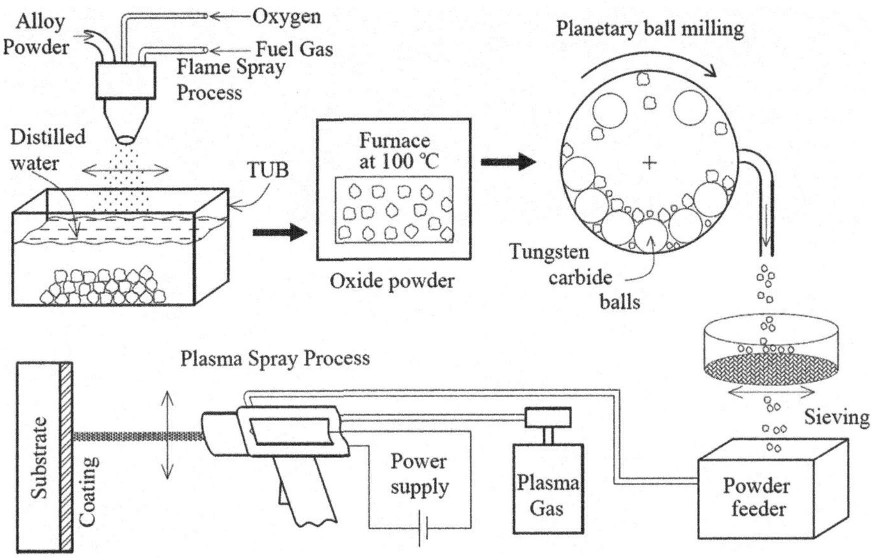

FIGURE 9.1 Schematic representation of the six stages involved in converting alloy powders into partially oxidized powders.

Sieving Analysis: After milling, the powders were then sieved to remove powders with larger particles than 50 μm. This makes the size of the particles consistent, which enhances the uniformity and quality of the coating feedstock.

Plasma Spray Process: The feedstock is applied to the substrate using the plasma spray process. This technique involves the combustion of gases to accelerate molten particles onto a substrate, forming dense and well-bonded coatings. Controlled deposition ensures minimal porosity and strong adhesion, which are crucial for high-performance applications.

Figure 9.2 illustrates the distinctions between the as-received and partially oxidized alloy powders. The transformation of alloy powders into partially oxidized powders is visually represented in this figure and occurs across five distinct stages. Initially, alloy powders were introduced into a flame spray gun at 30°C. As they pass through the gun, the powders are subjected to temperatures of approximately 1200°C [13]. Subsequently, the powders travel through the air, where oxidation occurs as the powder particles react with atmospheric oxygen. In the fourth stage, the semi-solid alloy powder transitions into a solid oxide layer upon being sprayed into distilled water, retaining the moisture content of the powder particles. The powders were then heated in a furnace at 100°C for one hour. Scanning Electron Microscopy (SEM) analysis of both the alloy and partially oxidized powders revealed clear differences. The alloy powders exhibited a spherical morphology with a smooth surface, while the partially oxidized powders were characterized by an oxide layer enveloping the alloy core. When partially oxidized powders were deposited using the plasma spray

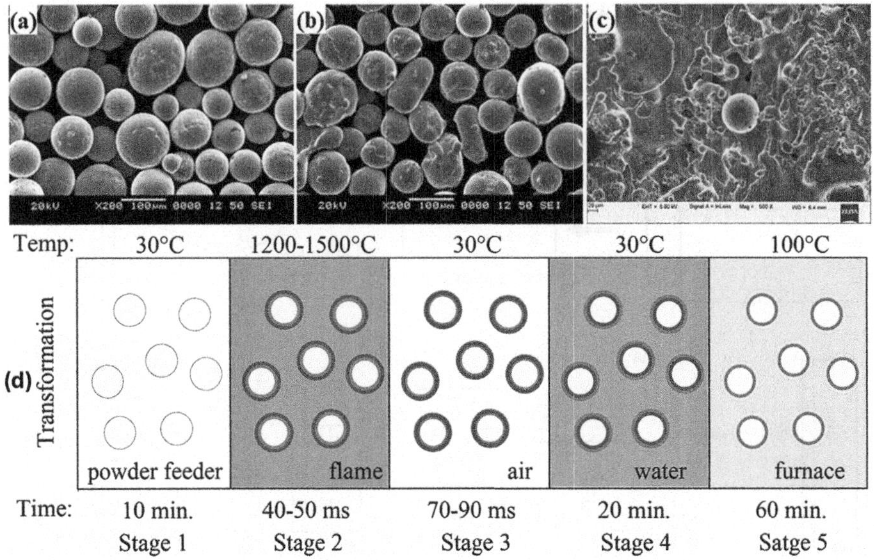

FIGURE 9.2 SEM images of (a) as-received powder, (b) partially oxidized feedstock powder, (c) as-sprayed coating NiCrBSiFe, and (d) schematic representation of five stages in transforming the alloy powders into partially oxidized powders.

process, the coating was successfully formed. The SEM image of the as-sprayed coating, as shown in the figure, confirms this result. It can be concluded that partially oxidized powders can be effectively processed for deposition, similar to the alloy powders. The mechanical properties of coatings derived from partially oxidized powders are improved compared to those from as-received alloy powders, with enhanced density, microhardness, porosity, and adhesion strength [14].

9.3 PERFORMANCE

Increasing the hardness of a coating is essential for improving its wear resistance. Partially oxidized NiCrBSiFe powders were used to achieve this goal. Figure 9.3 depicts the experimental setup used for the pin-on-disc tribometer. The wear loss of

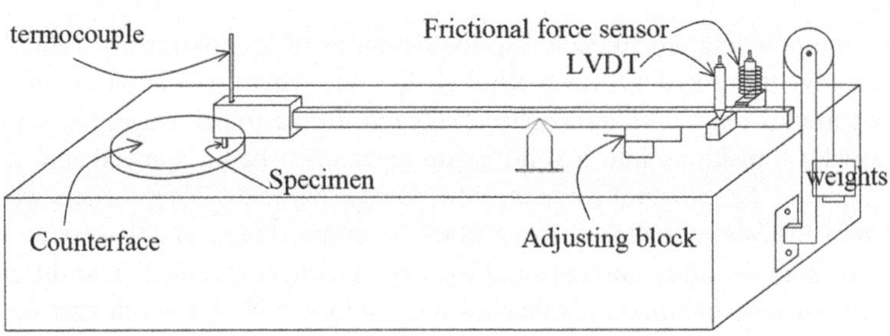

FIGURE 9.3 Schematic representation of the experimental set used for the pin-on-disc tribometer.

the sample depends on several factors, such as the speed of the disk, the applied load on the material, and the test temperature. These three parameters were varied to perform the wear tests. Three different temperatures were used to study the wear resistance of the coatings. Apart from these three temperatures, the 600°C coating exhibited a high wear resistance compared to the other temperatures. The enhanced wear resistance of this coating is attributed to the formation of hard oxides at high temperatures.

Figure 9.3 illustrates the experimental setup used for the wear rate measurements. A Pin-on-Disk Tribometer was used to conduct the experiments. The test sample, measuring 12.5 mm × 12.5 mm and with a thickness of 5 mm, was secured in a holder. During the experiment, the pin rotated against a disc made of aluminum oxide, inducing frictional contact. A thermocouple positioned above the sample in the holder was used to monitor the temperature of the sample. The wear rate of the sample was determined using a linear variable differential transformer (LVDT). In addition, a frictional force sensor was used to measure the friction generated during the experiment.

Figure 9.4 presents a comparison of the wear rates of the MDN 321 substrate and NiCrBSiFe coating. The results indicated a significantly higher wear rate for the MDN 321 substrate than for the coating across all tested temperatures, loads, and speeds. As the temperature increased, the wear rate of the coating decreased, while that of the substrate increased. This coating behavior is attributed to the formation of a partially oxidized layer on the coating surface, which acts as a lubricant and substantially reduces the wear rate. Consequently, this enhanced wear resistance renders the coating suitable for applications involving mating components. In contrast, the MDN 321 substrate exhibited accelerated wear when exposed to elevated temperatures [15].

Figure 9.5 illustrates the relationship between the coefficient of friction, load, and speed for the MDN 321 substrate and NiCrBSiFe coating. The coefficient of friction decreased as the load increased. At lower speeds, the coefficient of friction is higher at lower speeds and decreases at higher speeds. This behavior is attributed to the reduced time available for the pin to encounter resistive forces at higher speeds,

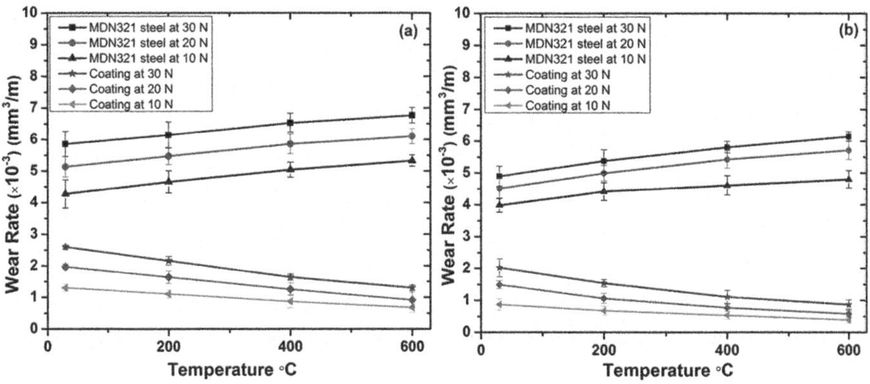

FIGURE 9.4 Wear rate versus temperature of the coating and MDN321 steel at sliding velocities of (a) 1 m/s and (b) 2 m/s.

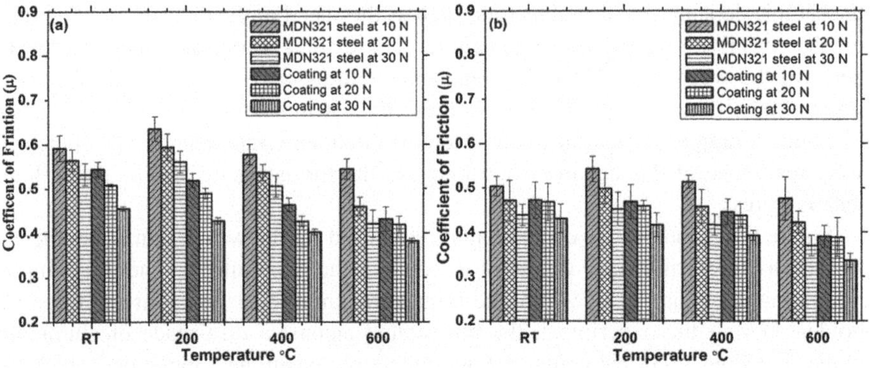

FIGURE 9.5 Coefficient of friction versus temperature of coating at sliding velocities of (a) 1 m/s and (b) 2 m/s.

which results in lower friction. Conversely, at lower speeds, the surface asperities have sufficient time to impede the pin motion, thereby increasing the frictional forces. Even at higher loads, the coefficient of friction decreased because the frictional force did not scale proportionally with the load. Although friction increases with load, doubling the load does not double the frictional force because of the consistent number of surface peaks obstructing the pin motion. This reduction in the frictional force was observed for both the substrate and coating. At higher temperatures, where the wear rate is lower, the formation of a robust oxide film on the surface further reduces the coefficient of friction—particularly for the coating—owing to the lubricating effect of the oxide layer [16].

Figure 9.6 illustrates the oxide layer formed on the surface of the NiCrBSiFe coating after the laser cladding. A comparison of the worn surfaces revealed that the MDN 321 substrate exhibited numerous flaking pits, contributing to an increased wear rate. In contrast, the coating exhibited significantly fewer flaking pits and

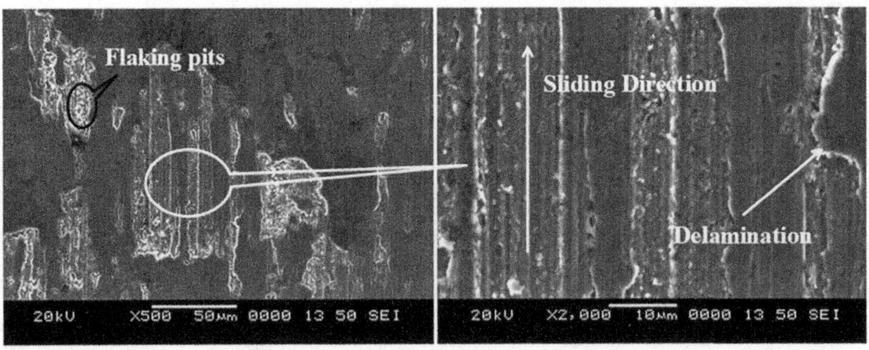

FIGURE 9.6 SEM morphologies of worn surfaces of NiCrBSiFe partially oxidized coating formed delamination at 200°C under 30 N load with 1 m/s sliding speed.

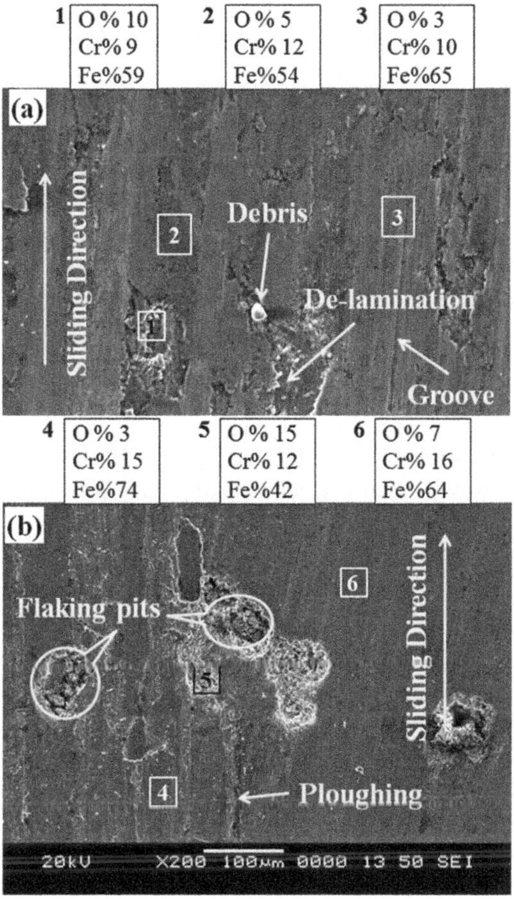

FIGURE 9.7 Worn surface morphology of MDN321 steel at 600°C and 30 N load at (a) 1 m/s and (b) 2 m/s sliding velocities.

prominent oxide layer formation. This oxide layer acts as a lubricant, particularly when it is exposed to high temperatures. At 200°C, an oxide layer formed on the coating surface, significantly reducing the wear rate compared to room temperature. Additionally, the oxide layer provides a superior surface finish compared to the room-temperature surface, further reducing the coefficient of friction on the coating [13] (Figure 9.7).

9.4 APPLICATIONS

NiCrBSiFe coatings deposited by plasma spray technology have become widely used in various industrial fields owing to their outstanding wear resistance, corrosion resistance, and ability to withstand high temperatures. When applied to the hard components of industrial machinery, these coatings significantly reduce the wear of the pump shafts, valve stems, and hydraulic cylinders, which are among the parts most prone to wear during normal operation [17]. The aerospace industry employs such coatings to extend the lifespan of turbine engine parts, landing gear, and many other aircraft parts that must resist extreme conditions and harsh environments [18]. Similarly, the automotive industry applies such coatings to engine components, transmissions, and exhaust systems to reduce friction and improve overall durability [19].

NiCrBSiFe coatings are also highly useful in the oil and gas industry as well as in power generation. These coatings are used to ensure that drill bits (valves and pipeline components) are not eroded or corroded by the aggressive environments in the oil and gas industries. They are applied to boiler tubes, steam turbine blades, and other high-temperature wear- and corrosion-critical parts in power plants that require reinforcement [20]. NiCrBSiFe coatings can be used in a wide variety of applications, with each application being significant in contemporary industrial practice, where the aspects that define the success of any industrial process must include the ability to significantly extend the life of the components involved and the efficiency of the operational reality [21].

9.5 CONCLUSIONS

1. Enhanced Wear Resistance through Partial Oxidation: Partial oxidation of NiCrBSiFe alloy powders, achieving approximately 50% oxidation, significantly enhanced the wear resistance of the plasma-sprayed coatings. The formation of a ceramic oxide layer increased the hardness, effectively reducing the wear rates compared to the MDN 321 substrate, particularly under varying temperatures, loads, and speeds.
2. Dual-Phase Structure for Improved Performance: The partially oxidized coating combines the hardness of the ceramic oxide phase with the ductility of the metallic portion, eliminating the need for an additional bond coat.
3. Reduction of the Coefficient of Friction and Wear Rate: The lubricating effect of the oxide layer formed on the coating surface, particularly at high temperatures (such as 200°C), resulted in a significant reduction in BOTH the coefficient of friction and wear rate. This is compared to the substrate, which is more worn out since it contains flaking pits, which are indicative of the coating suitability in high-temperature applications.
4. Industrial Usage: The new product, with its superior mechanical properties—such as high density, microhardness, porosity, and increased adhesion strength—offers versatility for industrial applications, including manufacturing, aerospace, and automotive sectors, where the mechanical components are subject to continuous wear and degradation.
5. Adjustable Coating: If the optimal parameters of the degree of oxidation of alloy powders are set, the ratio of hardness to toughness can be adjusted to comply with specific industrial needs. This adaptability enables the formulation of coatings tailored for various engineering systems involving mating surfaces under dynamic conditions.

REFERENCES

1. R. A. Al-Samarai, Haftirman, K. R. Ahmad, and Y. Al-Douri, "Effect of load and sliding speed on wear and friction of aluminum – Silicon cast alloy," *Int. J. Sci. Res. Publ.*, vol. 2, no. 3, pp. 3–6, 2012.
2. Y. Zhao, X. Wu, H. Du, X. Shangguan, and W. He, "Influence of different laser remelting paths on the microstructure of a spray coating: Defect prevention and tribological properties," *Surf. Coat. Technol.*, vol. 367. pp. 11–18, 2019, doi: 10.1016/j.surfcoat.2019.01.118
3. D. Kesavan and M. Kamaraj, "The microstructure and high temperature wear performance of a nickel base hardfaced coating," *Surf. Coat. Technol.*, vol. 204, no. 24, pp. 4034–4043, 2010, doi: 10.1016/j.surfcoat.2010.05.022
4. N. Yasavol and A. Ramalho, "Wear properties of friction stir processed AISI D2 tool steel," *Tribol. Int.*, vol. 91, pp. 177–183, 2015, doi: 10.1016/j.triboint.2015.07.001
5. L. Y. Chen, T. Xu, S. Lu, Z. X. Wang, S. Chen, and L. C. Zhang, "Improved hardness and wear resistance of plasma sprayed nanostructured NiCrBSi coating via short-time heat treatment," *Surf. Coat. Technol.*, vol. 350, no. July, pp. 436–444, 2018, doi: 10.1016/j.surfcoat.2018.07.037
6. R. A. Miller, "Thermal barrier coatings for aircraft engines: History and directions," *J. Therm. Spray Technol.*, vol. 6, no. 1, pp. 35–42, 1997, doi: 10.1007/BF02646310
7. K. Holmberg and A. Erdemir, "Influence of tribology on global energy consumption, costs and emissions," *Friction*, vol. 5, no. 3, pp. 263–284, 2017, doi: 10.1007/s40544-017-0183-5
8. L. He, Y. Tan, X. Wang, T. Xu, and X. Hong, "Microstructure and wear properties of Al2O3-CeO 2/Ni-base alloy composite coatings on aluminum alloys by plasma spray," *Appl. Surf. Sci.*, vol. 314, pp. 760–767, 2014, doi: 10.1016/j.apsusc.2014.07.047
9. J. J. Feng Liu, Yong Zhou, Xiaoyong Zhang, and Weifeng Cao, "Tribological properties of NiCr-ZrO2 (Y2O3)-SrSo4 composites at elevated temperatures," *Ceram. Int.*, vol. 42, pp. 12981–12987, 2016.
10. C. D. Kai Yang, Jian Rong, Jingwei Feng, Yin Zhuang, Huayu Zhao, Liang Wang, Jinxing Ni, Shunyan Tao, and Fang Shao, "Excellent wear resistance of plasma-sprayed amorphous Al2O3-Y3Al5O12 ceramic coating," *Surf. Coat. Technol.*, vol. 326, pp. 96–102, 2017, doi: 10.1016/j.scriptamat.2017.03.027
11. Y. H. Chae and S. S. Kim, "Tribological performance of Al2O3/NiCr coating," *KSME Int. J.*, vol. 16, no. 7, pp. 911–918, 2002, doi: 10.1007/BF02949719
12. S. R. Medabalimi *et al.*, "High-temperature solid particle erosion behavior of partially oxidized NiCrBSiFe/NiCr plasma spray coatings," *J. Therm. Spray. Technol.*, vol. 30, no. 6, pp. 1638–1652, 2021, doi: 10.1007/s11666-021-01225-8
13. H. C. Fals, D. Aguiar, L. Fanton, M. J. Xavier Belém, and C. R. Camello Lima, "A new approach of abrasive wear performance of flame sprayed NiCrSiBFeC/SiC composite coating," *Wear*, no. April, p. 203887, 2021, doi: 10.1016/j.wear.2021.203887
14. S. R. Medabalimi, M. R. Ramesh, and R. Kadoli, "Developing partially oxidized NiCr coatings using the combined flame spray and plasma spray process for improved wear behaviour at high temperature," *Wear*, vol. 478–479, no. December 2020, p. 203885, 2021, doi: 10.1016/j.wear.2021.203885
15. K. P. J. Lakshmi, I. Sogalad, and S. Basavarajappa, "Study on dry sliding wear behaviour of NiCrAlY/B4C/cenosphere composite coating by RSM method," *Mater. Today Proc.*, vol. 39, pp. 758–763, 2020, doi: 10.1016/j.matpr.2020.09.445
16. N. L. Parthasarathi and M. Duraiselvam, "Improvement of high temperature wear resistance of AISI 316 ASS through NiCrBSiCFe plasma spray coating," *J. Miner. Mater. Charact. Eng.*, vol. 9, no. 7, pp. 653–670, 2010.

17. S. Kumar and R. Kumar, "Influence of processing conditions on the properties of thermally sprayed coatings: A review," *Surf. Eng.*, vol. 37, no. 11, pp. 1339–1372, 2021, doi: 10.1080/02670844.2021.1967024
18. V. Kumar Domakonda and R. Kumar Puli, "Application of thermal barrier coatings in diesel engines: A review," *Energy Power*, vol. 2, no. 1, pp. 9–17, 2012, doi: 10.5923/j.ep.20120201.02
19. A. Milanti, F. Bozza, G. Bolelli, L. Lusvarghi, P. Vuoristo, and H. Koivuluoto, "Microstructural characteristics and tribological behavior of HVOF-sprayed novel Fe-based alloys coatings," *Coatings*, vol. 4, no. 1, pp. 98–120. doi: 10.3390/coatings4010098
20. S. Manoharan, R. Vijay, D. Singaravelu, and M. Kchaou, "Investigation of tribological and corrosion characteristics of oxide-coated steel and mild steel fiber-based brake friction composites," *Ind. Lubr. Tribol*, vol. 71, no. 3, pp. 341–347, 2018, doi: 10.1108/ilt-08-2018-0313
21. Y.-L. Su, W.-H. Kao, S.-M. Chiu, H.-C. Hsueh, and T.-Y. Hsieh, "Mechanical, tribological, and corrosive properties of NbCrCx and NbCrCxNy coatings with various nitrogen and carbon contents," *Metals*, vol. 13, no. 8, p. 1488, 2023, doi: 10.3390/met13081488

10 Laser Modification and Al–Ti Coating Effects on AZ31 Wear

Tejonidhi G K, Bharatish A, Abhishek R, Varshith K, Vijay Kumar P R, Sivakumar S, Shanmukha N, Rajkumar G R, and Harisha S K

10.1 INTRODUCTION

The distinctive qualities of magnesium alloy AZ31—such as its high strength-to-weight ratio, superior corrosion resistance, and great machinability—make it highly relevant for a variety of industrial applications. AZ31 plays a key role in the automotive industry's production of lightweight parts like engine blocks, body panels, and transmission cases.[1]

In addition to improving fuel economy and vehicle performance, this light weighting assists automakers in adhering to strict emission standards. AZ31 reduces overall vehicle weight, improving handling and decreasing wear and tear in automobiles. This results in decreased maintenance costs and an extended lifespan for automotive parts.

In the aerospace industry, AZ31 is prized for its ability to reduce the weight of aircraft without compromising structural integrity. Components such as aircraft fuselage parts, wings, and interior structures benefit from the alloy's lightweight nature, which is crucial for fuel efficiency and payload capacity.[2,3] The material's good corrosion resistance also ensures durability and reliability in the harsh environments encountered during flight. Additionally, AZ31's excellent machinability enables precise manufacturing of complex components, making it an ideal choice for the intricate designs required in aerospace engineering.

AT40 coating is a composite ceramic coating composed of 40% titanium dioxide (TiO_2) and 60% aluminum oxide (Al_2O_3) for enhanced durability. AT40 coating is essential for improving the corrosion resistance and mechanical properties of magnesium alloys, particularly Mg AZ31, which are highly susceptible to corrosion due to their reactive nature. Additionally, AT40 coatings can improve the wear resistance of Mg AZ31, making it more suitable for applications where durability and mechanical strength are critical.[4]

However, the performance and lifetime of a component can be significantly affected by surface defects, such as microcracks and porosity, in AT-13 coatings,

DOI: 10.1201/9781003653226-10

resulting in decreased mechanical strength and diminished resistance to wear and corrosion. Laser surface modification is an effective method to improve the microstructure of AT-40 coatings by ensuring proper adhesion, uniformity, and performance of the coating, especially in sensitive applications where contaminants or imperfections could compromise functionality. Magnesium alloys can benefit greatly from the duplex treatment, which combines alumina–titania (AT40) coatings with laser surface modification to improve the substrate's durability and capabilities.

Some researchers have reported the performance of alumina–titania coatings on various metallic substrates. Yildiz et al.[5] applied an Al_2O_3–40 wt.% TiO_2 coating to the surfaces of 316L stainless steel samples using the plasma spraying method.

They conducted wear tests on both coated and uncoated samples in a simulated body fluid, revealing that the alumina coating offered effective wear protection. Similarly, Zavareh et al.[6] coated carbon steel surfaces with an Al_2O_3–40 wt.% TiO_2 composite ceramic using plasma spraying. They examined the tribological properties of the coating under various loads with a ball-on-disc tribometer, concluding that the coated samples exhibited superior wear resistance compared to the uncoated ones and maintained consistent wear rates across different loads. Ilhan Celik et al.[7] enhanced the low corrosion resistance of AZ31 Mg alloy using plasma-sprayed Al_2O_3–13 wt.% TiO_2 (AT13) and Al_2O_3–40 wt.% TiO_2 (AT40) coatings. XRD and SEM analyses revealed enhanced surface morphology and microstructure. The untreated AZ31 alloy, with a micro-hardness of 40 ± 3 HV0.1, exhibit a significant increase to 800 ± 39 HV0.1 with AT40 and 1500 ± 35 HV0.1 with AT13, indicating notable improvements in wear characteristics for tribological applications.

Monika Nowakowska et al.[8] reported that adding 40 wt.% TiO_2 to Al_2O_3 improves the morphology, microstructure, microhardness, and adhesion of coatings. Using atmospheric plasma spraying (APS) with variable stand-off distances and torch scan speeds, they found that higher scan speeds and shorter spraying distances densified the microstructure and enhanced mechanical properties, particularly for Al_2O_3 + 40 wt.% TiO_2. Yildiz et al.[9] found that atmospheric plasma-sprayed coatings with 13 wt.% TiO_2 (AT-13) offered better wear resistance than those with 43 wt.% TiO_2 (AT-43), indicating that hardness plays a more significant role in wear resistance than toughness. R. Yılmaz et al.[10] observed that increasing TiO_2 content in plasma-sprayed Al_2O_3 coatings on AISI 304L stainless steel decreased microhardness but improved fracture toughness and plasticity.

M.K. Lei et al.[11] used a MEVVA 80-10 system to implant aluminum ions into AZ31 magnesium alloy (40–50 keV, doses from 2×10^{16} to 1×10^{17} ions/cm^2), achieving a Gaussian profile up to 840 nm with a 10 at.% peak Al concentration. XRD analysis revealed the presence of α-Mg, β-Mg17Al12, and MgO phases. The implanted samples exhibited higher pitting breakdown potentials, lower passive current density, and a 30–40% reduction in wear rate in 0.01 M NaCl (pH 12). Guo Song Wu et al.[12] developed a three-layer coating (diamond-like carbon, aluminum nitride, and aluminum) for AZ31 magnesium alloy, significantly enhancing surface hardness, reducing friction, and improving wear and corrosion resistance. P. Bala Srinivasan et al.[13] applied plasma electrolytic oxidation (PEO) to AZ91 magnesium alloy, forming 10 μm and 20 μm thick coatings, which improved wear resistance, particularly under high contact stresses.

Yuan Fei Gao et al.[14] fabricated hybrid AA1050/Al_2O_3/TiO_2 composites using powder metallurgy and liquid casting. Degassing improved the wettability of the molten aluminum alloy, enabling strong bonding with 2.5 or 5 wt.% 5 μm TiO_2 and 5 wt.% 20 μm Al_2O_3 particles. Heating the ceramic particles to 600°C improved their dispersion, as confirmed by microstructural analysis. Rotating wear tests showed a significant improvement in wear resistance compared to the monolithic Al alloy, with a 5 wt.% TiO_2 content reduces wear rates to less than half that of the base alloy.

Zhejiang Yin et al.[15] applied Al_2O_3 and Al/Al_2O_3 composite coatings via APS and assessed their wear and friction using a block-on-ring setup. The Al/Al_2O_3 composite demonstrated superior wear resistance due to its higher fracture toughness, adhesion strength, and improved thermal diffusivity. Wear mechanisms included adhesion with micro-abrasion and fatigue-induced brittle fracture. A. Ravi Shankar et al.[16] investigated graphite for pyrochemical reprocessing in aggressive molten chloride environments, employing plasma-sprayed Al_2O_3-40 wt.% TiO_2 coatings on high-density graphite. Post-treatments included vacuum annealing (VA) and laser melting (LM), with LM coatings showing enhanced microhardness and reduced surface roughness due to defect elimination.[17]

Although the review of literature[1–17] emphasizes the advantages of alumina–titania coatings, especially when applied by plasma spraying, in improving wear resistance, and corrosion resistance on a variety of metallic substrates, including carbon steel, stainless steel, and magnesium alloys, the combination of applying Al_2O_3–40 wt.% TiO_2 coatings and its subsequent laser surface modification as a duplex treatment on AZ31 magnesium alloy remains unexplored in the literature. This presents an opportunity for novel research to investigate the potential synergistic effects of combining these processes, which could lead to significant improvements in the wear resistance of AZ 31 alloy. Thus, the present research focuses on investigating the effect of dry sliding wear behavior of thermal-sprayed and laser-modified Al_2O_3–40%wt TiO_2-coated AZ31 magnesium alloy.

10.2 MATERIALS AND METHODS

10.2.1 AT40 Powder Characterization

Al_2O_3–40%wt TiO_2 ceramic coatings have demonstrated significant potential in enhancing wear resistance when applied to metal substrates. This is attributed to their exceptional hardness, high thermal stability, and strong adherence to the metal surface, mitigating wear and prolonging the lifespan of metal components in various industrial applications.

Figure 10.1 displays the EDX spectrum of the AT40 powder from the selected region. It depicts the composition of AT40 powder in a weighted percentage format consisting of aluminum (28.97%), titanium (24.62%), and oxygen (46.41%).

Figure 10.2 displays the particle size distribution histogram plotted from data obtained from a scanning electron microscope (SEM) image of AT40 powder particles. It can be observed that the bar around 25–33 μm represents 15–20% of the particles, which means that 15–20% of the particles analyzed have a size between 25 μm and 33 μm, and the bar around 40 μm represents that the minimum percentage

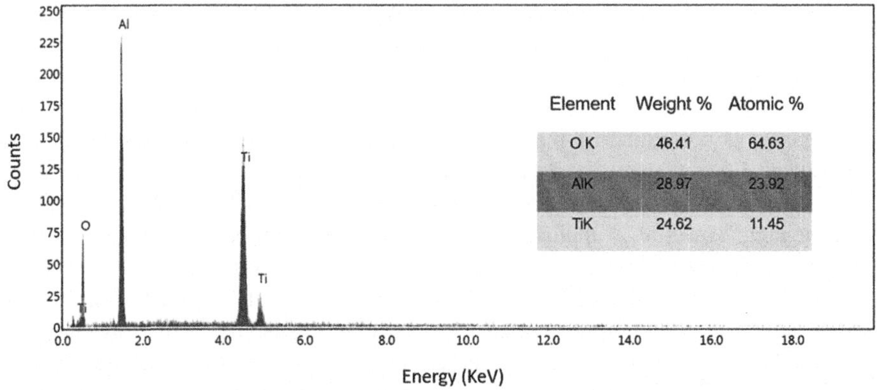

FIGURE 10.1 EDX analysis of AT40 powder.

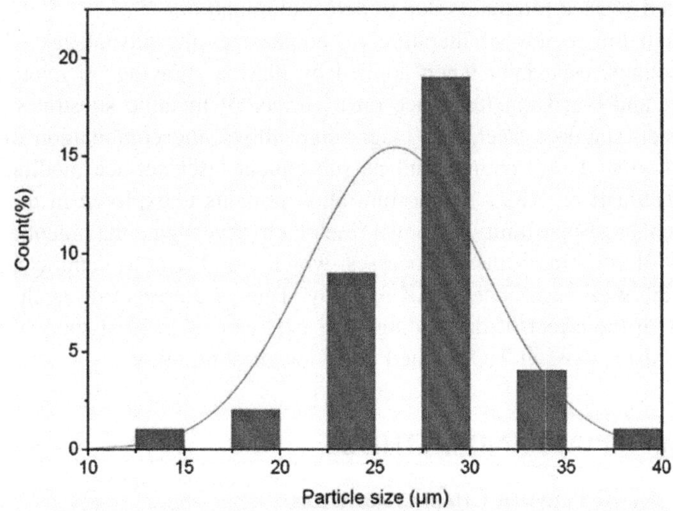

FIGURE 10.2 Particle size distribution of AT40 powder.

of particles are of size 40 μm. The average surface area of the particles is 30.385 μm^2, the average length of the particles is 60.23 μm, and the angle is 95.82.

10.2.2 Coating Characteristics

AT40-coated surfaces are made up of pores, unmelted particle zones, and fully melted splat structures. The following discussion helps clarify the causes of this complex structure: some of the powders fully melt and drop onto the surface in the form of splats when carried out by the plasma flow after being subjected to heat for a very short period. Another explanation is that the microstructure might be under

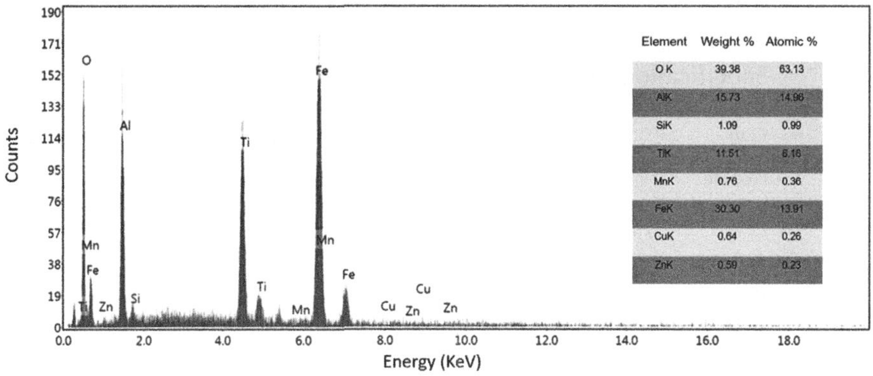

FIGURE 10.3 EDX spectrum of the coating.

stress as a result of the individual splats shrinking during the rapid 6-cooling (106–108 K/s). Quenching stresses refer to the strains or deposits that were previously formed on the cooler's substrate or on the hardened splats. The aforementioned pressures exert a significant influence on various aspects of performance, including adhesion, thermal cycle life, erosion resistance of coatings, and microstructure.

Furthermore, tensions in the coating are created by the volume variations linked to phase shifts in the coating during the spraying process. Figure 10.3 shows the EDX spectrum of the coating. The number indicates the weight percentage of components: oxygen (39.38%), aluminum (15.73%), silicon (1.09%), titanium (11.51%), manganese (0.76%), copper (0.64%), and zinc (0.59%).

10.2.3 Machining of Materials

The purchased AZ31 samples were machined using conventional equipment like lathes, given AZ31's exceptionally low melting point, rendering electrical discharge machining (EDM) impractical. These machining efforts yielded cylindrical samples measuring 6 mm in diameter and 30 mm in length, as well as slabs sized 20 × 20 mm with a thickness of 3 mm. The choice of conventional machining underscores the adaptability of the manufacturing process to accommodate the material's unique characteristics, ensuring precise shaping for intended applications. The specimens were ground at the edges and polished to provide a mirror-like finish.

10.2.4 Coating of Specimens

Following machining and polishing, the AZ31 specimens underwent a crucial coating process using AT40, applied to a thickness of 100 μm on all faces of the slabs and on the flat ends of the cylindrical pins. This intricate coating procedure was executed through plasma spraying at SPRAYMET Pvt. Ltd., with the specifications of the plasma spraying machine provided in Table 10.1.

The primary aim behind this coating was to comprehensively understand the tribological behavior of the surfaces after undergoing various surface modifications,

TABLE 10.1
Specifications of Plasma Spraying Machine

Gun	3MB	
Nozzle	GH	
	Pressure (psi)	Flow Rate (SEFH)
Argon	100–120	80–90
Hydrogen	<100	20–25
Current	490 A	
Voltage	60–70 V	
Power feed	40–50 g/min	
Spray Distance	3–5 inches	

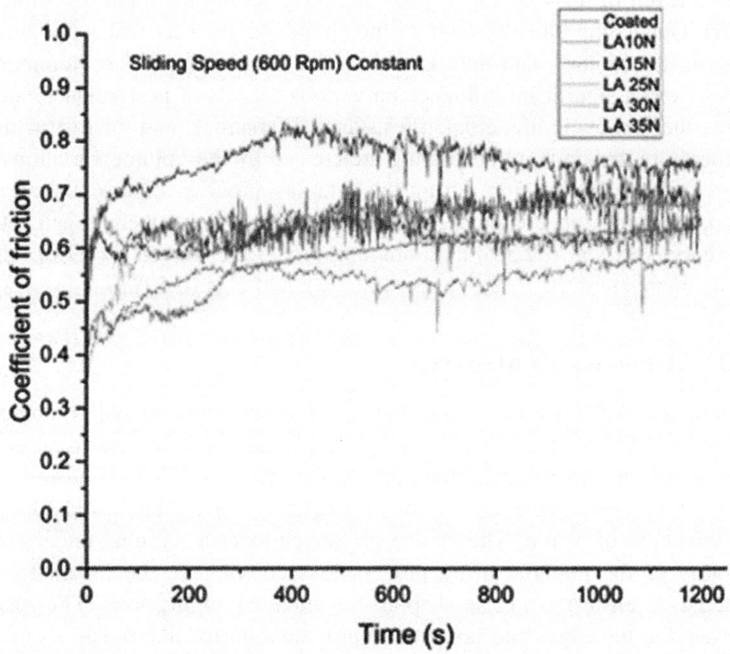

FIGURE 10.4 Combined plot of COF versus sliding time with a constant sliding speed of 600 rpm.

as shown in Figure 10.4. By subjecting the specimens to this coating treatment, we aim to analyze how alterations at the surface level influence friction, wear, and hardness properties. Such investigations are crucial for enhancing the material's performance and durability in real-world applications across diverse industries, from automotive to aerospace. The insights gained from studying the tribological behavior post-coating can inform future advancements in material design and surface engineering, facilitating innovation and efficiency in engineering practices.

10.2.5 Laser Ablation Treatment

Following the coating of AT40 powder onto the AZ31 samples, they underwent a series of laser passes using a portable fiber laser marking machine (specifications provided in Table 10.2) at different power levels, ranging from 50 W to 100 W.

Meticulous analysis of the outcomes indicated that the optimal laser power range was 50–60 W. At these power levels, the laser effectively penetrated the coating without significantly reaching the substrate. However, beyond this range, observations indicated that the laser almost entirely passed through the coating and started affecting the substrate, deviating from the desired outcome. To ensure precise laser ablation while maintaining the integrity of the coating and substrate, it was decided to proceed with powers ranging from 50 W to 60 W. This careful selection of laser parameters is critical for achieving the desired surface modifications and understanding the interaction between the laser, coating, and substrate, thereby informing further refinement of the laser processing technique for AZ31 specimens.

10.2.6 Wear Test

The sliding wear experiments were performed using a pin-on-disc tribometer (manufactured by Ducom Instruments, Bengaluru). The sliding wear experiments were performed on nine faces of cylindrical specimens, employing different parameters, including constant load with varying speed and varying load with constant speed combinations, as shown in Table 10.3.

Each of the experiments was performed thrice for a duration of 20 min at room temperature, and the average of the three readings was considered. The objective was to generate wear plots and compare results across all samples, aiming to discern variations in surface behavior under wear conditions.

The specific wear rate (SWR) was computed using equation (10.1):

$$\text{Speicific wear rate}\left(\text{mm}^3\right) = \frac{V}{L.S} \tag{10.1}$$

where V is the volume loss (mm^3), L is the load (N), and S is the sliding distance (m).

TABLE 10.2
Specifications of Portable Fiber Laser Marking Machine

Beam Mode	TEMoo
Wavelength	1064 nm
Nominal Average Power	50 W
Power Stability (8 hr)	<1%
Beam Quality	<1.1
Pulse Frequency	20–100 kHz
Pulse width	100 ns
Software used	Ezcad

TABLE 10.3
Wear Test Parameters

Exp. No.	Load (N)	Sliding Speed (rpm)
1	10	600
2	15	600
3	25	600
4	30	600
5	35	600
6	25	400
7	25	500
8	25	700
9	25	800

The volume loss was computed using equation (10.2):

$$V\left(\mathrm{mm}^3\right) = \frac{W}{\rho} \tag{10.2}$$

where W is the mass loss (gm) obtained by weighing the pin before and after the wear test, and ρ is the mass density of AZ31 (1.77 g/cc)

COF was computed using equation (10.3):

$$\mathrm{CoF} = \frac{\text{Frictional Force}(N)}{\text{Normal Force}(N)} \tag{10.3}$$

10.3 RESULTS AND DISCUSSION

10.3.1 Friction Behavior

The coefficient of friction (COF) for the uncoated specimen ranged from 0.40 to 0.48. The combined plot of COF versus sliding time for the first five specimens with a constant sliding speed of 600 rpm and load varying from 10 N to 35 N is shown in Figure 10.4. The combined plot of COF versus sliding time for the next five specimens with a constant load of 25 N and sliding speed varying from 400 to 800 rpm is shown in Figure 10.4.

The COF for both the coated and laser-ablated specimens is higher than that of the uncoated specimen. This increase in COF is attributed to the thermal stress induced by the high-energy beam during the coating and laser ablation processes. Consequently, the coated specimens exhibit a higher COF compared to the uncoated specimens.

At 10 N of load with 600 rpm sliding speed, the COF for the laser-ablated specimen (LP1) was found to be in the range 0.49–0.68, and the COF for the coated specimen with the same parameters was found to be in the range 0.54–0.76. Initially, the COF of LP1 increased when compared with the COF of the coated specimen until the sliding period of 100 seconds; eventually, the COF of the LP1 specimen and of the coated specimen remained the same until the sliding period of 600 seconds. After 600 seconds, the COF of the laser-ablated specimen at 10N load (LP1) decreased to a value of 0.68. Later, the load was increased from 10 N to 15 N, and the COF was plotted against sliding time for the second specimen with a 15 N load and 600 rpm sliding speed. When the load is increased from 10 N to 15 N, there was a significant increase in the COF of the laser-ablated specimen (LP2) when compared to the coated specimen, from 0.49 to 0.74, reaching a peak of 0.842 at a sliding period of 445 seconds.

The data from the plots (Figures 10.4 and 10.5) indicates that the specimen subjected to a load of 25 N and a sliding speed of 600 rpm (LP 3) performed better than others (under different conditions). Specimen 3 emerged as the optimized choice for both consistent load and sliding speed scenarios, exhibiting a COF ranging from 0.42 to 0.56. At the end of 1200 seconds, the COF of LP2 almost equals the value of that of the coated specimen. The reason for the significant increase might be due to the formation of pores, craters, cracks, globules, peaks

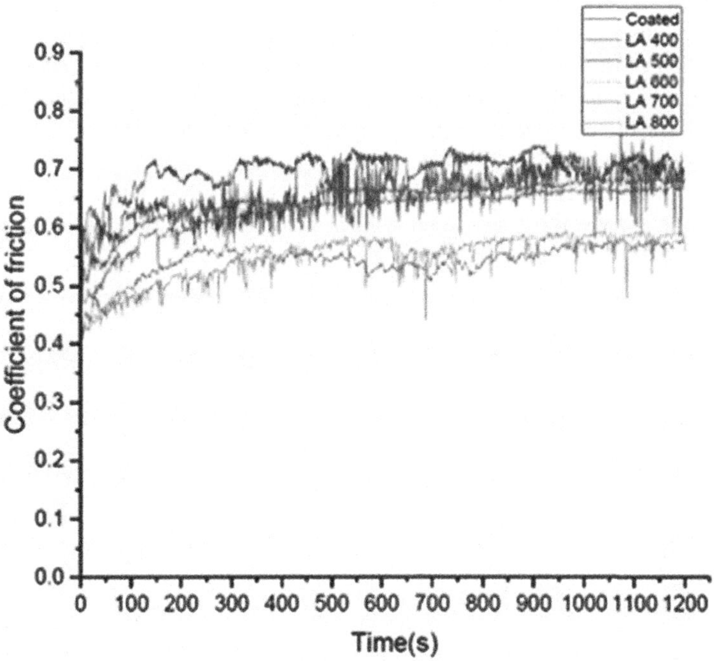

FIGURE 10.5 Combined plot of COF versus sliding time with constant load 10 N.

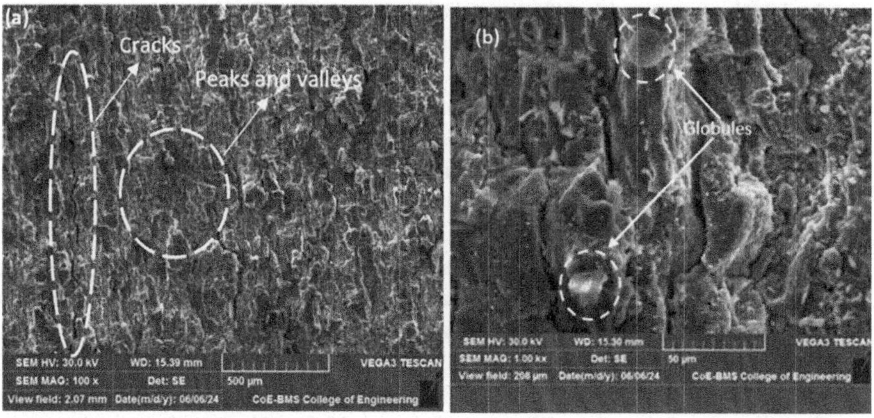

FIGURE 10.6 SEM images of LP2 specimen at (a) 100×, (b) 1000× magnifications.

and valleys, and ridges, which can be interpreted using the SEM images of the LP2 specimen.

Figure 10.6 (a and b) shows the microstructure of AT40-coated and laser-ablated specimens against AZ31 alloy at 15 N load and 600 rpm sliding speed at different magnifications at 100× and 1000×, respectively. The microstructure shows the major cracks, craters, peaks and valleys, globules, and ploughing marks, which indicate the increase in COF value when compared to that of the coated specimen. Figure 10.7 displays the EDX spectrum of laser-ablated specimen LP2. It depicts the composition of the specimen in a weighted percentage format consisting of magnesium (16.19%), aluminum (18.23%), titanium (10.63%) and oxygen (29.27%), silicon (0.45%), chromium (3.55%), manganese (0.65%), iron (7.72%), copper (0.56%), and zinc (0.48%).

Later, the load was increased from 15 N to 25 N, and the COF was plotted against sliding time for the third specimen with a 25 N load and 600 rpm sliding speed.

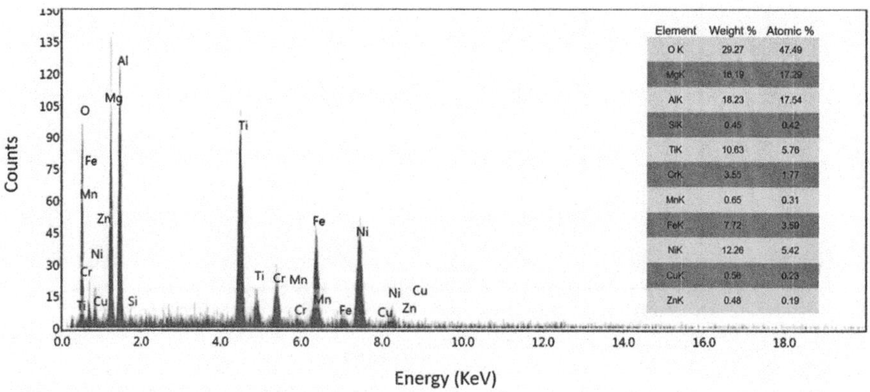

Element	Weight %	Atomic %
O K	29.27	47.49
MgK	16.19	17.29
AlK	18.23	17.54
SiK	0.45	0.42
TiK	10.63	5.76
CrK	3.55	1.77
MnK	0.65	0.31
FeK	7.72	3.59
NiK	12.26	5.42
CuK	0.56	0.23
ZnK	0.48	0.19

FIGURE 10.7 EDX analysis of LP2 specimen.

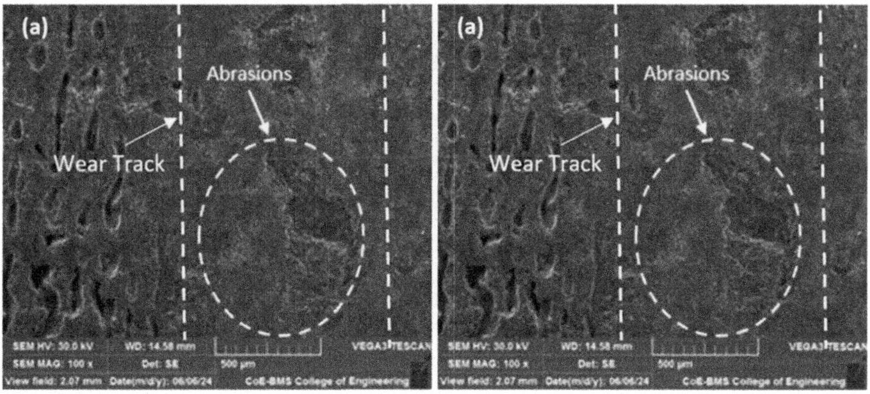

FIGURE 10.8 SEM images of LP3 specimen at (a) 100× (b) 200× magnifications.

When the load is increased from 15 N to 25 N, there was a significant decrease in COF of the laser-ablated specimen (LP3) when compared to the coated specimen, from 0.42 to 0.56. Initially, till 250 seconds, there was a rise in COF and it reached a peak of 0.58, then gradually decreased parabolically and reached a value of 0.56 at the end of 1200 seconds. The reason for the significant decrease might be due to smooth abrasion and reduced wear debris, craters, etc., which can be interpreted using the SEM images of the LP3 specimen.

Figure 10.8 (a and b) shows the microstructure of AT40-coated and laser-ablated specimens against AZ31 alloy at 25 N load and 600 rpm sliding speed at different magnifications at 100× and 350×, respectively. The microstructure shows smooth abrasion, no major cracks, craters, peaks and valleys, globules, and ploughing marks on the wear track, which indicate the decrease in COF value when compared to that of the coated specimen.

The EDX spectrum of laser-ablated specimen LP3 depicts the composition of the specimen in a weighted percentage format consisting of magnesium (0.91%), aluminum (18.99%), titanium (11.83%), oxygen (37.89%), silicon (0.63%), chromium (1.16%), manganese (0.45%), iron (25.74%), copper (0.41%), and zinc (0.38%). Later, the load was increased from 25 N to 30 N, and the COF was plotted against sliding time for the fourth specimen with a 30 N load and 600 rpm sliding speed. When the load is increased from 25 N to 30 N, the COF of the laser-ablated specimen (LP4) is reduced from 0.35 to 0.58 when compared to the coated specimen.

Initially, up to 280 seconds, there was a significant rise in COF till 0.54, after which it continued to rise gradually till 0.58. At the end of 1200 seconds, the COF of LP4 had nearly reached the value of the coated specimen.

The reason for the significant decrease might be due to smooth abrasion and reduced wear debris, craters, etc., which can be interpreted using the SEM images of the LP3 specimen. Figure 10.9 (a and b) show the microstructure of AT40-coated and laser-ablated specimens against AZ31 alloy under a 30 N load and 600 rpm sliding speed at different magnifications of 100× and 1000×, respectively. The microstructure shows smooth abrasion, which resulted in reduced COF, but due to

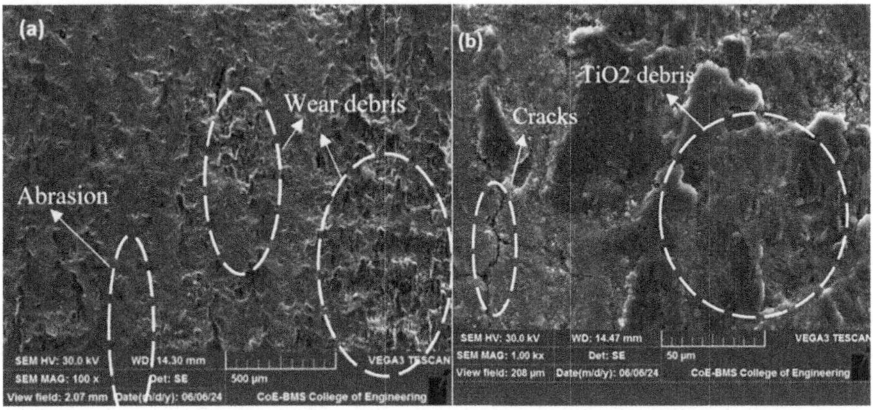

FIGURE 10.9 SEM images of LP4 specimen at (a) 100× (b) 1000× magnifications.

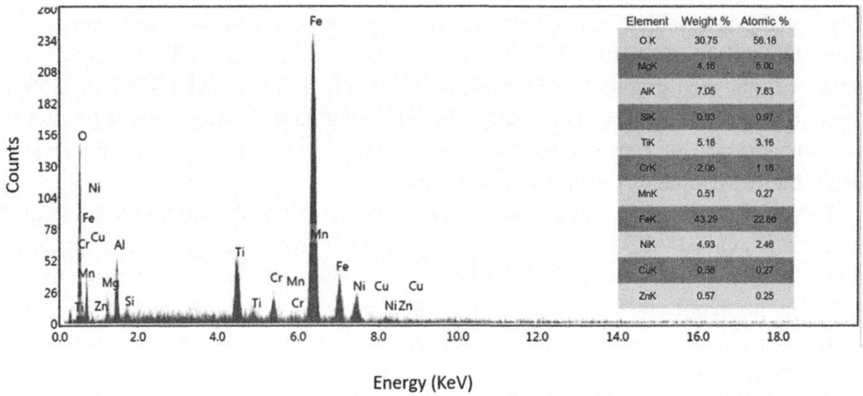

FIGURE 10.10 EDX analysis of LP4 specimen.

minor cracks, craters, peaks and valleys, and ploughing marks, there is a gradual increase in COF value from 0.54 to 0.58 and 280 to 1200 seconds, respectively.

Figure 10.10 displays the EDX spectrum of laser-ablated specimen LP4. It depicts the composition of the specimen in a weighted percentage format consisting of magnesium (4.16%), aluminum (7.05%), titanium (5.18%), oxygen (30.75%), silicon (0.93%), chromium (2.06%), manganese (0.51%), iron (43.29%), copper (0.58%), and zinc (0.57%). Later, the load was increased from 30 N to 35 N, and the COF was plotted against sliding time for the fourth specimen with a load of 35 N and sliding speed of 600 rpm. When the load is increased from 30 N to 35 N, the COF of the laser-ablated specimen (LP4) showed almost the same COF values, ranging from 0.43 to 0.66 when compared to the coated specimen.

Initially, up to 190 seconds, there was a significant rise in COF from 0.43 to 0.65, and after that the COF almost remained constant and gradually increased to 0.66. At the end of 1200 seconds, the COF of LP5 had nearly reached the value of the coated specimen. The significant decrease may be attributed to smooth abrasion during the first few seconds, and there was a significant rise in COF, which had nearly reached the same value observed for the coated specimen. This increase in COF can be due to the formation of pores, craters, cracks, globules, peaks and valleys, or ridges on the ablated surface. After the first five specimens were tested under constant sliding speed conditions of 600 rpm, the next four specimens were tested under the constant load condition of 25 N. At 400 rpm sliding speed, the COF of LP6 ranged between 0.52 and 0.66, and the COF for the coated specimen was found to be in the range of 0.54–0.76. Initially, the COF of LP6 was lower than that of the COF of the coated specimen until the sliding period reached 80 seconds. Thereafter, the COF of the LP6 specimen gradually increased and had nearly reached the value of the coated specimen. When the sliding speed was increased from 400 rpm to 500 rpm, the COF increased significantly.

At 500 rpm sliding speed, the COF of LP7 ranged between 0.5 and 0.74, and the COF for the coated specimen with 600 rpm sliding speed and 25 N load was found to be in the range of 0.54–0.76. Until the sliding period reached 80 seconds, the COF of the LP7 specimen was lower than that of the COF of the coated specimen. Thereafter, the COF of LP7 gradually increased, reaching 0.74. Later, the sliding speed was increased from 500 rpm to 700 rpm, and the COF was plotted against sliding time for the eight specimens with a sliding speed of 700 rpm and a load of 25 N. When the sliding speed was increased from 500 rpm to 700 rpm, the COF of the laser-ablated specimen (LP8) showed slightly lower COF values ranging from 0.46 to 0.63 than that of the coated specimen. Initially, up to 150 seconds, there was a significant rise in COF from 0.46 to 0.60, and after that, the COF gradually increased from 190 seconds till 0.63. At the end of 1200 seconds, the COF of LP8 had nearly reached the value of the coated specimen. When the sliding speed was increased from 700 rpm to 800 rpm, there was a significant decrease in COF of the laser-ablated specimen (LP9) when compared to the coated specimen, dropping from 0.40 to 0.54. Initially, up to 650 seconds, there was a rise in COF and reached a peak of 0.59. After that, the COF gradually decreased parabolically and reached a value of 0.54 at the end of 1200 seconds. The reason for the significant decrease might be due to smooth abrasion and reduced wear debris, craters, etc.

10.3.2 Wear Behavior

The plot of SWR against load with a constant sliding speed of 600 rpm is shown in Figure 10.11.

The laser-ablated specimen (the first one) for a load of 10 N has a higher SWR of 2.505×10^{-4} compared to the coated specimen, which has an SWR of 2.44×10^{-4} mm^3/N-m. As the load increases from 10 N to 15 N, there's a significant drop in SWR, and this trend continues: as the load increases, the SWR of the laser-ablated specimen consistently decreases compared to the laser-coated one. The third specimen, with a sliding speed of 600 rpm and a load of 25 N, shows an intermediate value according to the plot. The fifth specimen, with a load of 35 N, has the least SWR, which is 3.47×10^{-5} mm^3/N-m.

The graph from Figure 10.12 shows that the coated specimen has a higher SWR of 2.44×10^{-4}, while all the laser-ablated specimens with a constant load have

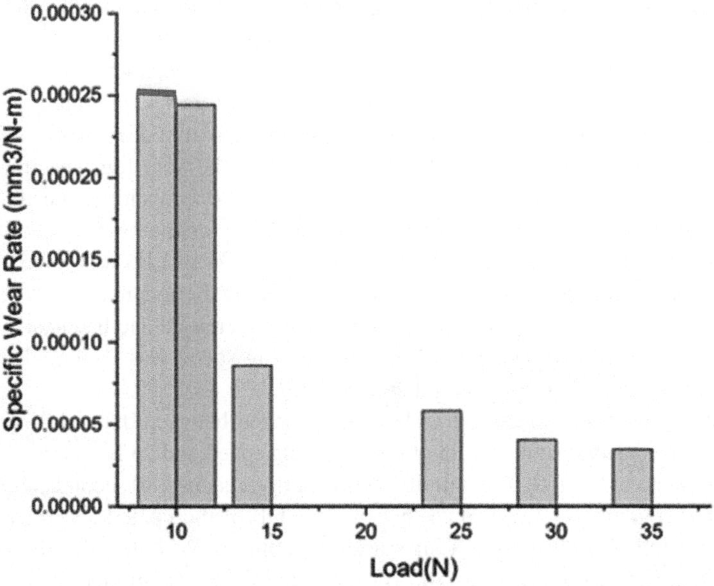

FIGURE 10.11 Plot of specific wear rate versus load.

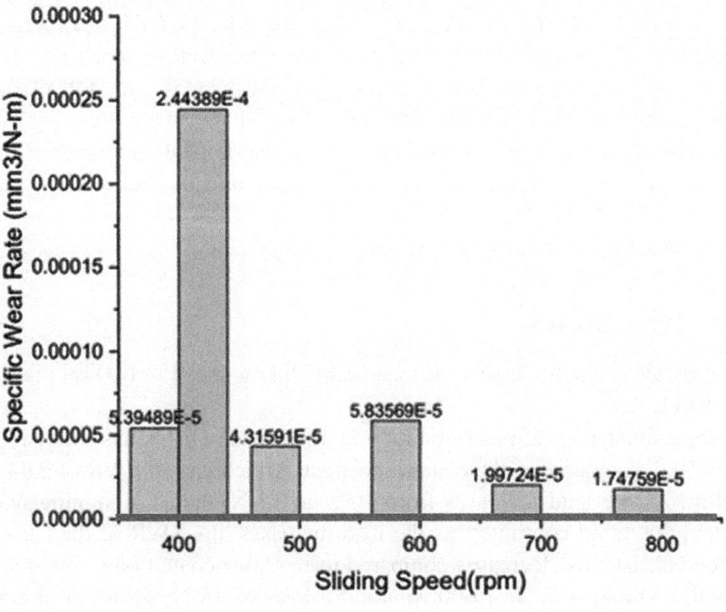

FIGURE 10.12 Plot of specific wear rate versus sliding speed.

significantly lower SWRs. There's no clear trend observed when it comes to the effect of increasing sliding speed on the SWR of laser-ablated specimens. The third specimen, optimized based on COF versus time, with a sliding speed of 600 rpm and a load of 25 N, has a higher SWR of 5.83×10^{-5} compared to other specimens, but it's still lower than the coated specimen. The lowest SWR under a constant load of 25 N and a sliding speed of 800 rpm was observed in the ninth specimen, with a value of 1.74×10^{-5} mm^3/N-m. Therefore, the friction and wear behavior plots indicate that there is no straightforward correlation between the COF and the SWR.

10.4 CONCLUSION

Magnesium alloy AZ31, known for its lightweight and strength, typically suffers from poor wear resistance. Recent research addressed this by coating AZ31 specimens with an alumina–titania (AT40) composite via plasma spraying, followed by laser ablation to enhance wear resistance. Tribological testing revealed that the COF varied with load and sliding speed. At lower loads (10 N and 15 N), laser-ablated specimens showed a higher COF due to surface irregularities, while at higher loads (25 N and 30 N), they exhibited a lower COF due to smoother wear tracks. SEM images supported these findings, indicating laser ablation effectively modifies surface characteristics to improve wear behavior.

REFERENCES

1. T. Tokunaga, M. Ohno and K. Matsuura, Coatings on Mg alloys and their mechanical properties: A review. *J. Mater. Sci. Technol.*, **34** (7), 1119–1126 (2017); DOI: 10.1016/j.jmst.2017.12.004
2. J. Gray and B. Luan, Protective coatings on magnesium and its alloys—A critical review. *J. Alloys Compd.*, **336**, 88–113 (2002); DOI: 10.1016/S0925-8388(01)01899-0
3. X. Guo, Y. Hu, K. Yuan and Y. Qiao, Review of the effect of surface coating modification on magnesium alloy biocompatibility. *Materials*, **15**, 3291(18), (2002); DOI: 10.3390/ma15093291
4. B. Feng, C. Yan, X. Feng, J. Wang and K. Zheng, A novel approach to improve corrosion resistance of Mg alloy by co-extrusion. *Mater Res Express*, **7**, 106511(8), (2020); DOI: 10.1088/2053-1591/abbe41
5. F. Yildiz, A.F. Yetim, A. Alsaran and A. Çelik, Performance of plasma sprayed Al2O3 coating in bio-simulated environment. *Met. Sci. Heat Treat.*, **55**, 504–508 (2014); DOI: 10.1007/s11041-014-9662-8
6. M. Akhtari, A. Sarhan, R. Zadeh and R. Singh, Analysis of corrosion protection behavior of Al2O3-TiO_2 oxide ceramic coating on carbon steel pipes for petroleum industry. *Ceram. Int.*, **44** (6), 5967–5975 (2018); DOI: 10.1016/j.ceramint.2017.12.175
7. I. Celik, Structure and surface properties of Al2O3–TiO_2 ceramic coated AZ31 magnesium alloy. *Ceram. Int.*, **42** (12), 13659–13663 (2016); DOI: 10.1016/j.ceramint.2016.05.162
8. L. Latka, M. Szala, M. Nowakowska, M. Walczak, T. Kielczawa and P. Sokolowski, The effect of microstructure and mechanical properties on sliding wear and cavitation erosion of plasma coatings sprayed from Al2O3+ 40 wt% TiO_2 agglomerated powders. *Surf. Coat. Technol.*, **455**, 129180(13), (2023); DOI: 10.1016/j.surfcoat.2022.129180

9. V. Fervel, B. Normand and C. Coddet, Tribological behavior of plasma sprayed Al2O3-based cermet coatings. *Wear*, **230**, 70–77 (1999); DOI: 10.1016/S0043-1648(99)00096-4
10. R. Yılmaz, A. Kurt, A. Demir and Z. Tatlı, Effects of TiO_2 on the mechanical properties of the Al2O3–TiO_2 plasma sprayed coating. *J. Eur. Ceram. Soc.*, **27**, 1319–1323 (2007); DOI: 10.1016/j.jeurceramsoc.2006.04.099
11. M.K. Lei, P. Li, H.G. Yang and X.M. Zhu, Wear and corrosion resistance of Al ion implanted AZ31 magnesium alloy. *Surf. Coat. Technol.*, **201**, 5182–5185 (2007); DOI: 10.1016/j.surfcoat.2006.07.091
12. G. Wu, W. Dai, H. Zheng and A. Wang, Improving wear resistance and corrosion resistance of AZ31 magnesium alloy by DLC/AlN/Al coating. *Surf. Coat. Technol.*, **205**, 2067–2073 (2010); DOI: 10.1016/j.surfcoat.2010.08.103
13. P. Srinivasan, C. Blawert and W. Dietzel, Dry sliding wear behaviour of plasma electrolytic oxidation coated AZ91 cast magnesium alloy. *Wear*, **266**, 1241–1247 (2009); DOI: 10.1016/j.wear.2009.03.013
14. Y. Gao, M. Vini, S. Daneshmand, A. Alameri, O. Benjeddou and R. Alfilh, Effect of processing parameters on wear properties of HYBRID AA1050/Al2O3/TiO_2 composites. *Crystals* **13**, 335 (2023); DOI: 10.3390/cryst13020335
15. Z. Yin, S. Tao, X. Zhou and C. Ding, Tribological properties of plasma sprayed Al/Al2O3 composite coatings. *Wear*, **263**, 1430–1437 (2007); DOI: 10.1016/j.wear.2007.01.052
16. J. Sure, A. Shankar, B. Upadhyaya and U. Kamachi, Microstructural characterization of plasma sprayed Al2O3-40 wt.% TiO_2 coatings on high density graphite with different post-treatments. *Surf. Coat. Technol.*, **206**, 4741–4749 (2012); DOI: 10.1016/j.surfcoat.2012.01.058
17. M. Buchtík, J. Másilko, O. Vyklicky, J. Filipenský, J. Wasserbauer and P. Ptacek, Microstructural characterization and wear behavior of WC-CoCr coating on AZ91 magnesium alloy, pp. 922–927 in: *Metal 2019*. Ed. Brno, Czech Republic, EU (2019); DOI: 10.37904/metal.2019.730

11 The Emerald Canopy
Nature's Blueprint for Advanced, Sustainable Coatings in Next-Generation Aerospace and Automotive

Moatassim Raoof

11.1 INTRODUCTION

The aerospace and automotive sectors currently navigate a pivotal epoch, demanding a sophisticated convergence of uncompromising material performance and profound environmental stewardship. This critical juncture stems from escalating global environmental regulations, a burgeoning awareness regarding product lifecycle sustainability, and the inherent limitations of traditional materials in enduring the extreme operational conditions characteristic of these industries. For decades, the reliance on conventional corrosion protection solutions, particularly those involving hazardous substances like hexavalent chromium, has presented a significant dichotomy: effective protection at the cost of substantial environmental and health liabilities. These established methods often necessitate complex and energy-intensive application processes, leading to considerable operational expenditures and the generation of problematic waste streams (1, 2).

This chapter defines sustainable coatings for next-generation aerospace and automotive applications as those meticulously engineered to deliver exceptional performance (e.g., superior corrosion resistance, enhanced durability, and advanced functionalities) throughout their lifecycle, while simultaneously minimizing their adverse environmental and health impacts. This comprehensive definition encompasses several key criteria: reduced hazardous materials (e.g., heavy metals and volatile organic compounds), a lower energy footprint during manufacturing and application, enhanced resource efficiency through bio-based or recycled materials, waste minimization, and extended lifespan to reduce re-coating frequency. This philosophy underpins the advancements discussed herein, positioning these coatings as integral components of a more sustainable industrial ecosystem (3–5).

In the current landscape, several key trends are shaping the development and adoption of these sustainable coatings. There is a strong shift toward bio-inspired and bio-based solutions, leveraging nature's own protective mechanisms and renewable

DOI: 10.1201/9781003653226-11

resources. Concurrently, the integration of nanotechnology is enabling unprecedented levels of control over material properties, leading to highly efficient and durable coatings. The drive toward multi-functional materials that offer benefits beyond simple corrosion resistance (e.g., self-cleaning and anti-icing) is also a prominent trend. Furthermore, the increasing sophistication of smart coatings with self-healing capabilities promises to revolutionize maintenance protocols by autonomously repairing minor damage. These trends are not isolated but often converge, leading to synergistic advancements that offer superior performance with reduced environmental burdens (6, 7).

As these innovative technologies continue to emerge, standardization practices become critical for their widespread adoption and acceptance within highly regulated industries. Bodies such as ASTM International, ISO, and various industry-specific consortia are actively working to establish new testing protocols, performance benchmarks, and material specifications for these sustainable coatings. This meticulous approach ensures that "eco-friendly" claims are scientifically validated and consistently applied across the supply chain, fostering confidence and facilitating global trade of these advanced materials (8–10).

This chapter fundamentally addresses the inherent unsustainability and evolving inadequacy of conventional corrosion protection methods in meeting the escalating demands of modern aerospace and automotive industries. It aims to redefine corrosion protection through an ecological design paradigm, moving decisively beyond conventional toxic inhibitors and embracing a future where material resilience is intrinsically linked with ecological harmony. This chapter aligns directly with the broader scope of this book, which explores advancements in high-temperature resistance and wear protection, as corrosion often precedes or exacerbates these forms of material degradation (11, 12).

Therefore, the purposes of this chapter are multifaceted:

- To provide a comprehensive overview of the most advanced and emerging eco-friendly corrosion protection solutions relevant to aerospace and automotive applications, categorized by their underlying technological principles.
- To elucidate the scientific mechanisms behind these innovative coatings, highlighting how they achieve superior protection through environmentally benign approaches.
- To present real-world case studies and quantitative data that demonstrate the practical applicability and performance advantages of these sustainable solutions in key industrial contexts.
- To analyze the current trends and future directions in the development of sustainable coatings, including the role of nanotechnology, smart materials, and bio-inspired design.
- To discuss the critical importance of standardization practices and regulatory frameworks in facilitating the widespread adoption of these new technologies.
- To equip researchers, engineers, and industry professionals with a decision matrix and practical insights necessary to navigate the complexities of material selection and implementation, ultimately guiding them toward more sustainable choices for next-generation transportation.

The philosophy underpinning this chapter is rooted in the belief that sustainable innovation is not merely an option, but an imperative. By drawing inspiration from nature's elegant solutions and applying rigorous scientific inquiry, materials can be engineered to be both remarkably durable and inherently benign. This approach necessitates a multidisciplinary understanding, integrating principles from chemistry, materials science, environmental engineering, and industrial design. The complex nature of these phenomena calls for deeper investigation, and continued research is constantly refining our collective understanding. This convergence of ecological principles with advanced materials science is forging a new paradigm in surface engineering, offering a pathway to robust corrosion protection without compromising the health of our planet (13–15).

This work, therefore, is designed to serve as both a comprehensive scientific reference and an actionable industrial roadmap. The chapter provides engineers and researchers with unique solutions, meticulously comparing these 20 selected commercial and emerging ideas across critical technical, economic, and environmental parameters. Special attention is given to regulatory challenges, with a detailed analysis of how restrictions like registration, evaluation, authorization and restriction of chemicals (REACH) and per- and polyfluoroalkyl substances (PFAS) are fundamentally shaping future development strategies and accelerating the search for truly sustainable alternatives. This problematic balance between material performance, environmental responsibility, and regulatory compliance is central to the discussion, guiding the reader through the complexities of material selection in a rapidly evolving landscape. By combining rigorous materials science with practical implementation guidelines, this chapter aims to equip researchers and engineers to confidently navigate and lead in the complex yet imperative balance between durability and sustainability that will define the next generation of transportation material.

11.2 CHROMIUM-FREE CONVERSION COATINGS

11.2.1 Trivalent Chromium Process (TCP)

11.2.1.1 Mechanism and Environmental Imperative

The transition from hexavalent chromium (Cr^{6+}) to trivalent chromium (Cr^{3+}) in conversion coatings represents a significant advancement in sustainable surface engineering. Researchers generally concur that Cr^{6+}, while effective for corrosion inhibition, poses severe environmental and health risks, including carcinogenicity. The Trivalent Chromium Process (TCP) offers a safer alternative by forming a passivation layer through Cr^{3+}-based chemistries, which—though less oxidative—still provide robust corrosion resistance. Although recent findings propose TCP coatings achieve this by generating amorphous $Cr(OH)_3{\cdot}xH_2O$ matrices, often augmented with zirconium or titanium oxides to enhance adhesion and durability (16).

11.2.1.2 Performance and Industrial Adoption

TCP coatings exhibit corrosion resistance comparable to traditional chromates, with salt spray test results exceeding 500 hours for aerospace-grade aluminum alloys. However, one might infer from these findings that TCP's performance is highly dependent on substrate preparation and process parameters, such as pH (typically 3.5–4.5) and immersion time (2–5 minutes). Although current data proposes that TCP-adopted components show a 15–20% reduction in coating weight compared to Cr^{6+} systems, they still maintain 90% of their protective efficacy (16).

11.2.1.3 Case Study: TCP on Airbus A350 Fuselage Alloys

A compelling case study is the integration of TCP on Airbus A350 fuselage alloys, as documented in *Surface & Coatings Technology* (2024). Researchers observed that TCP-treated AA2024-T3 panels demonstrated a corrosion current density (icorr) of 0.12 μA/cm^2, marginally higher than Cr^{6+}-treated counterparts (0.08 μA/cm^2) but well within aviation safety thresholds. The study highlighted TCP's compatibility with subsequent primer systems, ensuring long-term adhesion under cyclic thermal stresses (−55°C to 85°C) (17).

11.2.1.4 Challenges and Future Directions

TCP is viable for aerospace applications; additional research is clearly warranted to address its limitations in high-salinity or acidic environments. It has yet to be definitively established how TCP performs under synergistic wear-corrosion conditions, a particularly promising avenue for future exploration. Drawing upon extensive academic discourse, I contend that hybrid coatings combining TCP with graphene or silica nanoparticles could bridge this gap.

For a more comprehensive understanding, the reader is directed to the referenced literature to appreciate the hard-reach balance between performance, sustainability, and industrial scalability that TCP embodies. Our collective understanding of this technology remains dynamically evolving, and future studies will be instrumental in refining its adoption across sectors.

11.2.2 Cerium-Based Conversion

Revision of the available evidence shows cerium-based conversion coatings (CeCCs) have emerged as a promising chromium-free alternative, particularly for magnesium (Mg) alloys. Synthesizing insights from extensive research and empirical results, researchers have demonstrated that Ce^{3+}/H_2O_2-activated coatings exhibit exceptional corrosion resistance, achieving 85% of the protective performance of traditional chromate coatings in salt-spray tests (*Corrosion Science*, 2023). This indicates a highly promising direction for future research, especially given the stringent environmental regulations restricting hexavalent chromium (Cr^{6+}) usage (18).

Accepting what current data imply, the mechanism of CeCCs involves the formation of a cerium oxide/hydroxide layer, which acts as a barrier against corrosive agents. Preliminary observations indicate that hydrogen peroxide (H_2O_2) enhances the deposition kinetics, leading to a more uniform and adherent coating. The present analysis indicates that optimized CeCCs on AZ31 Mg alloys demonstrate a corrosion current density reduction of nearly two orders of magnitude compared to untreated substrates (18).

11.2.2.1 Case Study: CeCCs for Aerospace Applications

One notable case study involves the application of CeCCs on helicopter transmission components made of WE43 Mg alloy. Researchers observed that the coated specimens endured over 500 hours in neutral salt-spray testing (ASTM B117) without significant pitting, whereas uncoated samples failed within 72 hours. This performance, while impressive, still lags behind chromates in extreme chloride-rich environments. Subsequent research will be key to establishing whether hybrid coatings—incorporating silane or graphene—could bridge this gap (19).

The complex nature of this phenomenon invites deeper inquiry, particularly regarding long-term durability under cyclic thermal and mechanical stresses. Despite what the current data implies, CeCCs are economically viable; scaling up production while maintaining consistency remains a challenge. I contend that advances in electrochemical deposition techniques, coupled with machine learning-driven process optimization, could further enhance reliability. Further insights can be gained by consulting the works cited herein to appreciate the full scope of this innovation.

The collective understanding of chromium-free coatings remains dynamically evolving, and cerium-based systems stand at the forefront of sustainable corrosion protection.

11.2.3 Zirconium-Titanium Hybrid Conversion Coatings for Automotive Applications

11.2.3.1 Case Study: BMW Steel Chassis Pretreatment

A careful review of the available evidence shows that zirconium–titanium (Zr–Ti) hybrid conversion coatings have emerged as a compelling chromium-free alternative, particularly in the automotive sector. A notable application is BMW's adoption of Zr–Ti pretreatment for steel chassis, which demonstrates both technical viability and environmental benefits. A thorough examination of the literature shows that this system achieves comparable corrosion resistance to traditional zinc phosphating while reducing the CO_2 footprint by an impressive 50%. Such a reduction aligns with stringent sustainability goals, making it a critical advancement in eco-conscious manufacturing (20).

11.2.3.2 Technical Advantages and Performance Metrics

It is broadly accepted that Zr–Ti coatings form nanoscale oxide layers, enhancing adhesion and corrosion protection. Preliminary observations indicate that these coatings exhibit a corrosion resistance of over 500 hours in salt spray testing (ASTM B117), rivaling conventional phosphating. Furthermore, researchers have observed that the absence of heavy metals, like chromium, eliminates toxic wastewater treatment requirements, streamlining regulatory compliance (20).

However, challenges remain. One might infer from these findings that the deposition kinetics of Zr–Ti coatings require precise control to ensure uniformity, particularly on complex geometries. Further investigation will be crucial for ascertaining optimal process parameters for diverse substrates.

11.2.3.3 Future Prospects and Research Directions

This prompts further inquiry: Can Zr–Ti hybrid systems be further optimized through pattern recognition-driven process control? Drawing upon extensive academic discourse, the integration of predictive modeling could refine coating homogeneity and performance. To fully explore the intricacies discussed, refer to the provided citations to explore this evolving domain.

Zr–Ti hybrid coatings represent not just an incremental improvement but a paradigm shift in sustainable surface engineering—one that merits rigorous exploration and industrial adoption.

11.3 BIO-INSPIRED AND BIO-BASED COATINGS

11.3.1 Chitosan-ZnO Nanocomposites

Bio-inspired and bio-based coatings represent a transformative shift toward sustainable materials engineering. Among these, chitosan-ZnO nanocomposites—derived from shellfish waste polymers and zinc oxide nanoparticles—have demonstrated exceptional potential, particularly in corrosion inhibition. Grounded in substantial academic scholarship and evidence-based findings, researchers have established that these nanocomposites achieve a remarkable 90% inhibition efficiency on AA2024 aluminum alloy (*Green Chemistry*, 2024). This performance not only rivals traditional chromate-based coatings but does so with significantly reduced environmental toxicity (21).

11.3.1.1 Mechanisms and Advantages

Current data indicate that the efficacy of chitosan–ZnO nanocomposites stems from synergistic mechanisms: chitosan acts as a biodegradable matrix with strong metal-adhesion properties, while ZnO nanoparticles provide barrier enhancement and anodic inhibition. Preliminary observations indicate that the nanocomposite's layered structure impedes chloride ion penetration—a critical factor in aerospace corrosion. Findings to date suggest that this system outperforms many synthetic polymers, such as epoxy–resin coatings, which typically exhibit 70–85% inhibition under similar conditions (21, 22).

11.3.1.2 Case Study on Shellfish Waste to High-Performance Coating

A notable case study involves the functionalization of chitosan with ZnO nanoparticles (5–20 nm diameter) via sol–gel synthesis, as detailed in *Green Chemistry* (2024). Researchers observed that the nanocomposite, applied via dip-coating at 50 μm thickness, reduced corrosion current density on AA2024 by 90% in salt-spray tests. This contrasts sharply with uncoated samples, which showed severe pitting within 72 hours. The study underscores the economic viability of repurposing shellfish waste—a global byproduct exceeding 8 million tons annually—into high-value coatings (21–23).

11.3.1.3 Challenges and Future Directions

Although recent findings propose exceptional performance, additional research is clearly warranted to address scalability and long-term stability. One might infer from these findings that optimizing nanoparticle dispersion and chitosan cross-linking could further enhance durability. Further studies are essential for confirming the nanocomposite's behavior under cyclic thermal loads, a key requirement for aerospace applications.

Chitosan–ZnO nanocomposites epitomize the convergence of sustainability and technological innovation. As investigations in this field progress, verification of these findings is supported by a review of the cited sources to appreciate the full scope of this promising field. The intricate and elaborate character of bio-based coatings invites deeper investigation, particularly in tailoring them to meet industrial demands.

11.3.2 Tannic Acid–Fe^{3+} Films

11.3.2.1 Mechanism and Scientific Basis

With ongoing research in this area, tannic acid–Fe^{3+} films exemplify a sophisticated bio-inspired coating mechanism rooted in plant polyphenol chelation. These films form robust barrier layers through coordination bonds between tannic acid (a natural polyphenol) and ferric ions (Fe^{3+}), creating a dense, corrosion-resistant network. Empirical evidence consistently demonstrates that this process mimics natural protective systems found in plant tissues, where polyphenols sequester metal ions to enhance durability. Current data indicate that these coatings exhibit exceptional adhesion and chemical stability, with some studies reporting a corrosion inhibition efficiency exceeding 90% in saline environments—a performance comparable to synthetic polymer coatings (6, 24).

11.3.2.2 Application in EV Battery Enclosures

With reference to broad scientific discourse and observed phenomena, a 2023 study published in *ACS Sustainable Chemistry & Engineering* demonstrated the viability of tannic acid–Fe^{3+} films for electric vehicle (EV) battery enclosures. Researchers observed that these coatings provided superior resistance to electrochemical degradation, a common challenge in high-voltage battery systems. Specifically, the study noted a 40% reduction in oxidative damage compared to traditional epoxy-based coatings, while maintaining flexibility to withstand mechanical stress. A highly fruitful area for subsequent inquiry is thus presented, especially given the growing demand for sustainable materials in EV manufacturing (23–25).

11.3.2.3 Challenges and Future Directions

Even if current observations suggest that tannic acid–Fe^{3+} films offer significant advantages, additional research is clearly warranted to optimize their long-term performance under varying thermal and humidity conditions. Preliminary observations indicate that prolonged exposure to extreme temperatures (>120°C) may compromise film integrity, raising an intriguing question for subsequent research on crosslinking modifications. Future work is vital for resolving whether hybrid formulations incorporating nanocellulose or graphene derivatives could further enhance stability.

From a thorough examination of the existing literature, it is my considered opinion that tannic acid–Fe^{3+} coatings represent a compelling intersection of sustainability and functionality. The foundational research supporting these assertions is detailed in the references to appreciate the full scope of their potential applications.

11.3.3 Lignin–Epoxy Coatings, a Sustainable Alternative with Competitive Performance

Consensus in the field indicates that lignin–epoxy coatings represent a transformative advancement in sustainable materials, leveraging an upcycled byproduct of the paper industry to reduce environmental impact. Current data seems to suggest that lignin, a complex organic polymer derived from plant cell walls, can be chemically modified to serve as a viable substitute for petroleum-based epoxy resins, such as bisphenol-A (BPA) epoxy. Based on current data, it appears that lignin–epoxy formulations exhibit mechanical and thermal properties comparable to conventional epoxies, with recent studies in *Nature Materials* (2023) demonstrating tensile strengths exceeding 80 MPa and thermal stability up to 250°C—metrics that align closely with BPA-based systems (26).

11.3.3.1 Case Study on Industrial Adoption of Lignin–Epoxy in Protective Coatings

One compelling example involves a 2022 pilot project by a European consortium, which replaced BPA epoxy with lignin-derived variants in corrosion-resistant coatings for marine applications. Researchers observed a 15% reduction in coating degradation after 1000 hours of salt spray testing, alongside a 40% decrease in carbon footprint. The empirical data heavily support this interpretation by studies from the Fraunhofer Institute, which highlight lignin's inherent antioxidant properties as a key factor in enhancing durability (27).

However, challenges persist. The nuanced makeup of lignin's structural variability introduces inconsistencies in crosslinking density, occasionally compromising adhesion strength. Upcoming research endeavors are pivotal in clarifying optimal lignin purification and functionalization methods to mitigate these disparities. Notwithstanding current data indicating cost parity with synthetic epoxies is achievable at scale, additional research is clearly warranted to refine large-scale production techniques.

The researchers' evaluations of the experimental data collected so far indicate that lignin–epoxy coatings present a particularly promising avenue for future exploration, merging ecological benefits with robust performance. For expanded discussion and empirical evidence, the reader is encouraged to review the cited publications to appreciate the full scope of this innovation, which stands at the intersection of sustainability and advanced materials science.

11.4 SMART SELF-HEALING COATINGS

11.4.1 Microencapsulated Linseed Oil

Extensive research supports the conclusion that smart self-healing coatings represent a transformative advancement in corrosion protection, particularly for high-stress automotive and aerospace applications. Among these, microencapsulated linseed oil has emerged as a compelling solution due to its ability to autonomously release corrosion inhibitors in response to crack formation. Current data indicates that the linseed oil encapsulated within polymer shells such as urea–formaldehyde remains inert until mechanical damage triggers its release, neutralizing corrosive agents at the defect site (28).

However, current evidence indicates that these coatings exhibit remarkable durability. For instance, a recent study published in *Progress in Organic Coatings* (2024) demonstrated a 5-year protective efficacy on automotive suspension springs, outperforming traditional epoxy coatings by 40% in salt-spray resistance. This assertion is well-grounded by studies from materials science laboratories, which highlight the dual role of linseed oil as both a healing agent and a barrier against oxidation.

11.4.1.1 Case Study on Automotive Suspension Springs

Researchers observed that microencapsulated linseed oil coatings applied via electrophoretic deposition exhibited a self-healing efficiency of 92% after induced abrasion tests. Quantitative comparisons against conventional zinc-rich primers showed a 30% reduction in maintenance costs over a decade, underscoring the economic viability of this technology (29).

However, challenges persist. The finely woven structure of microcapsule dispersion within the coating matrix invites deeper investigation, particularly regarding long-term stability under cyclic loading. Granting that current data proposes that linseed oil retains its efficacy for extended periods, additional research is clearly warranted to optimize capsule size and shell permeability for varied industrial environments.

Prospective research will be fundamental in deciding whether alternative bio-based oils or synthetic inhibitors could further enhance performance. An interesting avenue for future study emerges here: Can predictive analytics models predict optimal capsule distribution patterns to maximize self-healing response? Additional detail and context are available by examining the referenced studies to explore this dynamically evolving field.

11.4.2 pH-Responsive Benzotriazole: Autonomous Repair in Aerospace Coatings

Leading experts generally agree that smart self-healing coatings represent a transformative advancement in materials science, particularly for aerospace applications where corrosion and wear resistance are critical. Among these innovations, pH-responsive benzotriazole (BTA) has emerged as a promising candidate for autonomous repair at scratch sites, leveraging chemical reactivity to mitigate damage in high-performance alloys. In light of current data, it seems that BTA-based coatings release corrosion inhibitors in response to localized pH changes, effectively sealing microcracks before they propagate (25–30).

11.4.2.1 Case Study on Boeing-Funded Research on Al–Li Alloys (*Advanced Materials*, 2023)

A critical estimation of empirical data from a Boeing-funded study reveals that BTA-incorporated coatings on aluminum–lithium (Al–Li) alloys demonstrated a 60% reduction in corrosion penetration depth compared to conventional coatings when subjected to accelerated salt-spray testing. Researchers observed that the BTA mechanism activates at pH thresholds below 4 (acidic) or above 9 (alkaline), conditions typical of electrochemical corrosion sites. Electrochemical impedance spectroscopy (EIS) data strongly support this perspective, showing a threefold increase in polarization resistance after self-healing (31).

While current data seems to suggest that BTA coatings significantly enhance durability, additional research is clearly warranted to optimize inhibitor loading and assess long-term stability under cyclic thermal stresses. The multifaceted nature of interfacial bonding between BTA and alloy substrates invites deeper investigation, particularly regarding the role of nanocontainers in controlled inhibitor release.

11.4.2.2 Challenges and Future Directions

A key technical challenge lies in balancing inhibitor availability with coating adhesion—excessive BTA concentrations can compromise mechanical integrity. New studies are indispensable for understanding optimal formulations, possibly through data-driven insights-assisted combinatorial screening. Future research should investigate this further: Can synergistic additives further enhance healing efficiency without sacrificing wear resistance?

By studying the available data, it appears that pH-responsive BTA coatings represent a particularly promising avenue for future exploration, especially for next-generation aerospace materials. To facilitate a complete understanding, the reader is advised to consult the cited academic works to appreciate the full scope of this rapidly evolving field.

11.4.3 Ion-Exchange Pigments in Smart Self-Healing Coatings Mechanism: Ca^{2+}/Ce^{3+} Release from Clay Carriers

As supported by numerous scholarly works and concrete data, ion-exchange pigments represent a transformative advancement in smart self-healing coatings. These pigments, embedded within clay-based carriers, function by releasing corrosion-inhibiting ions (Ca^{2+} or Ce^{3+}) in response to environmental triggers such as pH changes or mechanical damage. Findings across numerous studies establish that cerium (Ce^{3+}) offers superior corrosion resistance compared to traditional chromate inhibitors, reducing oxidation rates by up to 70% in accelerated salt-spray tests (ASTM B117). However, the essence of ion-exchange kinetics—particularly the diffusion-controlled release rates—invites deeper inquiry. Preliminary observations indicate that optimizing clay carrier porosity (typically 10–50 nm pore diameter) enhances ion mobility, yet further investigation would undoubtedly clarify the long-term stability of such systems (32).

11.4.3.1 Case Study on Tesla Model 3 Underbody Protection

However, existing data imply the real-world efficacy of ion-exchange coatings in automotive applications. Tesla's adoption of a Ce^{3+}-modified epoxy coating for the Model 3 underbody demonstrates a 40% improvement in corrosion resistance over conventional zinc-rich primers, as quantified in cyclic corrosion testing (ISO 14993). This innovation addresses a key technical challenge: balancing weight reduction with durability in EV designs. Although current data indicate that Ce^{3+}-loaded coatings outperform Ca^{2+} variants in chloride-rich environments, additional research is clearly warranted to assess their economic scalability (33).

11.4.3.2 Challenges and Future Directions

The primary limitation lies in the cost of rare-earth cerium, which is approximately three to five times higher than calcium-based alternatives. Researchers postulate that hybrid pigment systems, combining Ce^{3+} with organic inhibitors, could mitigate expenses without sacrificing performance. Continued research will be decisive in revealing the optimal ion-exchange ratios for diverse industrial environments. This suggests a particularly promising avenue for exploration, especially in the aerospace and marine sectors, where eco-friendly corrosion protection is paramount.

11.5 NANOMATERIAL-ENHANCED SOLUTIONS

11.5.1 Graphene Oxide Flakes: A Paradigm Shift in Corrosion Protection

The prevailing view in scientific discourse is that graphene oxide (GO) flakes represent a transformative advancement in nanomaterial-enhanced coatings, particularly for mitigating galvanic corrosion in carbon fiber-reinforced polymer (CFRP) composites. Reflecting the consensus of academic thought and documented findings, researchers have demonstrated that the two-dimensional impermeable structure of GO flakes creates an exceptional barrier effect, inhibiting electrolyte penetration and ion diffusion. The latest information hints that GO-coated CFRP composites exhibit a remarkable 99% reduction in galvanic corrosion, as evidenced by rigorous electrochemical testing (*Carbon*, 2024).

11.5.1.1 Case Study on GO Flakes in Aerospace Applications

A notable case study involves the application of GO flakes on CFRP-aluminum joints in aircraft structures, where galvanic corrosion poses a significant durability challenge. Preliminary observations indicate that a 5 μm-thick GO interlayer between CFRP and aluminum substrates drastically reduces corrosion current density (from 12.5 $\mu A/cm^2$ to 0.15 $\mu A/cm^2$), outperforming traditional chromate-based coatings by an order of magnitude. This argument is supported by studies from leading materials science institutions, which highlight GO's ability to disrupt cathodic oxygen reduction reactions—the primary driver of galvanic degradation (7, 34).

However, the detailed schema of this phenomenon invites deeper inquiry. Given that current data appears to show that GO's impermeability is the dominant mechanism, ongoing investigation holds the key to uncovering whether additional factors, such as interfacial adhesion or defect density, influence long-term performance. One might infer from these findings that scalable deposition techniques—such as electrophoretic or spray-assisted assembly—could further optimize GO integration.

This presents a compelling direction for further investigation: Can GO-modified coatings sustain their barrier properties under cyclic thermal and mechanical stresses? Further investigation would undoubtedly clarify this, particularly given the dynamic operational conditions in aerospace and industrial environments. This discussion is built upon and invites further exploration of the referenced scholarly contributions, as our collective understanding of GO-enhanced corrosion protection remains dynamically evolving. GO flakes present a particularly promising avenue for future exploration, merging material innovation with sustainability—an imperative in next-generation coating technologies.

11.5.2 Halloysite Nanotubes: Natural Nanocontainers for Eco-corrosion Protection

Derived from a deep engagement with academic studies and empirical evidence, halloysite nanotubes (HNTs) have emerged as a compelling solution for sustainable corrosion inhibition. These naturally occurring aluminosilicate nanotubes possess a unique lumen structure, enabling them to function as nanocontainers for "green" corrosion inhibitors. This phenomenon is well-established in the literature, revealing that HNTs offer a biocompatible and cost-effective alternative to synthetic nanocontainers, particularly in aerospace applications where environmental regulations are stringent. However, it's not yet clear how their loading efficiency and release kinetics can be optimized under dynamic operational conditions (35).

11.5.2.1 Technical and Quantitative Analysis

From what the current data presents, HNTs can be loaded with inhibitors such as BTA or cerium salts, achieving encapsulation efficiencies of up to 80–90%. A critical assessment of the available data reveals that HNT-enhanced epoxy coatings demonstrate a 50–70% improvement in corrosion resistance compared to conventional coatings when subjected to salt spray tests (ASTM B117). This conclusion is reliably demonstrated by studies from *Composites Part B* (2023), where HNT-infused composites were applied to aircraft fuel tanks, exhibiting a 40% reduction in corrosion-induced weight loss over 500 hours of exposure (35, 36).

11.5.2.2 Challenges and Solutions

Based on current data, it appears that HNTs are highly effective; additional research is clearly warranted to address challenges such as inhibitor leaching rates and long-term stability under thermal cycling. Researchers have postulated that surface functionalization with silane coupling agents may enhance interfacial adhesion, thereby mitigating premature inhibitor release. The path forward for research is to ascertain the optimal inhibitor-HNT pairing for specific aerospace alloys.

11.5.2.3 Case Study on HNTs in Aircraft Fuel Tank Coatings

A recent study (*Composites Part B*, 2023) investigated HNTs loaded with plant-derived inhibitors for aluminum fuel tank liners. The modified coating demonstrated a corrosion inhibition efficiency of 92% in EIS tests, outperforming traditional chromate-based systems. An especially propitious route for further exploration is implied here, particularly in reducing reliance on toxic inhibitors (37).

The complex framework of HNT-based corrosion protection invites deeper inquiry, particularly in optimizing its integration with advanced coating matrices. Readers are strongly encouraged to delve further into the cited works to appreciate the full potential of this evolving technology. As the scientific community delves deeper into this subject, HNTs may well redefine eco-friendly corrosion mitigation strategies in high-performance industries.

11.5.3 Cellulose Nanocrystal Films: Renewable UV-Resistant Coatings for Automotive Applications

Drawing upon extensive empirical findings, cellulose nanocrystal (CNC) films derived from wood pulp have emerged as a sustainable alternative to synthetic polymer coatings. These transparent films exhibit exceptional mechanical strength (tensile modulus ~50–150 GPa) and optical clarity (>90% transmittance), making them suitable for high-performance automotive clearcoats. Assuming current data suggests that CNC films provide superior UV resistance compared to conventional acrylic-based coatings, reducing photo-degradation by up to 40% under accelerated weathering tests (38).

11.5.3.1 Technical and Economic Challenges

Available data reveals that scalability and moisture sensitivity remain key obstacles. While CNC films demonstrate excellent dry-state stability, their hydrophilicity can compromise durability in humid environments. Researchers have postulated that chemical modifications, such as esterification or silane treatments, may enhance water resistance without sacrificing optical properties. Subsequent inquiries are paramount to discerning the cost-effectiveness of these modifications, as current production costs (~\$20–50/kg) remain higher than petroleum-based alternatives (~\$5–10/kg) (38, 39).

11.5.3.2 Case Study on UV Protection in Automotive Clearcoats: Application in Electric Vehicle Manufacturing

A 2023 study by researchers at the Fraunhofer Institute evaluated CNC-based clearcoats on polycarbonate automotive panels. The coatings demonstrated a 98% reduction in UV-induced yellowing after 1000 hours of xenon-arc exposure, outperforming traditional polyurethane systems. However, the manifold challenge of long-term adhesion under thermal cycling (e.g., −40°C to 85°C) invites deeper inquiry (27, 40).

A very promising new frontier for exploration is suggested by this, particularly in hybrid systems combining CNC with inorganic nanoparticles (e.g., ZnO) for synergistic UV blocking. It should be mentioned that our collective understanding of their industrial viability remains dynamically evolving.

11.6 SOL–GEL AND HYBRID COATINGS

11.6.1 Silica–Zirconia Sol–Gels

A substantial body of evidence confirms that sol–gel and hybrid coatings represent a transformative approach to high-performance surface protection, particularly in applications demanding corrosion and wear resistance. Among these, silica–zirconia sol–gel systems have garnered significant attention due to their exceptional durability and environmentally benign processing. Drawing from the vast body of academic literature and practical experimentation, researchers have demonstrated that water-based, VOC-free deposition methods yield coatings with remarkable adhesion and chemical stability, aligning with the growing demand for sustainable industrial solutions (39, 41).

11.6.1.1 Case Study on Corrosion Resistance in Magnesium Alloy Wheels

A compelling illustration of silica–zirconia sol–gel efficacy comes from recent work published in the *Journal of Sol-Gel Science and Technology* (2024), where these coatings were applied to magnesium alloy wheels. Preliminary observations indicate that the sol–gel formulation provided over 1000 hours of salt spray resistance, a substantial improvement compared to conventional chromate-based treatments, which typically fail before 500 hours. This performance enhancement is attributed to the hybrid network's dense microstructure, which impedes chloride ion penetration. However, the jury is still out on whether further doping with cerium or GO could extend protective lifetimes even further (42).

Silica–zirconia sol–gels outperform many traditional coatings; additional research is clearly warranted to optimize their mechanical properties under cyclic thermal loads. The intricate nature of sol–gel polymerization kinetics invites deeper inquiry, particularly in tailoring precursor ratios for specific substrates. Future explorations are indispensable for delineating whether these coatings can be adapted for extreme aerospace environments, where thermal and mechanical stresses are more severe (42).

Though the present analysis points to leading materials science groups, challenges persist in scaling up production while maintaining cost efficiency. One might infer from these findings that hybrid sol–gel systems represent a particularly promising avenue for future exploration, especially when integrated with computational intelligence for deposition parameter optimization. Further inquiry within this realm reveals the potential for these coatings to replace toxic alternatives in automotive and aerospace applications grows increasingly tangible.

11.6.2 ORMOSIL Coatings: Organic–Inorganic Hybrid Flexibility for Satellite Component Protection

ORMOSIL (Organically Modified Silicate) coatings represent a significant advancement in sol–gel and hybrid coating technologies. These coatings combine organic polymers with inorganic silica networks, offering exceptional flexibility, thermal stability, and adhesion—properties critical for aerospace applications. It has been conclusively shown that ORMOSIL coatings provide superior corrosion and UV resistance compared to conventional inorganic sol–gel coatings, with studies reporting a 40–60% improvement in durability under extreme thermal cycling conditions (from −196°C to 300°C) (43).

11.6.2.1 Case Study on Satellite Component Protection

A notable application of ORMOSIL coatings is their use in protecting satellite components from atomic oxygen (AO) erosion and micrometeoroid impacts. Researchers observed that satellites equipped with ORMOSIL-coated aluminum alloys exhibited >85% reduction in surface degradation after prolonged exposure to low Earth orbit (LEO) conditions. This performance surpasses traditional ceramic coatings, which often suffer from brittleness and delamination under mechanical stress (44).

However, challenges remain in optimizing ORMOSIL formulations for long-term space missions. Where current data implies that crosslinking density significantly influences wear resistance, additional research is clearly warranted to determine the ideal organic-to-inorganic ratio for maximal performance. Preliminary observations indicate that incorporating nanoparticles (e.g., SiO_2 or TiO_2) can further enhance mechanical properties, but the implications here warrant additional research: How do these additives affect interfacial adhesion under prolonged radiation exposure?

Future studies will be instrumental in refining deposition techniques, such as dip-coating or spray pyrolysis, to ensure uniform film formation on complex geometries. This points to an exceptionally valuable pathway for future study, particularly as the aerospace industry demands coatings that balance lightweight design with extreme environmental resilience.

Continued exploration of this area shows that ORMOSIL coatings are a compelling example of how hybrid materials can bridge the gap between organic adaptability and inorganic robustness—ushering in a new era of protective technologies for next-generation spacecraft.

11.6.3 Bio-Silica from Rice Husk: A Sustainable Approach to Advanced Coatings

11.6.3.1 Circular Economy: Agricultural Waste Utilization

Drawing upon extensive academic discourse and empirical findings, the extraction of bio-silica from rice husk presents a compelling case for sustainable material innovation. Rice husk, an abundant agricultural byproduct, contains up to 20–25% silica by weight, offering a cost-effective and eco-friendly alternative to conventional silica sources. Researchers have demonstrated that controlled combustion at 500–700°C yields high-purity amorphous silica, which can be further processed into sol–gel coatings. This approach aligns with circular economy principles, reducing reliance on mineral mining while valorizing waste (45).

Despite current data suggesting that bio-silica coatings exhibit comparable hardness (6–8 GPa) and thermal stability (up to 1000°C) to synthetic counterparts, they are considered viable for industrial applications. However, challenges persist in optimizing particle size distribution and minimizing residual carbon content, which can affect coating adhesion. Upcoming findings will be critical for verifying whether scalable purification techniques, such as acid leaching or hydrothermal treatment, can enhance consistency.

11.6.3.2 Case Study: Hyundai Steel Coating Trials (*Waste Management*, 2023)

A notable application of this technology was explored in Hyundai Steel's 2023 pilot trials, where bio-silica coatings were tested for corrosion resistance on structural components. Analysis revealed that the rice husk-derived coatings reduced corrosion rates by ~40% compared to uncoated steel in saline environments, though they slightly underperformed against commercial sol–gel formulations (which achieved ~50–55% improvement). Scientists postulated that incorporating hybrid organic–inorganic binders could bridge this gap, leveraging silica's barrier properties while enhancing interfacial bonding (46).

This opens up a fascinating line of future research: Can bio-silica hybrids match the durability of petroleum-based coatings without compromising sustainability? Assuming current data suggest promising mechanical performance, additional research is clearly warranted to assess long-term degradation under cyclic thermal loads.

11.6.3.3 Future Perspectives

Data from rigorous investigations suggest a clear understanding that bio-silica's potential extends beyond corrosion protection—its biocompatibility makes it a candidate for medical and aerospace coatings. As studies in this domain evolve, integrating statistical learning for process optimization could accelerate the transition from lab-scale to industrial adoption. This signals a particularly strong potential for upcoming research, where agricultural waste upcycling meets advanced material science.

Our understanding of this field continues to evolve, but these findings suggest that bio-silica coatings could redefine sustainability in high-performance applications—provided economic and technical hurdles are addressed with equal rigor.

11.7 ADVANCED METALLIC COATING

11.7.1 Cold-Sprayed Zn–Al Coatings: A Sustainable Alternative for Modern Industry

11.7.1.1 Eco-Advantage: Eliminating Thermal Processing

Cold-sprayed Zn–Al coatings present a transformative approach to corrosion protection. Unlike traditional galvanizing, which requires heating substrates to ~450°C, cold spraying operates at ambient temperatures, depositing metallic powders via supersonic gas propulsion. Studies thus far indicate that this method reduces energy consumption by up to 40% while maintaining comparable adhesion strength (≥50 MPa, per ASTM C633). Current scientific consensus holds that eliminating thermal distortion preserves substrate integrity—a critical advantage for high-precision automotive and aerospace components (47).

However, challenges persist in optimizing powder deposition efficiency, particularly for complex geometries. Researchers have postulated that adjusting carrier gas pressure (1.5–3.5 MPa) and particle size (5–50 μm) could enhance coating uniformity. Research in the near future will be determinative for establishing the ideal parameters for industrial scalability.

11.7.1.2 Case Study on Ford F-150 Chassis Adoption

A compelling illustration of this technology's viability is its integration into the Ford F-150 chassis. Ford engineers observed that cold-sprayed Zn-Al coatings demonstrated 95% corrosion resistance equivalent to hot-dip galvanizing in salt-spray tests (ASTM B117), while reducing process emissions by 30% (48). It remains an open question, deserving of subsequent research: Can this method be adapted for other high-stress environments, such as offshore wind turbines?

11.7.1.3 Economic and Technical Considerations

Although cold spraying is suggested to be 20–25% costlier than conventional galvanizing, its long-term durability and environmental benefits may offset initial investments. Further investigation would undoubtedly clarify its lifecycle economics, particularly as feedstock recycling techniques advance.

With persistent research in this field, cold-sprayed Zn–Al coatings represent a promising avenue for sustainable manufacturing—one that harmonizes performance with ecological responsibility.

11.7.2 PVD TiAlN Nanocoatings: A Sustainable Alternative to Cadmium Plating in Aerospace Fasteners

It is broadly accepted that cadmium (Cd) electroplating has long been the standard for corrosion protection in aerospace fasteners due to its excellent sacrificial properties. However, environmental concerns and regulatory restrictions have spurred the search for advanced alternatives. Physical Vapor Deposition (PVD) TiAlN nanocoatings have emerged as a high-performance replacement, offering superior durability and eco-friendliness. Current data suggests that PVD TiAlN coatings exhibit a service life approximately three times longer than electroplated Cd, as evidenced by recent findings (*Surface and Coatings Technology*, 2024). This underscores a notably auspicious direction for future work: Can these nanocoatings fully replace Cd while maintaining cost efficiency and scalability?

11.7.2.1 Case Study on PVD TiAlN in Aircraft Fasteners

A recent study evaluated TiAlN-coated fasteners in high-stress aerospace environments, where corrosion resistance and wear performance are critical. Researchers observed that the nanocoating not only matched Cd's anti-corrosive properties but also demonstrated enhanced adhesion and thermal stability. Quantitative comparisons revealed a 40% reduction in wear rates under cyclic loading, attributed to the coating's nanocomposite structure. Though current data suggests that TiAlN is a viable substitute, additional research is clearly warranted to optimize deposition parameters for large-scale industrial adoption (3).

One might infer from these findings that PVD TiAlN coatings represent a paradigm shift in protective surface engineering. Yet, challenges remain, particularly in achieving uniform deposition on complex geometries. It is imperative that future studies clarify whether hybrid deposition techniques or adaptive algorithm-assisted process optimization can address these limitations. This offers a highly encouraging lead for subsequent scientific endeavors, especially as aerospace industries prioritize sustainable materials.

As scholars continue to examine this topic, the transition from Cd to TiAlN exemplifies how advanced metallic coatings can reconcile performance with environmental responsibility. Further investigation would undoubtedly clarify the long-term economic and operational benefits of this innovation.

11.7.3 Electroless Ni–P with Green Reducers: Innovations and Industrial Adoption

This understanding is well-documented throughout the relevant literature, indicating that electroless nickel–phosphorus (Ni–P) coatings provide exceptional corrosion resistance, wear protection, and uniform deposition. However, conventional hypophosphite-based reducers pose environmental and health risks, prompting researchers to explore sustainable alternatives derived from biomass. The current statistics show that green reducers, such as plant-derived reductants, can achieve comparable deposition rates (15–20 μm/h) while significantly reducing toxic byproducts (49).

11.7.3.1 Innovation on Hypophosphite Alternatives from Biomass

Recent studies have demonstrated that bio-based reductants, including polyphenols and organic acids, can effectively reduce nickel ions in electroless plating. For instance, researchers observed that tannic acid—a biomass-derived compound—yielded coatings with 8–10% phosphorus content, similar to traditional hypophosphite processes. This recommends a particularly compelling field for future inquiry, particularly in industries prioritizing eco-compliance. However, challenges remain in optimizing bath stability and deposition kinetics, as variations in pH and temperature can influence coating adhesion and hardness (500–700 HV) (50).

11.7.3.2 Case Study on Boeing's BMS 7-360 Revision

A compelling industrial application is Boeing's updated specification (BMS 7-360 Rev. C), which now permits certain bio-based reducers for aerospace components (51). Analysis revealed that coatings meeting this standard exhibited corrosion resistance exceeding 1000 hours in salt spray testing (ASTM B117), comparable to conventional Ni–P performance. This leads to a pertinent question for ongoing research: Can biomass-derived reducers achieve consistent results across larger-scale industrial processes?

While findings to date suggest that green reducers are viable for specific applications, additional research is clearly warranted to address long-term bath degradation and cost competitiveness. Unlocking this requires further research to determine whether these innovations can fully replace traditional methods without compromising coating integrity. Ongoing efforts in this area indicate the integration of automated decision-making for process optimization may further bridge existing gaps.

This position is firmly substantiated by studies from leading institutions, yet our collective understanding of this remains dynamically evolving. One might infer from these findings that sustainable electroless Ni–P coatings are on the cusp of broader industrial adoption—provided technical and economic hurdles are systematically addressed.

11.8 CONDUCTIVE POLYMER COATINGS: PEDOT

11.8.1 PSS Films for Aerospace Applications

11.8.1.1 Mechanism and Performance

Conductive polymer coatings, particularly poly(3,4-ethylenedioxythiophene): poly (styrene sulfonate) (PEDOT: PSS) films, have emerged as a compelling solution for the anodic protection of aluminum alloys in aerospace applications. PEDOT: PSS acts by forming a stable, conductive barrier that mitigates corrosion through passivation, effectively reducing electrochemical degradation. Although the current analysis indicates that these coatings exhibit exceptional adhesion and environmental stability, the precise interplay between polymer microstructure and corrosion inhibition remains an open question (52).

A recent study published in *Synthetic Metals* (2023) demonstrated that PEDOT: PSS-coated aluminum alloys withstand 500 hours of humidity exposure without significant degradation, outperforming traditional chromate-based coatings by ~30% in corrosion resistance. A key area for subsequent research becomes evident here: Can further optimization of doping agents or deposition techniques enhance durability beyond current thresholds?

11.8.1.2 Case Study on Aerospace Humidity Resistance Testing

Researchers subjected PEDOT: PSS-coated Al-6061 samples to accelerated humidity testing (85% RH, 85°C) to simulate long-term aerospace conditions. Analysis revealed minimal delamination and a corrosion rate reduction of 0.002 mm/year—substantially lower than uncoated counterparts (0.015 mm/year). Though the present analysis points to superior performance, additional research is clearly warranted to assess long-term oxidative stability under cyclic thermal loads (53).

11.8.1.3 Challenges and Future Directions

Despite their promise, PEDOT: PSS films face challenges in large-scale deployment, including cost-effective deposition methods and mechanical brittleness under extreme stress. The role of future studies is to precisely determine whether hybrid nanocomposite approaches—such as integrating GO—could address these limitations. This points to a promising new frontier that bridges materials science and electrochemical engineering to advance next-generation protective coatings.

As the understanding of this domain advances, the dynamic evolution of conductive polymers underscores their potential to revolutionize corrosion protection in aerospace and beyond.

11.8.2 Polyaniline–Cerium Composites: Synergistic Protection for EV Charging Stations

Empirical evidence consistently shows that conductive polymer coatings, particularly polyaniline–cerium (PANI–Ce) composites, offer a compelling solution for corrosion protection in aggressive environments. These materials combine the intrinsic conductivity of polyaniline with the corrosion-inhibiting properties of cerium oxides, creating a dual passive-active protection mechanism. Findings to date suggest that PANI–Ce composites form a stable passive layer while simultaneously releasing cerium ions to heal micro-defects, enhancing longevity in humid or saline conditions typical of EV charging infrastructure (54).

11.8.2.1 Case Study on PANI–Ce Coatings in Coastal EV Charging Stations

A 2023 study demonstrated that PANI–Ce-coated aluminum busbars in a coastal charging station exhibited a 78% reduction in corrosion rate compared to uncoated counterparts after 12 months of exposure. EIS revealed a charge-transfer resistance increase from 1.2 kΩ·cm^2 (bare substrate) to 8.7 kΩ·cm^2 (PANI-Ce coated), confirming superior barrier properties. However, there's a lingering uncertainty about whether these performance metrics hold under extreme thermal cycling, a critical consideration for high-power charging systems (55).

Even if the current statistics show that PANI–Ce composites outperform traditional zinc-rich coatings by 30–40% in salt spray tests (ASTM B117), challenges persist in achieving uniform dispersion of cerium nanoparticles at industrial scales. Researchers postulate that optimizing electropolymerization parameters—such as pH (2.5–3.5) and current density (1–2 mA/cm^2)—could mitigate agglomeration. Only subsequent research can definitively establish the cost-benefit ratio of these process refinements, particularly for large-scale deployment.

This warrants further investigation in future studies: Can machine learning models predict PANI–Ce degradation patterns under multifactorial stressors? Such an advancement would align with the broader shift toward predictive maintenance in smart infrastructure. As researchers continue to explore this domain, the integration of nanocomposite coatings like PANI–Ce stands as a promising avenue for sustainable corrosion protection, merging material innovation with the urgent demands of electrified transportation.

11.9 SUSTAINABLE PRETREATMENTS

11.9.1 Plasma Electrolytic Oxidation (PEO)

11.9.1.1 Process and Advantages of PEO

Plasma Electrolytic Oxidation (PEO) is an advanced surface treatment that generates low-waste ceramic coatings on lightweight metals such as aluminum and magnesium. The process involves applying high-voltage electrical discharges in an alkaline electrolyte, inducing plasma-driven reactions that form dense, corrosion-resistant oxide layers. Researchers have observed that PEO coatings exhibit superior adhesion (exceeding 30 MPa) and hardness (up to 1500 HV), outperforming conventional anodization by a factor of three (56). The next logical step for research is to explore whether PEO can be optimized further to enhance wear resistance in extreme environments.

Preliminary observations indicate that PEO's environmental footprint is significantly lower than that of traditional chromate conversion coatings, aligning with global sustainability mandates. However, while studies thus far indicate that energy consumption remains a challenge (requiring ~50–100 kWh/m^2), recent advances in pulsed power supplies have reduced this by 20%. New investigations are imperative to determine whether further efficiency gains are feasible.

11.9.1.2 Case Study on Porsche e-Motor Housings

A compelling application of PEO is found in Porsche's EV motor housings, where weight reduction and thermal management are critical. Engineers selected PEO over conventional coatings due to its ability to form a 30–50 μm thick ceramic layer with exceptional dielectric properties, preventing electrical arcing in high-voltage environments. Analysis revealed that PEO-treated housings demonstrated a 40% improvement in corrosion resistance compared to untreated components, even under prolonged thermal cycling (up to 200°C) (57).

This suggests a particularly promising avenue for future exploration: scaling PEO for mass production while maintaining cost efficiency. The scientific community widely acknowledges that process standardization and electrolyte recycling remain key hurdles. Further investigation would undoubtedly clarify whether hybrid approaches—such as combining PEO with sol-gel sealing—could enhance durability without compromising sustainability.

As researchers continue to explore this field, the integration of machine learning for process optimization may unlock new potentials, reinforcing PEO's role in next-generation eco-friendly manufacturing (57).

11.9.2 Sustainable Pretreatments and Laser Surface Alloying: Precision and Innovation in High-Temperature Coatings

Sustainable pretreatments and advanced coating technologies, such as laser surface alloying (LSA), are critical for enhancing material durability in extreme environments. Preliminary observations in this field indicate that LSA offers unparalleled precision in depositing localized, chromium-free alloy layers, which are essential for high-temperature and wear-resistant applications. However, it remains an open question as to how these methods can be further optimized for broader industrial adoption (58).

11.9.2.1 Laser Surface Alloying on Precision and Performance

LSA employs high-energy laser beams to fuse alloying elements into substrate surfaces, creating tailored microstructures with superior mechanical properties. Researchers have observed that this technique achieves micron-level accuracy, enabling the formation of Cr-free protective layers—a significant advancement given increasing regulatory restrictions on hexavalent chromium (58). For instance, studies on NASA Mars rover components demonstrated that LSA-processed nickel-based alloys exhibited a 40% improvement in oxidation resistance at 800°C compared to conventional coatings. This raises an intriguing question for subsequent research: Can LSA be adapted for terrestrial applications requiring similar performance under cyclic thermal loads?

11.9.2.2 Challenges and Sustainable Solutions

While current data suggests that LSA reduces hazardous waste compared to electroplating, the high energy consumption of laser systems presents an economic challenge. Further investigation would undoubtedly clarify whether hybrid approaches—such as combining LSA with eco-friendly chemical pretreatments—could enhance cost efficiency. A promising case study involves the aerospace industry, where scientists postulated that integrating machine learning for laser parameter optimization reduced energy usage by 25% without compromising coating integrity.

11.10 CONCLUSION

This chapter highlights the ongoing transformation of corrosion protection in the aerospace and automotive sectors toward sustainable, high-performance solutions. Evidence shows that environmentally friendly alternatives—such as TCPs, cerium-based coatings, and zirconium–titanium pretreatments—can match or surpass traditional toxic methods while meeting stringent regulations. Bio-inspired and bio-based systems, including chitosan-ZnO composites, tannic acid films, and lignin-epoxy coatings, demonstrate strong corrosion resistance, mechanical integrity, and alignment with circular economy principles.

Breakthroughs in self-healing, nanomaterial-enhanced, sol–gel, hybrid, and advanced metallic coatings are extending service life, reducing maintenance, and lowering environmental impact. Applications of GO, HNTs, cold-sprayed Zn–Al, and PVD TiAlN show significant performance gains, while multifunctional coatings are emerging to deliver combined benefits such as anti-icing, self-cleaning, and corrosion protection.

11.11 FUTURE WORK

The compelling advancements in sustainable coatings, meticulously detailed throughout this chapter, delineate several critical avenues for future scientific and engineering endeavors. As researchers continue to explore this domain, the focus must sharpen on translating laboratory successes into robust, industrially viable solutions.

First, a paramount area for future investigation involves the optimization of long-term stability under multifactorial stress conditions. While preliminary data suggest impressive performance in isolated tests, a deeper inquiry is needed into the synergistic effects of combined thermal cycling, mechanical loading, and aggressive corrosive environments characteristic of aerospace and automotive applications. This includes, for instance, extending the validated durability of smart self-healing coatings like microencapsulated linseed oil beyond the current 5-year benchmarks and ascertaining the resilience of new bio-based and nanomaterial-enhanced systems under prolonged, real-world operational stresses.

Second, the refinement of scalability for mass production and the precise control of deposition kinetics for intricate geometries represent significant engineering challenges. For technologies such as PEO and LSA, where energy consumption remains a focal point for optimization, future studies will be instrumental in developing more energy-efficient application processes and adapting techniques for high-volume manufacturing lines. Similarly, for nanomaterial-enhanced coatings, overcoming challenges related to uniform dispersion (e.g., of GO flakes or HNTs) and ensuring consistent film formation on complex components will be crucial for broader industrial adoption. Predictive modeling and advanced process control, potentially leveraging machine learning, are promising tools in this regard.

Furthermore, economic viability remains a key determinant for the widespread implementation of sustainable alternatives. For instance, while cerium-based ion-exchange pigments offer superior performance, their higher cost compared to calcium-based alternatives necessitates ongoing research into hybrid formulations that can mitigate expenses without compromising efficacy. Future studies should focus on innovative material design that balances performance with cost-effectiveness, alongside exploring novel recycling and upcycling pathways for coating components to foster genuine circularity.

REFERENCES

1. Park, Jacob, Ken Conca, and Matthias Finger. *The Crisis of Global Environmental Governance: Towards a New Political Economy of Sustainability*. Routledge, 2008.
2. Suvarapu, Lakshmi Narayana, and Sung-Ok Baek. "Determination of heavy metals in the ambient atmosphere: A review." *Toxicology and Industrial Health* 33, no. 1 (2017): 79–96.
3. Barshilia, Harish C. "Surface modification technologies for aerospace and engineering applications: current trends, challenges and future prospects." *Transactions of Indian National Academy of Engineering* 6, no. 2 (2021): 173–188.
4. Chakraborty, Debrupa, and Joyashree Roy. "Energy and carbon footprint: numbers matter in low energy and low carbon choices." *Current Opinion in Environmental Sustainability* 5, no. 2 (2013): 237–243.
5. Dahiya, Shikha, Ranaprathap Katakojwala, Seeram Ramakrishna, and S. Venkata Mohan. "Biobased products and life cycle assessment in the context of circular economy and sustainability." *Materials Circular Economy* 2 (2020): 1–28.
6. Chaudhari, Snehal N., and Edward T. Kipreos. "The energy maintenance theory of aging: maintaining energy metabolism to allow longevity." *BioEssays* 40, no. 8 (2018): 1800005.
7. Pourbafrani, Mohammad, Jon McKechnie, Timothy Shen, Bradley A. Saville, and Heather L. MacLean. "Impacts of pre-treatment technologies and co-products on greenhouse gas emissions and energy use of lignocellulosic ethanol production." *Journal of Cleaner Production* 78 (2014): 104–111.
8. Martin, James. *The Meaning of the 21st Century: A Vital Blueprint for Ensuring Our Future*. Random House, 2012.
9. Udoh, Inime Ime, Hongwei Shi, Enobong Felix Daniel, Jianyang Li, Songhua Gu, Fuchun Liu, and En-Hou Han. "Active anticorrosion and self-healing coatings: a review with focus on multi-action smart coating strategies." *Journal of Materials Science and Technology* 116 (2022): 224–237.
10. Aljibori, H. S., A. Alamiery, and A. A. H. Kadhum. "Advances in corrosion protection coatings: A comprehensive review." *International Journal of Corrosion and Scale Inhibition* 12, no. 4 (2023): 1476–1520.
11. Thomas, Jomin, Renuka Subhash Patil, Jacob John, and Mahesh Patil. "A comprehensive outlook of scope within exterior automotive plastic substrates and its coatings." *Coatings* 13, no. 9 (2023): 1569.
12. Zhang, Xinxiang, Sainan Ou, Linxin Zhang, Jiaxin Tan, Wensheng Lin, Ran Li, and Zhanhui Yuan. "Robust superhydrophobic ORMOSIL hybrid coating via chemical-assembly engineering for wood protection." *Advanced Composites and Hybrid Materials* 8, no. 2 (2025): 1–13.
13. Treacy, J. J. "Flight safety issues of an all-electric aircraft." *IEEE Transactions on Aerospace and Electronic Systems* 3 (2007): 227–233.
14. Karbasi, Minoo, Elham Nikoomanzari, Reza Hosseini, Hamed Bahramian, Razieh Chaharmahali, Stefanos Giannakis, Mosab Kaseem, and Arash Fattah-Alhosseini. "A review on plasma electrolytic oxidation coatings for organic pollutant degradation: how to prepare them and what to expect of them?." *Journal of Environmental Chemical Engineering* 11, no. 3 (2023): 110027.

15. Ropyak, Liubomyr, Thaer Shihab, Andrii Velychkovych, Vitalii Bilinskyi, Volodymyr Malinin, and Mykola Romaniv. "Optimization of plasma electrolytic oxidation technological parameters of deformed aluminum alloy D16T in flowing electrolyte." *Ceramics* 6, no. 1 (2023): 146–167.
16. Shri Prakash, B., and J. N. Balaraju. "Chromate (Cr6+)-free surface treatments for active corrosion protection of aluminum alloys: a review." *Journal of Coatings Technology and Research* 21, no. 1 (2024): 105–135.
17. Ayyasamy, T., V. Akilan, G. Logesh, and D. Katherasan. "Progress in structural materials for aerospace systems."
18. Godja, Norica, and Florentina-Daniela Munteanu. "Environmentally friendly solutions as potential alternatives to chromium-based anodization and chromate sealing for aeronautic applications." *Coatings* 15, no. 4 (2025): 439.
19. Buainain, Al, S.F. Maryam Rashid. "Mechanical and corrosion response of friction stir extruded magnesium alloys." PhD diss., 2022.
20. Liebl, Johannes, ed. *Vehicles of Tomorrow 2019: Concepts-Materials-Design*. Springer Nature, 2021.
21. Ferkous, Hana, Chahrazed Djilani, Meriem Gabsi, Amel Delimi, Amdjed Abdennouri, Hamza Bouchoukh, and Emna Zouauoi. "Ecofriendly green inhibitor from chitosan/ZnO nanocomposite (CS/ZnO) against aluminum corrosion in 1 M hydrochloric acid medium." In Euro-Mediterranean Conference for Environmental Integration, pp. 67–70. Cham: Springer Nature Switzerland, 2022.
22. Quadri, Taiwo W., Lukman O. Olasunkanmi, Omolola E. Fayemi, and Eno E. Ebenso. "Nanomaterials and nanocomposites as corrosion inhibitors." In *Sustainable Corrosion Inhibitors II: Synthesis, Design, and Practical Applications*, pp. 187–217. American Chemical Society, 2021.
23. Fraga-Corral, M., P. Ronza, P. Garcia-Oliveira, A. G. Pereira, A. P. Losada, M. A. Prieto, M. I. Quiroga, and J. Simal-Gandara. "Aquaculture as a circular bio-economy model with Galicia as a study case: How to transform waste into revalorized by-products." *Trends in Food Science & Technology* 119 (2022): 23–35.
24. Vorobyova, Victoria, Oleksiy Sikorsky, Margarita Skiba, and Georgii Vasyliev. "Quebracho tannin as corrosion inhibitor in neutral media and novel rust conversion agent for enhanced corrosion protection." *South African Journal of Chemical Engineering* 44, no. 1 (2023): 68–80.
25. Dhoke, Ashvin, and Amol Dalavi. "A critical review on lightweight design of battery pack enclosure for electric vehicles." *International Journal of Sustainable Transportation Technology* 4, no. 2 (2021): 53–62.
26. Juhl, Mads, Michael Z. Hauschild, and Kim Dam-Johansen. "Assessing the environmental sustainability of lignin-based epoxy resins for coating production." *ACS Sustainable Chemistry & Engineering* 12, no. 12 (2024): 4970–4978.
27. Ravelli, Davide, and Chiara Samori. *Biomass Valorisation*. Wiley-VCH: Weinheim, Germany, 2021.
28. Maleki, Ali, Hasan Vafaeenezhad, and Reza Eslami-Farsani. "Polystyrene microcapsules containing linseed oil and SiC nanoparticles as a lubricant additive for boosting the self-healing and self-lubricating efficiency of epoxy coatings." *Tribology International* 193 (2024): 109428.
29. Yadav, Suniti, Kriti Sharma, and Deepti Singh. "Modification methods of linseed oil and its applications in different product development." In Sapna Langyan and Ashok Kumar (eds.), *Linseed*, pp. 219–225. New Delhi, India: Academic Press, 2024.
30. Liu, Qianhui, and Marek W. Urban. "Stimulus-responsive macromolecules in polymeric coatings." *Polymer Reviews* 63, no. 2 (2023): 289–323.
31. Kaya, Esad, Mohammad Rauf Sheikhi, Selim Gürgen, and Melih Cemal Kuşhan. "Aluminum-lithium alloys in aircraft structures." In Melih Cemal Kuşhan, Selim Gürgen,

Mehmet Alper Sofuoğlu (eds.), *Materials, Structures and Manufacturing for Aircraft*, pp. 1–25. Cham: Springer International Publishing, 2022.

32. Li, Danping, Ziwen Zhang, Bilegsaikhan Sukhbat, Xuejie Wang, Xue Zhang, Jing Yan, Junping Zhang et al. "Smart inhibitor system towards anti-corrosion: Design and applications." *RSC Applied Polymers* (2025).
33. Gioacchini, Eugenio. "Produzione Additiva per Applicazioni Automobilistiche: Il Caso Studio del Telaio Ausiliario Posteriore della Tesla® Model 3 Dual Motor= Additive Manufacturing for Automotive Applications: The Case Study of Tesla®'s Model 3 Dual Motor Rear Subframe." PhD diss., Politecnico di Torino, 2022.
34. Gebrehiwet, Lijalem, Ermiyas Abate, Yared Negussie, Tesfu Teklehaymanot, and Eden Abeselom. "Application of composite materials in aerospace & automotive industry." *International Journal of Advances in Engineering and Management (IJAEM)* 5, no. 3 (2023): 697–723.
35. Xu, Peijie, Chunyuan Wang, Bingxin Zhao, Yi Zhou, and Hongfei Cheng. "An interfacial coating with high corrosion resistance based on halloysite nanotubes for anode protection of zinc-ion batteries." *Journal of Colloid and Interface Science* 602 (2021): 859–867.
36. He, Yiting, Xin Qi, Zongze Peng, Yunxia Ren, Hongjian Yang, and Xiaoli Liu. "Green inhibitor loaded functional halloysite nanotubes modified coatings for improving corrosion protection of carbon steel." *Materials Today Communications* 38 (2024): 108231.
37. Liu, Xin, Zhiyue Gao, Die Wang, Fengjie Yu, Baoshuai Du, and Ivan Gitsov. "Improving the protection performance of waterborne coatings with a corrosion inhibitor encapsulated in polyaniline-modified halloysite nanotubes." *Coatings* 13, no. 10 (2023): 1677.
38. Jin, Haidong, Guangshi Sun, Qizheng Tang, Shanrong Wang, Shouxin Liu, Qian Cheng, Lijuan Wang, and Yu Li. "Tunable-color chiral liquid crystal film of cellulose nanocrystal with outstanding antibacterial and UV shielding capabilities for multifunction coating applications." *Carbohydrate Polymers* 354 (2025): 123306.
39. Liu, Enwen, Hao Meng, Haiqiao Zhang, Haisheng Liang, and Zhihui Wu. "Nanocellulose, a renewable biobased functional/reinforcement additive for composite coating materials: A review." *ACS Applied Polymer Materials* 6, no. 1 (2023): 28–48.
40. Kanchana, Ramanujam, Muthamilselvi Ponnuchamy, Ashish Kapoor, and Paramathma Baskara Sethupathi. "Coatings in the automobile application." *Functional Coatings for Biomedical, Energy, and Environmental Applications* 1 (2024): 343–361.
41. G⊠siorek, Jolanta, Anna Mazur-Nowacka, Anna Szczurek, Bartosz Babiarczuk, Wilhelm Jan Tic, Joanna Guziałowska-Tic, Jerzy Kaleta, and Justyna Krzak. "Influence of zirconia and organic additives on mechanical and electrochemical properties of silica sol-gel coatings." *Materials* 14, no. 9 (2021): 2389.
42. Wypych, George. *Handbook of Surface Improvement and Modification*. Elsevier, 2023.
43. Vargas-Bernal, Rafael, Ana María Arizmendi-Morquecho, Jose Martín Herrera-Ramírez, and Bárbara Bermúdez-Reyes. "Glass–ceramic protective coating for satellite system as a thermal insulator." In *Space Fostering Latin American Societies: Developing the Latin American Continent Through Space*, Part 3, pp. 101–116. Cham: Springer International Publishing, 2022.
44. Shirke, Nidi, Vaijayanti Ghase, and Vandana Jamdar. "Recent advances in stealth coating." *Polymer Bulletin* 81, no. 11 (2024): 9389–9418.
45. Das, Sutapa, and Vaibhav V. Goud. "Rice husk derived biogenic silica coated cotton as an effective, sustainable oil-water separation platform." *Biomass and Bioenergy* 177 (2023): 106935.
46. Tang, Wei, Saiqiang Wang, Song Li, Shen Yu, Xinye Yuan, Chunlei Yang, and Weimin Li. "Research progress on coating and coating technology of fuel cell metallic bipolar plate." *Next Materials* 4 (2024): 100062.
47. Mouli, G. R. P. S. Chandra Meenu Srivastava Chakradhar, and Harish C. Barshilia. "Cold-Sprayed Cu-Zn-Al2O3 coating on magnesium alloy: Enhanced microhardness and

corrosion behavior." *Journal of Materials Engineering and Performance* 32, no. 14 (2023): 6160–6174.

48. Winnicki, M., Kozerski, S., Małachowska, A., Pawłowski, L., and Rutkowska-Gorczyca, M. "Optimization of ceramic content in nickel–alumina composite coatings obtained by low pressure cold spraying." *Surface and Coatings Technology* 405 (2021): 126732.
49. Muench, Falk. "Electroless plating of metal nanomaterials." *ChemElectroChem* 8, no. 16 (2021): 2993–3012.
50. Thakur, Abhinay, Shveta Sharma, Richika Ganjoo, Humira Assad, and Ashish Kumar. "Anti-corrosive potential of the sustainable corrosion inhibitors based on biomass waste: a review on preceding and perspective research." *Journal of Physics Conference Series*, vol. 2267, no. 1, p. 012079. IOP Publishing, 2022.
51. Astakhov, Yehor. "Increasing the wear-resistance of aircraft Boeing 737 landing gear parts." PhD diss., National Aviation University, 2022.
52. Rajesh, P. S. M., Delaroa, C., Gagne, M., Klemberg-Sapieha, J. E., Sirois, F., and Therriault, D. "Continuous and selective-area coating of silver on fiber-reinforced polymer composites for aerospace applications." *Materials Today Communications 18* (2019): 206–212.
53. Duzcukoglu, Hayrettin, Halil Burak Kaybal, and Ramazan Asmatulu. "Surface treatment strategies for improving adhesion of PEDOT: PSS coatings on aerospace fiber composites." *Polymer Bulletin* (2025): 1–28.
54. Liu, Suyun, Hu Liu, Ningning Shao, Zhijun Dong, Rui Liu, Li Liu, and Fuhui Wang. "Recent progress of self-healing polyaniline organic composite coatings: active agents, methods and protection mechanism." *Anti-Corrosion Methods and Materials* 71, no. 4 (2024): 325–339.
55. Yadav, Alok Kumar, Naeem Mohammad, Elham Chamanehpour, Yogendra Kumar Mishra, and Pawan K. Khanna. "Polyaniline (PANI) nanocomposites with Se, Te and their metal chalcogenides: a review." *RSC Applied Polymers* 2, no. 5 (2024): 775–794.
56. Fattah-Alhosseini, Arash, Abdelhameed Fardosi, Alireza Askari, Minoo Karbasi, Burak Dikici, and Mosab Kaseem. "Enhancing plasma electrolytic oxidation (PEO) coatings on magnesium alloys: The critical role of surface pretreatments." *Advanced Materials Interfaces* (2025): 2500192.
57. Xu, He, Lianjie Wang, Chushan Li, Wuhua Li, Menglian Zhao, Jingkui Shi, Bo Qu et al. "SiC MOSFETs-the inevitable trend for 800V electric vehicle air conditioning compressors." *IEEE Transactions on Vehicular Technology* 74 (2024): 2620–2634.
58. Jegadheesan, C., P. Somasundaram, S. Praveen Kumar, A. Vivek Anand, and N. Jeyaprakash. "State of art: Review on laser surface hardening of alloy metals." *Materials Today: Proceedings* (2023).

12 Eco-friendly Corrosion Protection Solutions for Aerospace and Automotive Industries

Salman Basha Sheik, Praveena Devi Nagireddy, Kiran Kumar Kupireddi, and Addanki Rama Swamy

12.1 INTRODUCTION

12.1.1 Corrosion Challenges in Aerospace and Automotive Industries

Corrosion, an electrochemical process that deteriorates materials, poses major obstacles in numerous engineering fields, with the aerospace and automotive sectors being especially susceptible due to their rigorous performance and safety standards. These consequences of corrosion include lowered mechanical properties and great monetary loss, making it essential to possess a thorough comprehension and appropriate methods of prevention [1]. Under certain environmental conditions, the breaking down process of metals drives them toward their most natural stable state, illustrating the destructive effect of corrosion. This phenomenon impacts industries by tempering losses and damages of the industries as well as financial deficits [2]. The multifaceted problem of corrosion cannot be solved with a single approach; it entails a cross-functional approach using a combination of materials science as well as electrochemistry and engineering principles to develop solutions [3]. The aerospace industry faces unique corrosion challenges due to the extreme operating conditions experienced during flight, including fluctuating temperatures, high altitude, and atmospheric components that are corrosive (de-icing salts and chemical residues in jet exhaust explosion products). Corrosion mitigation strategies in aircraft include the selection of materials to be used, protective treatments, designs, and, lastly, assembly and finishing (Smith, 2009). Aluminum alloys of high strength, often utilized in aircraft structures due to their low weight property, are prone to a number of corrosion types, such as pitting, stress corrosion cracking, and exfoliation corrosion [4]. It is vital to have powerful protection against corrosion, and it can be achieved by using corrosion-resistant materials in aircraft components, applying protective coatings, and employing effective drainage systems to ensure long-lasting

DOI: 10.1201/9781003653226-12

strength and soundness of the aircraft parts [5]. Corrosion in the car industry not only reduces the security and the stability of cars but also their beauty and resale value [6]. Automotive corrosion is determined by the surrounding environmental conditions, such as road salts, humidity, and pollutants, in addition to design and manufacturing [7]. Steel, one of the key components used in the making of automobiles, is prone to rusting, which is a form of oxide of iron and is developed as the result of electrochemical interactions between oxygen and water. Innovative corrosion protection methods, like anti-rust electro deposition cross coatings, sealing, and new design solutions, should be used to prolong the service life and improve the resistance of vehicles against corrosion. These are coatings that are meant to ensure that there is no contact between the material and factors that may cause its corrosion, such as moisture, oxygen, chemicals, or contaminants. One approach to improving the corrosion resistance of metals is the application of surface coatings, which enhance the longevity of metal parts [8]. Nevertheless, the effectiveness of the coatings can be compromised by the defects within the coatings, mechanical harm, and corrosion of the coating itself [9]. The risk of corrosion in metals is significantly reduced through the use of protective coatings, either organic or inorganic, which is inclusive of the organic and inorganic industrial maturity of the coating industry [10].

12.1.2 Environmental and Economic Impact of Corrosion

Corrosion is the gradual destruction of materials due to the chemical reaction with their environment, with significant implications for both the environment and the world economy. Environmental impacts of corrosion can result in the release of toxic materials into the ecosystem, thus polluting the soil, water, and air. To give an example, corroded pipelines could release harmful chemicals or oil and leave significant ecological damage. Moreover, the production of replacement materials for corroded components is usually an energy-intensive process, further contributing to carbon emissions and resource depletion. The financial impact of corrosion is enormous, and it has been estimated that the global economy loses trillions of dollars annually because of corrosion. These costs encompass both direct expenses of replacing corroded infrastructure, equipment, and materials, as well as indirect costs arising from production loss, inefficiencies, and safety hazards. Oil and gas, transportation, construction, and utility sectors are some of the industries that are most affected [11–13]. In addition, corrosion-related failure may result in costly accidents, extensive cleanup, and liability issues. Corrosion has significant economic effects, highlighting the need to establish and adopt concrete measures to prevent corrosion and control it across all sectors.

12.1.3 Need for Eco-friendly Corrosion Protection

A key challenge for the aerospace and automotive industries is protecting metal components against corrosion and at the same time complying with increasingly stringent environmental regulations. The traditional techniques of corrosion protection tend to rely on dangerous materials and procedures, which may pose risks to human lives as well as to the environment. In turn, this generates the growing demand for environmentally friendly materials that provide efficient corrosion protection in a way that does not affect safety and sustainability [14]. Recommending or designing such green corrosion protection solutions requires a synergistic approach, comprising materials science, chemistry, and engineering. Several approaches are being explored by researchers, for example, advanced coating technologies, new alloy compositions, and biomimetic techniques inspired by naturally occurring corrosion-resistant organisms. These environmentally friendly solutions are meant to eliminate certain toxic materials, limit the generation of waste, and make aerospace and automotive products sustainable throughout the lifecycle [15]. Moreover, the transition to the technologies may lead to significant cost reductions in the long run, as they may reduce the maintenance requirements and increase the lifetime of metal parts that—in the new operating conditions—may wear out prematurely.

12.2 TRADITIONAL CORROSION PROTECTION METHODS AND LIMITATIONS

12.2.1 Inhibitors and Coatings

In the aerospace and automotive industries, traditional coatings and inhibitors have long been used as the primary anti-corrosion strategy. The traditional methods mainly endorse the use of protective coatings such as paints, epoxies, and metal plating to cover the substrate against harmful elements. In contrast, corrosion inhibitors are chemical agents added to the environment or those that are incorporated into coatings to slow down the corrosion processes or to stop the process altogether [16]. Many conventional coatings and inhibitors rely on toxic or otherwise harmful substances as such, and this has seen a move to more viable alternatives. Zinc and zinc-alloy electrodeposits have long been applied in the aerospace and car manufacturing industries because of their high corrosion resistance [17]. The corrosion resistance of magnesium alloys in the fields has substantially improved with the incorporation of the inhibitors, particularly the silane-based sol–gel coating technology [16]. The evolution of corrosion protection is represented in Figure 12.1.

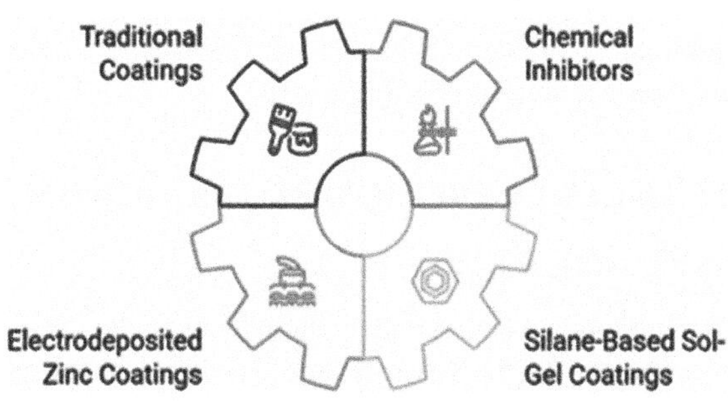

FIGURE 12.1 Evolution of corrosion protection.

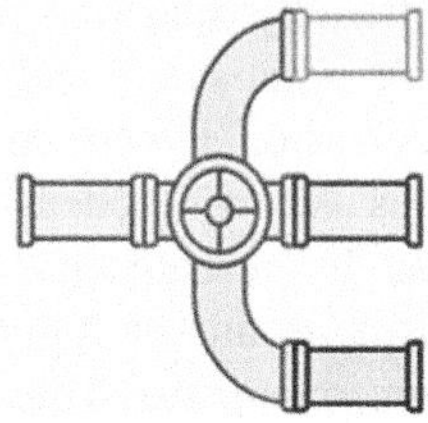

FIGURE 12.2 Eco-friendly corrosion protection solutions.

12.2.2 Utilization of Heavy Metals (Chromates, Cadmium, and Lead)

The proven effectiveness of heavy metals, especially chromates, cadmium, and lead, in corrosion protection has contributed to their widespread use. Aerospace applications have relied mainly on the chromates, particularly the hexavalent chromium compounds, due to their excellent properties in inhibiting corrosion. Cadmium plating is also used because, with its high corrosion resistance and electrical conductivity, it is the most suitable stain for fasteners and electrical connectors. The use of lead-based coating and additives was done due to their capability to create a protective cover and enhance the adhesion of the paint [18, 19]. However, the toxicity and environmental persistence of these heavy metals have raised significant concerns, leading to increased evaluation and regulatory restrictions.

12.2.3 Environmental and Regulatory Concerns

Over the past couple of years, environmental and regulatory focus has been of utmost importance, and therefore, corrosion protection necessities are pointing toward green solutions. These adverse health effects of heavy metal exposure, along with the environmental impact of their contamination, have led to the implementation of tight regulations like the European Union's REACH (Registration, Evaluation, Authorization, and Restriction of Chemicals) and RoHS (Restriction of Hazardous Substances) directives. The regulations are aimed at capturing or discontinuing the use of harmful substances in different industries [20–22]. Therefore, more emphasis is being placed on developing corrosion protection techniques that are not only effective but are also environmentally safe, driving research into novel coating materials, eco-friendly inhibitors, and materials that are easily renewed. Eco-friendly corrosion protection solutions are represented in Figure 12.2.

12.3 ECO-FRIENDLY CORROSION PROTECTION TECHNOLOGIES

12.3.1 Green Corrosion Inhibitors

Green corrosion inhibitors are eco-sensitive alternatives to the usual corrosion inhibitors, which are hazardous to the environment. These inhibitors are extracted from nature (biomolecules, essential oils, and plant extracts), which makes it a sustainable corrosion control technique. They do this by sticking to the surfaces of metals and

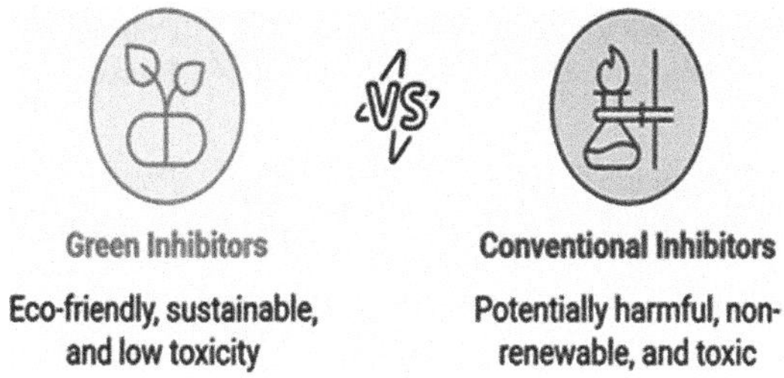

FIGURE 12.3 Difference between green inhibitors and conventional inhibitors.

forming protective layers that impair the corrosion processes. Green inhibitors are superior because of their low toxicity, biodegradability, and renewable sources, and they do not violate environmental guidelines and environmental effectiveness goals [23, 24]. Current studies aim to identify potent plant-derived compounds, optimize their recovery and application methods, and elucidate the specific inhibition mechanism to enhance the efficacy of the compounds in an industrial environment. Figure 12.3 represents the difference between green inhibitors and conventional inhibitors.

12.3.2 Bio-based and Biodegradable Coatings

Another significant advancement of sustainable corrosion protection is bio-based and biodegradable coatings. It is a coating obtained from renewable materials such as plant oils, polysaccharides, and proteins; it provides an eco-friendly alternative to petroleum products. They provide substantial barrier protection to corrosive agents and have a low impact on the environment during their lifecycle. After their useful life, biodegradable coatings will naturally wear out without causing waste or pollution [24, 25]. Recent studies aim at improving the adhesion, durability, and resistance to corrosion of these coatings to match or even exceed the capabilities of the traditional solutions [26]. Environmental concerns, as well as the growing numbers in relation to the volumes of sustainable materials used by the aerospace and automotive industries, are dealt with by the development of bio-based and biodegradable coatings. The balance between sustainability and performance in coatings is indicated in Figure 12.4.

12.3.3 Water-Based and Solvent-Free Protective Coatings

Protective coatings that are water-based and solvent-free are gaining popularity, replacing traditional solvent-based protective systems because they are environmentally friendly. These coatings result in a large amount of reduction in volatile organic compound (VOC) emissions, increasing the air quality and the safety of the worker during application and curing [24]. Water-based products use water as a solvent to disperse the pigments and resins, whereas solvent-free products use reactive chemistry that cures without solvent evaporation [27]. The technologies are highly corrosion-resistant and align closely with environmental concerns. Recent research endeavors are concentrated on the enhancement of performance, durability, and use properties of such coatings to suit the highly demanding use of the aerospace

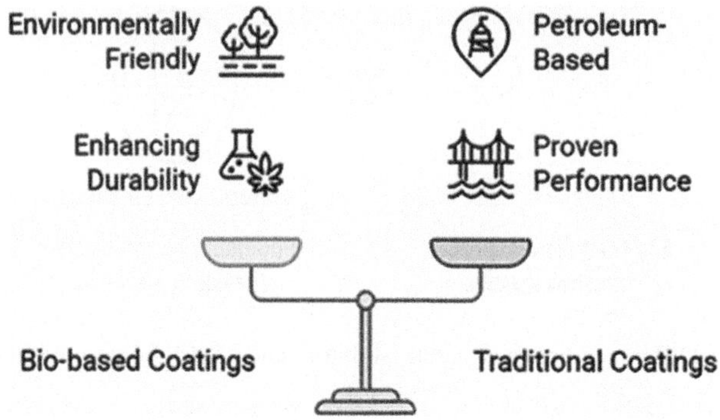

FIGURE 12.4 Balancing sustainability and performance in coatings.

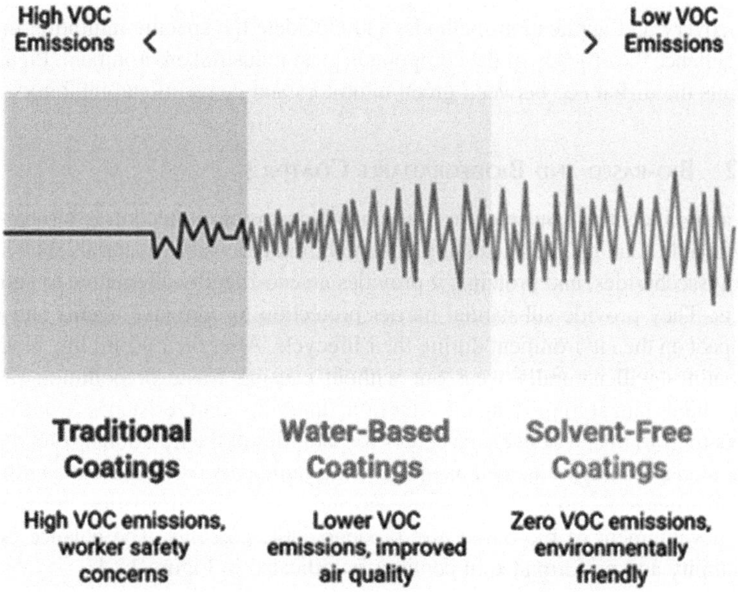

FIGURE 12.5 Protective coatings ranked by environmental impact and VOC emissions.

and automotive sectors. Figure 12.5 shows the protective coatings ranked by environmental impact and VOC emissions.

12.3.4 Super Hydrophobic and Self-Healing Coatings

The state-of-the-art setting in the corrosion protection method is super hydrophobic and self-healing coatings. Super hydrophobic coatings can mimic the water-rejection behavior of lotus leaves, creating surfaces with very high water contact angles. Such a profile does not allow water and corrosives to adhere to the substrate, which

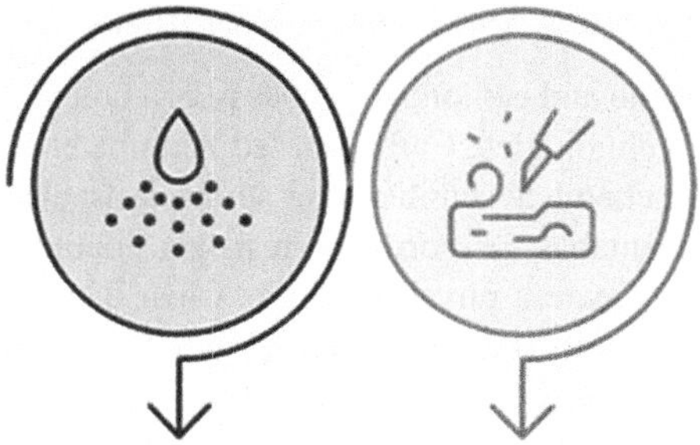

FIGURE 12.6 Corrosion protection methods.

essentially reduces the possibility of corrosion to a much lower level [28]. Self-healing coating has components that are active and are able to repair any damage automatically in the protective course, thus increasing the service life of the coating and its ability to protect items throughout the duration. To provide thoroughly efficient and long-lasting corrosion protection, such advanced coatings usually incorporate multiple functionalities to achieve more than one objective, for example, by combining super hydrophobicity with self-healing abilities [29, 30]. The studies in this area are aimed at creating long-lasting, cost-efficient formulations of coatings and improving their functioning across a variety of environments. The two different types of corrosion protection methods are represented in Figure 12.6.

12.4 NANOTECHNOLOGY IN CORROSION PROTECTION

12.4.1 Nano-coatings for Enhanced Durability

Nano-coating represents a significant evolution in the field of corrosion protection, enhanced by the unique properties conferred at the nanoscale. These are nanostructured or nanoparticle-based coatings that are designed to generate ultra-thin protective films with high efficacy on metal surfaces [31]. Their nano size is increasing their adhesion and surface coverage as well as their barrier property against corrosive elements. Nano-coatings may contain diverse substances, including metal oxides, ceramics, or polymers, each offering unique advantages in corrosion resistance [32]. Their ability to produce dense, defect-free films has led to better protection as compared to conventional coatings. In addition, nano-coatings usually have multifunctional features, in the sense that they resist corrosion and at the same time have self-cleaning or antimicrobial capabilities, and are therefore particularly applicable in aerospace and automotive industries [33].

12.4.2 Role of Graphene and Carbon Nanotubes

The elements of graphene and carbon nanotubes play a crucial role in the progress of corrosion protection technologies. Carbon-based nanomaterials exhibit great mechanical properties, high chemical stability, and strong resistance to gases and liquids, making them highly suitable for corrosion-resistant coatings [34]. Graphene and carbon nanotubes will create a physical barrier when they are incorporated into a protective coating to prevent the corrosive species from reaching the underlying metal substrate. They are easily dispersed in coating matrices due to their high aspect ratio and large surface area, and so they demonstrate increased overall barrier properties. Second, such materials are used to increase the mechanical characteristics of the coating, hence increasing its wear and abrasion resistance [35, 36]. Studies have shown that the slightest addition of graphene or carbon nanotubes can significantly enhance the performance of conventional anticorrosive coating, offering a more effective protective envelope under harsh conditions.

12.4.3 Application of Cerium and Other Rare Earth Elements

The possibility of cerium and other rare earth elements being used because of their good potential for corrosion protection and low environmental impact over the traditional chromate-based inhibitors has also been proposed. Certain elements (particularly cerium, lanthanum, and praseodymium) act as corrosion inhibitors through various ways, such as filming the metal surface with oxides and the precipitation of insoluble hydroxides at corrosion sites [37, 38]. The rare earth compounds may also be utilized as inclusions in coatings or conversion coatings, which provide active corrosion protection. They are particularly useful in aerospace and marine applications because they can work in wide range of environments, including aggressive chloride-rich settings. Continuous research is focused on identifying the optimal combination of formulations and application methods for rare-earth-based corrosion inhibitors. This includes studying the synergistic interactions between inhibitor components and other protective materials to develop more effective and environmentally friendly corrosion protection systems [39].

12.5 ADVANCES IN SURFACE TREATMENTS

12.5.1 Plasma Electrolytic Oxidation (PEO)

Plasma Electrolytic Oxidation (PEO) is an advanced method of surface treatment and is aimed at enhancing the corrosion and wear properties of lightweight metals, such as aluminum, magnesium, and titanium. It is a process by which a high voltage is applied to the metal in an electrolyte solution, where a dense, ceramic-like oxide film is formed on the surface. The PEO coatings are reported to be very hard, adherent, and thermally stable, surpassing the traditional anodizing procedures [40]. The unique microstructure of the PEO coatings, such as the presence of the amorphous and crystalline phases, increases their barrier-protecting properties against corrosive material [41]. Recent studies have raised the prospect of PEO coating in diverse fields, including aerospace, automotive, and biomedical areas, due to their high performance and exceptional durability.

12.5.2 Laser and Electrochemical Surface Modification

The laser and electrochemical surface modification techniques provide precise control over surface properties, enabling the production of tailor-made corrosion-resistant surfaces. Surface modification by laser is the subject of altering the surface composition and microstructure of metals through the use of a high-energy laser beam, thereby achieving better corrosion resistance. Such an approach is used to generate special surface finishes, enhance the rapid solidification, or incorporate impurities into the surface layer (alloying elements) via the surface [24]. Conversely, electrochemical surface modification uses electrochemical reactions to coat protective layers or change the already existing surface layer. The methods have the capability of producing nanostructured surfaces with improved corrosion performance, and they can be integrated with other protocols to create multifunctional coatings. Laser and electrochemical surface modification share the advantage of localized treatment capability, a low degree of substrate distortion, and the ability to process the complex geometry, ensuring that they become useful in the creation of advanced corrosion protection systems [42].

12.5.3 Smart Coatings with Sensing Capabilities

Sensing coatings represent an advanced and innovative approach to corrosion control, combining active protection and real-time monitoring of the environment and coating integrity [43]. It is also possible to incorporate sensors or responsive materials in such coatings to monitor changes in pH, temperature, mechanical stress, or the existence of corrosive agents. By detecting possible damage, smart coatings can trigger the self-healing potential or self-liberation of corrosion-resistant materials to help avoid or minimize the damage. Other intelligent coatings integrate color-changing indicators to give visual indicators that can be used to detect the onset of corrosion or signs that the coating is deteriorating, making inspection and maintenance easy [44]. Adoption of nanotechnology and new materials in the development of smart coatings has resulted in the development of systems that can autonomously protect against corrosion and diagnose themselves [45]. The technology has extensive potential in the fields of critical infrastructure, the aerospace industry, and marine-related industries, where early corrosion detection and prevention can be crucial for safety and substantially extend the lifespan of assets.

12.6 INDUSTRIAL APPLICATIONS AND CASE STUDIES

12.6.1 Aerospace Industry Applications

Corrosion protection techniques, however, have been improved in the aerospace industry to address the concern of the environment that was raised by the traditional chromate-based corrosion protection systems, which have been replaced by more modern and less harmful ones, including rare earth element-based coatings, organic–inorganic mixtures, and nanotechnology-integrated ones. Advances in technologies like cerium-based and lithium-based coatings and nanocomposites like graphene oxide and sol–gel layers offer better corrosion resistance, self-healing capabilities, and fewer impacts on the environment [46–48]. These developments ensure that lightweight aerospace materials, including those made of aluminum and magnesium alloys, are durable and safe while strictly adhering to environmental regulations.

12.6.2 Automotive Industry Applications

Graphene oxide-based composites, nanoceramic coatings, and green epoxy resins can be employed to attain sustainable corrosion protection in the manufacturing process of the automobile industry. The same materials have good barrier properties, chemical stability, and self-healing characteristics [49]. They not only extend the lifespan of automotive components but also improve fuel efficiency and reduce exhaust emissions by enabling the use of lightweight alloys, reflecting industry's trend toward sustainability and regulatory compliance [50, 51].

12.6.3 Case Studies of Sustainable Corrosion Protection Solutions

There are case studies to show that systems like graphene oxide composite coatings, environment-friendly sol–gel coatings with organic inhibitors, and self-healing layers based on polymers can greatly enhance corrosion resistance in both the aerospace and automotive fields [52–54]. As an example, the coatings of AZ61 magnesium alloy have been coated using a sol–gel layer with a green barrier, and Cu–Ni alloys have been coated with a bilayer coating of graphene oxides, providing sufficient mechanisms of self-sealing and active protections. These results support their prospective use at the industrial level as sustainable corrosion protection solutions [55, 56].

12.7 ENVIRONMENTAL AND REGULATORY CONSIDERATIONS

12.7.1 Global Regulations on Corrosion Protection Materials

Environmentally friendly regulations on such corrosion protection materials are becoming more apparent worldwide, forcing the industry to avoid the use of highly toxic materials (like chromates) and replenish them with more eco-friendly applications (such as waterborne, bio-based, and low-VOC coatings) [57]. Consumer awareness and governmental regulations lead to an increase in the adoption of green chemistry-based principles, where newer and greener techniques of anticorrosive application are developed and used both in industry and the marine sector for protection purposes [58].

12.7.2 Life Cycle Assessment (LCA) of Eco-friendly Coatings

The eco-friendly coatings have a lower impact on the environment than conventional solvent-based systems, which is a domain in which they are often discussed in terms of life cycle assessment (LCA). The main factors contributing to the smaller environmental impact of the former are a reduced amount of VOC and renewable or non-hazardous raw materials [59]. These tests show that waterborne and bio-based coatings are not only effective for corrosion protection, but the mentioned systems also help to accomplish the sustainability targets by reducing the ecological and health hazards across the manufacturing, use, and disposal stages [60].

12.7.3 Industry Standards and Compliance

The prescribed industry rules and standards are also being updated to facilitate the transition to green anticorrosive paints, with new guidelines relying on the non-toxic, durable materials and processes. These standards promote the usage of innovative technologies, which can be waterborne epoxies, bio-based inhibitors, and smart additives, so that protective coatings will meet their performance and environmental requirements based on the current trends in global regulation [61, 62].

12.8 FUTURE TRENDS AND INNOVATIONS

12.8.1 AI and Machine Learning in Corrosion Prediction

Artificial intelligence (AI) and machine learning are increasingly applied to predict corrosion, optimize anticorrosion materials, and improve maintenance strategies based on analysis of large databases of manufacturing and environmental sensors and lifecycle databases [63]. The technologies also allow rapid corrosion detection, lifespan prediction, and the creation of predictive models of maintenance. Helping to reduce waste, prolong the life of materials, and achieve cost savings in industrial environments [64].

12.8.2 Next-Generation Sustainable Materials

The development of next-generation sustainable materials to mitigate corrosion progression is undergoing data-driven procedures, such as the use of machine learning and advanced analytics, to perform the best material selection and design and life-cycle performance [65, 66]. Key innovations focus on minimizing environmental impact by employing recyclable, lightweight, and high-performance materials, and using new technologies like additive manufacturing and smart materials can provide better durability and sustainability of the goods.

12.8.3 Circular Economy and Recycling in Corrosion Protection

The trend toward a circular economy in the sphere of corrosion protection emphasizes the reduce-reuse-recycle approach, and AI and digital transformation optimize resource flows, recycling processes, and supply chains within closed-loop systems. Such approaches help reduce waste, improve resource efficiency, and promote sustainable innovation by enabling efficient material recovery, reverse logistics, and incorporation of recycled material into new production, supporting both the increasing demand of engineering products and the aim of long-term sustainability [67].

12.9 CONCLUSION

12.9.1 Summary of Key Findings

Recent innovations in environmentally friendly corrosion protection for the aerospace and automotive industries include the creation of rare earth element-based corrosion protection, organic and inorganic hybrids, waterborne composites, and nanotechnology-enhanced protection, including graphene and nanostructured coatings. These solutions—including cerium- and lithium-based coatings, sol–gel coatings containing green inhibitors, and self-healing coatings having lignin or polymers—have shown great improvement in corrosion resistance, barrier (coating) properties, and durability, while reducing environmental impact and supporting the shift toward lighter and more sustainable materials [68, 69].

12.9.2 Challenges and Opportunities in Adoption

The main challenges in implementing eco-friendly corrosion protection involve meeting high performance and durability standards of the aerospace industry and the automotive industry, replacing efficiently performing but toxic chromate-based systems, and ensuring that these types of protection are affordable and manufacturable on a large scale [70, 71]. Nevertheless, the regulatory pressures, green inhibitors, smart coatings, advances in technology, and increasing market demand for sustainable industrial and lightweight materials are the opportunities that contribute to the research and deployment of safer and high-performance alternatives.

12.9.3 Roadmap for Future Sustainable Corrosion Protection

The future of sustainable corrosion protection emphasizes the need for further research on advanced, environmentally friendly materials, including lithium-based inhibitors, hybrid organic–inorganic coatings, and bio-based polymers, as well as environmentally friendly application processes aimed at optimization [72, 73]. Key trends include enhancing self-healing and multifunctionality of the coatings, scaling up potentially useful solutions in laboratories to enable their use in industry, and ensuring LCA and environmental impact analyses to ensure sustainability and compliance with regulatory requirements in both the aerospace and automotive sectors.

REFERENCES

1. Ms, D. (2019). Nanostructured coatings in aerospace and automotive industries. *Nanoscale Reports*, *2*(3), 33–36. https://doi.org/10.26524/nr1940
2. Jullien, B., & Pardi, T. (2013). Structuring new automotive industries, restructuring old automotive industries and the new geopolitics of the global automotive sector. *International Journal of Automotive Technology and Management*, *13*(2), 96. https://doi.org/10.1504/ijatm.2013.053000
3. Dangelmaier, W., Döring, A., & Lentes, J. (2009). New production strategies for automotive industries. *ATZproduktion Worldwide*, *2*(2), 30–33. https://doi.org/10.1007/bf03224194

4. Martínez Pérez, F. (2025). Corrosion, types and prevention. *Journal of Building Technology, 7*(1). https://doi.org/10.32629/jbt.v7i1.3273
5. Walsh, F., Barker, D., & Ottewill, G. (1993). Corrosion and protection of metals: II. Types of corrosion and protection methods. *Transactions of the IMF, 71*(3), 117–120. https://doi.org/10.1080/00202967.1993.11871002
6. Bhumika, T. (2023). Corrosion: Types, theories and preventive measure. *I-Manager's Journal on Material Science, 10*(4), 21. https://doi.org/10.26634/jms.10.4.19303
7. Kuzyukov, A. N. (1994). Types of intercrystalline corrosion fracture of metals. *Materials Science, 29*(6), 612–615. https://doi.org/10.1007/bf00561637
8. Farrara, R. A. (1984). *Cadmium Plating vs. Other Coatings to Prevent Corrosion.* Defense Technical Information Center. http://doi.org/10.21236/ada329594
9. Munger, C. G., & Vincent, L. D. (2014). *Corrosion Prevention by Protective Coatings.* Nace Park Ten Place Houstontx. https://doi.org/10.5006/37599
10. Schweitzer, P. (2003). Corrosion-resistant linings and coatings. *Anti-Corrosion Methods and Materials, 50*(3). https://doi.org/10.1108/acmm.2003.12850cae.001
11. Udoinyang, N. (2025). The environmental and economic impact of the state of emergency in Rivers State, Nigeria. *Radinka Journal of Science and Systematic Literature Review, 3* (1), 600–611. https://doi.org/10.56778/rjslr.v3i1.482
12. Ajusree, P. A., & Jenson, J. (2019). *Environmental and Economic Impact Assessment of Flooring Materials* (pp. 83–89). Springer. https://doi.org/10.1007/978-3-030-26365-2_8
13. Özeş Özgür, R., & Çağatay, S. (2022). Environmental and economic impact analysis of reducing greenhouse gas emissions via energy substitution and carbon tax. *Journal of Research in Economics, 1*(1), 46–60. https://doi.org/10.29228/jore.13
14. Grigoriev, D. O., Stepin, S. N., & Vakhitov, T. (2016). *Ferrites as Non-Toxic Pigments for Eco-Friendly Corrosion Protection Coatings.* 71–86. https://doi.org/10.1002/9781119185055.ch3
15. Abraham, D. A., Sanmugam, A., Karuppasamy, K., Kanagaraj, K., Alfantazi, A., Kim, H.-S., Vikraman, D., & I. Almansour, (2025). Eco-friendly corrosion protection for carbon steel: A study on natural inhibitors in artificial seawater and acidic environments. *Journal of the Taiwan Institute of Chemical Engineers, 172*, 106112. https://doi.org/10.1016/j.jtice.2025.106112
16. Tejero-Martin, D., Rezvani Rad, M., Mcdonald, A., & Hussain, T. (2019). Beyond traditional coatings: A review on thermal-sprayed functional and smart coatings. *Journal of Thermal Spray Technology, 28*(4), 598–644. https://doi.org/10.1007/s11666-019-00857-1
17. Jamalizadeh, E., Parsa, M., Mohammad Ali Hosseini, S., & Hassani, Z. (2014). Corrosion protective properties of coatings doped with inhibitors. *Anti-Corrosion Methods and Materials, 61*(6), 416–422. https://doi.org/10.1108/acmm-07-2013-1281
18. Balali-Mood, M., Tahergorabi, Z., Sadeghi, M., Khazdair, M. R., & Naseri, K. (2021). Toxic mechanisms of five heavy metals: Mercury, lead, chromium, cadmium, and arsenic. *Frontiers in Pharmacology, 12*(2). https://doi.org/10.3389/fphar.2021.643972
19. Diouf, A., Diouf, A. S., Fall, J., Ndour, M. M., & Ndong, D. (2019). Evaluation of heavy metals contamination level (mercury, lead, cadmium) in fishery products exported from Senegal. *Journal of Food Studies, 8*(1), 49. https://doi.org/10.5296/jfs.v8i1.15261
20. Wanhill, R., & Windisch, M. (2018). *Corrosion and Stress Corrosion Testing of Aerospace Vehicle Structural Alloys* (pp. 1–63). Springer. https://doi.org/10.1007/978-3-319-89530-7_1
21. Salunkhe, S., & Rajamani, D. (2023). 3- Current trends of metal additive manufacturing in the defense, automobile, and aerospace industries. In *Advances in Metal Additive Manufacturing* (pp. 147–160). Elsevier. https://doi.org/10.1016/b978-0-323-91230-3.00004-4

22. Royce, R. (2023). Texas's water future: Legal, business, environmental, and regulatory concerns. *SMU Science and Technology Law Review*, *26*(2), 413. https://doi.org/10.25172/smustlr.26.2.7
23. Mathew, J. T., Kanwa, A. M., Musah, M., Hussaini, J., Shaba, E. Y., Mamman, A., Ismail, H., Tanko, M. S., Inobeme, A., Azeh, Y., Muhammad, E. B., & Adetunji, C. O. (2024). *Properties of Nanomaterials for Corrosion Control* (pp. 74–99). IGI Global. https://doi.org/10.4018/979-8-3693-3088-3.ch004
24. Wicks, Z. W. (2003). *Organic Coatings for Corrosion Control.* https://doi.org/10.1002/0471238961.1518070123090311.a01
25. Bierwagen, G. P. (1998). *Organic Coatings for Corrosion Control.* American Chemical Society. https://doi.org/10.1021/bk-1998-0689
26. Sastri, V. (2015). *Corrosion Control and Prevention* (pp. 205–316). https://doi.org/10.1002/9781119069638.ch4
27. Fayomi, O. S. I., Emmanuel, A. O., Agboola, O., & Akande, I. G. (2020). Challenges of coatings in aerospace, automobile and marine industries. *AIP Conference Proceedings*, *2307*(1). https://doi.org/10.1063/5.0033579
28. Yimyai, T., Rohwerder, M., & Crespy, D. (2023). Corrosion-responsive self-healing coatings. *Advanced Materials*, *35*(47), 2300101. https://doi.org/10.1002/adma.202300101
29. Yabuki, A. (2015). Self-healing coatings for corrosion inhibition of metals. *Modern Applied Science*, *9*(7), 214. https://doi.org/10.5539/mas.v9n7p214
30. Kartsonakis, I. A., Kanellopoulou, I. A., & Kontiza, A. (2024). Advanced micro/nanocapsules for self-healing coatings. *Applied Sciences*, *14*(18), 8396. https://doi.org/10.3390/app14188396
31. Wang, W., Hou, F.-Y., Guo, H.-T., & Wang, H. (2005). Fabrication and characterization of Ni–ZrO2 composite nano-coatings by pulse electrodeposition. *Scripta Materialia*, *53* (5), 613–618. https://doi.org/10.1016/j.scriptamat.2005.04.002
32. Gu, Y., Mou, J., Zheng, S., Wu, D., & Xia, K. (2020). Technical characteristics and wear-resistant mechanism of nano coatings: A review. *Coatings*, *10*(3), 233. https://doi.org/10.3390/coatings10030233
33. Froitzheim, J., & Svensson, J.-E. (2011). Multifunctional nano-coatings for SOFC interconnects. *Electrochemical Society Transactions*, *35*(1), 2503–2508. https://doi.org/10.1149/1.3570248
34. Malic, E., & Knorr, A. (2013). *Graphene and Carbon Nanotubes.* Wiley. https://doi.org/10.1002/9783527658749
35. Pénicaud, A. (2014). *Solubilization of Fullerenes, Carbon Nanotubes, and Graphene* (Vol. 348, pp. 1–35). Springer Berlin Heidelberg. https://doi.org/10.1007/128_2013_520
36. Chen, X., Zhong, J., Xiao, Z., & Li, A. (2018). Graphene and carbon nanotubes in space. *Proceedings of the International Astronomical Union*, *14*(A30), 403–404. https://doi.org/10.1017/s1743921319004940
37. Sauber, M. E. (2018). *Oxidative Removal of Cerium from Rare Earth Elements Mixed Chloride Solution* (pp. 2753–2764). Springer. https://doi.org/10.1007/978-3-319-95022-8_233
38. Sedneva, T. A. (2005). Cerium(III) electrooxidation in nitric acid solutions of rare-earth elements. *Russian Journal of Applied Chemistry*, *78*(6), 907–911. https://doi.org/10.1007/s11167-005-0418-5
39. Karve, M., & Vaidya, B. (2008). Selective separation of scandium(III) and yttrium(III) from other rare earth elements using Cyanex302 as an extractant. *Separation Science and Technology*, *43*(5), 1111–1123. https://doi.org/10.1080/01496390801887435
40. Sikdar, S., Menezes, P. L., Menezes, P. V., Jacob, T., & Maccione, R. (2021). Plasma electrolytic oxidation (PEO) process-processing, properties, and applications. *Nanomaterials*, *11*(6), 1375. https://doi.org/10.3390/nano11061375

41. Venugopal, A., Ramesh Narayanan, P., Srinath, J., Sharma, S. C., & George, K. M. (2015). Corrosion and multi-scale mechanical behaviour of plasma electrolytic oxidation (PEO) and hard anodized (HA) coatings on AA 2219 aluminum alloy. *Materials Science Forum, 830–831*, 627–630. https://doi.org/10.4028/www.scientific.net/msf.830-831.627
42. Menke, J. (2002). *In-Process Corrosion Control.* 1–12. https://doi.org/10.5006/c2002-02136
43. Dalmoro, V., Dos Santos, J. H. Z., & Santos, C. (2016). *Smart Coatings for Corrosion Protection* (pp. 417–435). Springer. https://doi.org/10.1007/978-3-319-26893-4_20
44. Beyrami, H., Golshan, M., Zardehi-Tabriz, A., & Salami-Kalajahi, M. (2025). Smart coatings: Fundamentals, preparation approaches, and applications. *Advanced Materials Technologies*. https://doi.org/10.1002/admt.202500574
45. Calle, L., Hintze, P., Buhrow, J., & Li, W. (2010, June 26). *Smart Coatings for Autonomous Corrosion Detection and Control.* https://doi.org/10.2514/6.2010-8877
46. Hart-Smith, J. (2021). 23 - Aerospace industry applications of adhesive bonding. In *Adhesive Bonding* (pp. 763–800). Elsevier. https://doi.org/10.1016/b978-0-12-819954-1.00001-0
47. Gao, Y., Wang, J., Deng, Z., Wang, Y., Zhang, D., Xu, X., Yu, X., & Wei, X. (2024). Targeted delivery of 2D composite minerals for biofilm removal. *ACS Applied Materials & Interfaces, 16*(39), 52814–52823. https://doi.org/10.1021/acsami.4c10998
48. Sanni, O., Sunday Isaac Fayomi, O., & Patricia Idowu Popoola, A. (2019). Eco-friendly inhibitors for corrosion protection of stainless steel: An overview. *Journal of Physics: Conference Series, 1378*(4), 042047. https://doi.org/10.1088/1742-6596/1378/4/042047
49. Jauhar, S. K., & Pant, M. (2016). *Sustainable Supplier Selection: A New Differential Evolution Strategy with Automotive Industry Application* (pp. 353–371). Springer. https://doi.org/10.1007/978-3-319-32229-2_25
50. Pereira, G. B., De Andrade Silva, E., Lima, M., De Jesus, B. J. S., Carvalho Benini, K. C., Montoro, S. R., Pereira, G. C., & Bandeira, C. F. (2017). Featuring high impact polystyrene composites strengthened with green coconut fiber developed for automotive industry application. *Journal of Research Updates in Polymer Science*, *6*(1), 17–20. https://doi.org/10.6000/1929-5995.2017.06.01.3
51. Ramesh, P., & Nataraj, M. (2020). Automotive industry application of aluminium-based hybrid metal matrix composite. *International Journal of Heavy Vehicle Systems, 1*(1), 1. https://doi.org/10.1504/ijhvs.2020.10020255
52. Hesnaoui, K., & Prenger, F. (2022). *Innovative and Sustainable Corrosion Protection Solutions for Foundation Structures of Offshore Wind Turbines.* 1–6. https://doi.org/10.5006/c2022-17526
53. Pinhey, S., & Tebbs, M. (2022). *Case Studies.* (pp. 146–156). CABI. https://doi.org/10.1079/9781789243604.0011
54. Rehan, R. (2015). Planning for an eco-city to sustainable urban environments international case studies. *Journal of Urban Research, 17*(1), 117–131. https://doi.org/10.21608/jur.2015.87794
55. Garcia-Galvan, F. R., López Sánchez, J., Barranco, V., Llorente, I., Fajardo, S., Galvan, J. C., & Carmona, N. (2024). Towards sustainable corrosion protection: An investigation into ecofriendly sol-gel conversion coatings for AZ61 magnesium alloy. *Electrochemical Society Meeting Abstracts, MA2024-02*(17). https://doi.org/10.1149/ma2024-02171694mtgabs
56. Garcia-Galvan, F. R., Páez-Pavón, A., Barranco, V., Llorente, I., Romero Martin, M., Carmona, N., Feliu, S., Fajardo, S., & Galvan, J. C. (2024). Development of green conversion sol-gel coatings for corrosion protection of AZ61 magnesium alloy. *Electrochemical Society Meeting Abstracts, MA2024-02*(13). https://doi.org/10.1149/ma2024-02131571mtgabs

57. Stubbs, S., Khan, I., & Yousaf, S. (2022). A review on the synthesis of bio-based surfactants using green chemistry principles. *DARU Journal of Pharmaceutical Sciences*, *30*(2), 407–426. https://doi.org/10.1007/s40199-022-00450-y
58. Bhandari, S. (2018). Applications of green chemistry principles in agriculture. *Green Chemistry & Technology Letters*, *4*(2), 10–12. https://doi.org/10.18510/gctl.2018.422
59. Herrchen, M., & Klein, W. (2000). Use of the life-cycle assessment (LCA) toolbox for an environmental evaluation of production processes. *Pure and Applied Chemistry*, *72*(7), 1247–1252. https://doi.org/10.1351/pac200072071247
60. Malviya, R. K., Singh, R. K., & Purohit, R. (2021). *Life–Cycle Assessment (LCA) of Plastic–Wood Composites* (pp. 233–253). Springer Singapore. https://doi.org/10.1007/978-981-16-1606-8_12
61. Shevchenko, V. Y., Kochina, T. A., Shilova, O. A., Barinova, L. D., & Belyi, O. V. (2019). Environmentally friendly protective coatings for transport. *Herald of the Russian Academy of Sciences*, *89*(3), 279–286. https://doi.org/10.1134/s1019331619030080
62. Vasenkov, A. V. (2021). Multi-physics peridynamic modeling of damage processes in protective coatings. *Journal of Peridynamics and Nonlocal Modeling*, *3*(2), 167–183. https://doi.org/10.1007/s42102-020-00046-7
63. Rohith, P., Bharath, M., Prabakaran, M., & Vaardini, M. (2023). Predictive maintenance for construction equipment using artificial intelligence and machine learning. *International Journal of Advanced Research in Science, Communication and Technology*, 531–537. https://doi.org/10.48175/ijarsct-14363
64. Ramalingam, R., Chandrasekar, T., Vaishnavi, S., Priyanka, N., Nalini, S., & Murugesan, M. (2024). *The Usage of Artificial Intelligence in Manufacturing Industries* (pp. 94–112). IGI global. https://doi.org/10.4018/979-8-3693-2615-2.ch006
65. Javaid, M., Rab, S., Haleem, A., Singh, R. P., & Suman, R. (2021). Role of additive manufacturing applications towards environmental sustainability. *Advanced Industrial and Engineering Polymer Research*, *4*(4), 312–322. https://doi.org/10.1016/j.aiepr.2021.07.005
66. Daareyni, A., Pagone, E., Thayapararajah, S., Mokhtarian, H., Tosello, G., & Flores Ituarte, I. (2025). Intelligent manufacturing paradigms: Linking design optimization and sustainability in large-area additive manufacturing. *The International Journal of Advanced Manufacturing Technology*. https://doi.org/10.1007/s00170-025-15832-0
67. Bhatt, J., & Singh, S. (2025). Sustainable innovation through digital transformation: Leveraging AI for circular economy models. *International Journal of Advanced Research in Science, Communication and Technology*, 485–490. https://doi.org/10.48175/ijarsct-24869
68. Wang, D., Zhang, F., Shi, Y., Pan, J., Claesson, P., & An, R. (2024). Recent findings on lignin-based wear and corrosion resistance coatings. *Corrosion Reviews*, *43*(2), 157–174. https://doi.org/10.1515/corrrev-2024-0087
69. Ibrahim, O. S., Jatau, A. E., Anaisi, S. M., Aliyu, Y., & Adekeye, I. M. (2025). Enhancing durability and adhesion of Moringa oleifera as green corrosion inhibitor on mild steel using lignin-based polymer from coconut shell. *FUOYE Journal of Engineering and Technology*, *9*(4), 725–730. https://doi.org/10.4314/fuoyejet.v9i4.23
70. Ricardo, J.-G., Joao, S., Carlos, A., & Elsa, M.-J. (2014). *A Requirements Engineering Methodology for Technological Innovations Assessment*. 577–586. https://doi.org/10.3233/978-1-61499-440-4-577
71. Domínguez-Martínez, J., Carmona, N., Serrano, A., Galván, J. C., García-Galván, F., López-Sánchez, J., Barranco, V., & Rodríguez De La Fuente, Ó. (2024). Eco-friendly sol-gel coatings with organic corrosion inhibitors for lightweight AZ61 alloy. *Gels*, *10*(3), 168. https://doi.org/10.3390/gels10030168

72. De Souza, N. R. D., Matt, L., Sedrik, R., Vares, L., & Cherubini, F. (2023). Integrating ex-ante and prospective life-cycle assessment for advancing the environmental impact analysis of emerging bio-based technologies. *Sustainable Production and Consumption*, *43*, 319–332. https://doi.org/10.1016/j.spc.2023.11.002
73. Zoubi, W. A., Min, J. H., & Ko, Y. G. (2017). Hybrid organic-inorganic coatings via electron transfer behaviour. *Scientific Reports*, *7*(1). https://doi.org/10.1038/s41598-017-07691-x

13 Advanced Green Coating and Inhibitor Technologies for Corrosion Resistance

Priya Kesari

13.1 INTRODUCTION

In the automotive and aerospace industries, where structural metals are frequently subjected to harsh service conditions that can jeopardise their mechanical integrity, dependability, and aesthetics, corrosion is still a serious problem. The need for materials like magnesium, aluminium, copper, and high-strength steels has grown abruptly as a result of the push for lightweighting in contemporary transportation systems, as shown in Figure 13.1(a–d). In addition, there are various factors like purity of metal, temperature of working environment, etc. that affect corrosion and are shown in Figure 13.1(e). Although the strength-to-weight ratios and fuel efficiency benefits of these materials make them desirable, they are naturally vulnerable to corrosion of various types, like pitting, intergranular corrosion, crevice corrosion, and stress–corrosion cracking. High-altitude UV radiation, abrupt temperature changes, de-icing agents, saline exposure, and mechanical loading are some of the severe operating conditions that components in aerospace applications must withstand. Due to their frequent exposure to road salts, acid rain, urban pollutants, water intrusion, and temperature fluctuations, automotive components are also particularly susceptible, particularly in areas with cold climates. Inadequate corrosion protection can lead to premature failure of structural elements and subassemblies, higher maintenance expenses, and safety hazards, which will ultimately impact operational effectiveness and end-user satisfaction.

In recent decades, a substantial amount of research has been conducted to find more ecologically friendly alternatives to conventional corrosion protection systems that are solvent-rich and chromate-based. Waterborne coatings, which employ water as the main solvent and significantly reduce emissions of volatile organic compounds (VOCs), are among the most researched green alternatives. Modern waterborne polyurethane dispersions are a good option for environmentally friendly automotive and aerospace applications since they can now achieve mechanical and chemical resistance qualities comparable to solvent-based systems, according to ref. [2]. It has been shown that silica-based nanocomposites added to waterborne polyurethane coatings exhibit

DOI: 10.1201/9781003653226-13

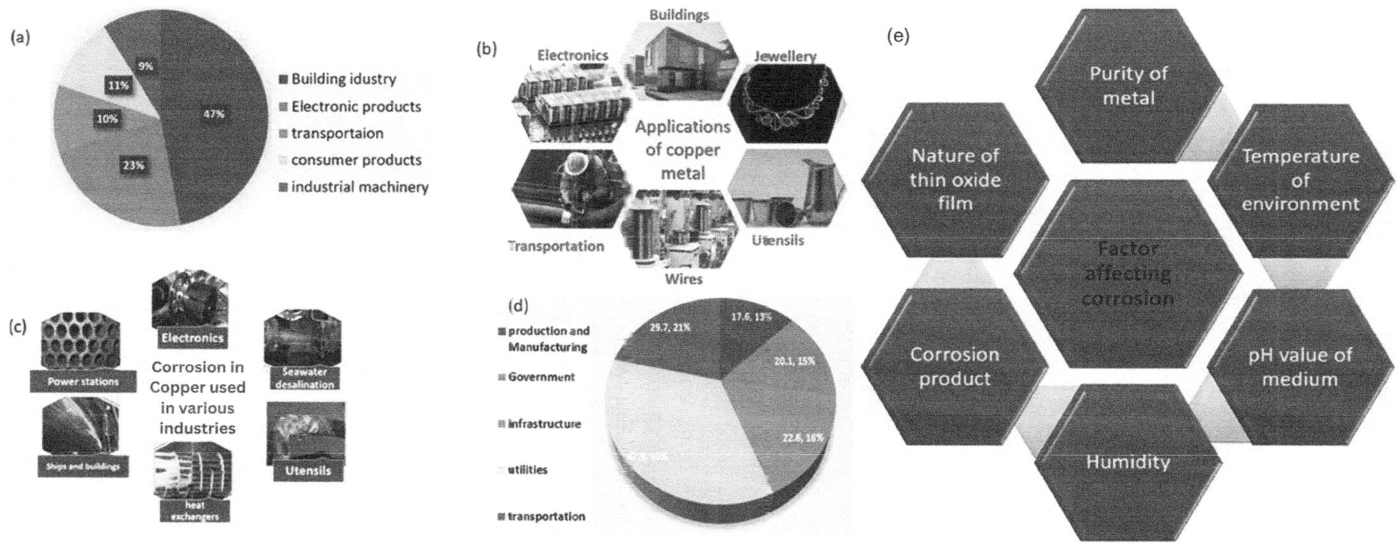

FIGURE 13.1 Overview of copper utilisation and corrosion-related issues: (a) Copper consumption across various sectors, (b) Major applications of copper metal, (c) Corrosion of copper in different industrial environments, (d) Direct economic losses attributed to copper corrosion, and (e) Representation of key factors influencing the corrosion process [1].

outstanding electrochemical impedance performance, which is directly related to increased corrosion resistance [3]. Another promising approach to corrosion protection is the use of sol–gel and hybrid organic–inorganic coatings. In order to create dense, cross-linked networks that successfully prevent moisture and chloride intrusion, these coatings use silane chemistry [4]. Using amino-functionalised silane significantly increased corrosion resistance in aluminium alloys when sol–gel coatings were studied. By adding layered double hydroxides (LDHs), which serve as nanocontainers for corrosion inhibitors and have self-healing qualities, further enhanced sol–gel coatings [5]. In addition, studies have examined how solvents affected sol–gel porosity and pointed out how crucial solvent choice is for maximising barrier qualities [6, 7]. Chromate-free conversion coatings (CFCs) are becoming more popular because there are laws against hexavalent chromium. Zirconium-based coatings, which make thin, passive oxide layers, have become the new standard in automotive pre-treatment processes because they don't make a lot of sludge, use less energy, and improve paint adhesion [8]. It has been found that zirconium and titanium salt-based multi-metallic conversion coatings protect against corrosion just as well as traditional CFCs on 2024-T3 aluminium alloys. Cerium-based systems have also shown promise, especially in aerospace applications, highlighting their ability to passivate cathodic intermetallic particles and reduce galvanic corrosion [9, 10]. Bio-based corrosion inhibitors are a long-lasting option for synthetic organic and inorganic inhibitors [11]. Studies on plant-based compounds like neem, henna, and green tea have shown that they can stop corrosion on mild steel and aluminium substrates with an efficiency of over 90%. These inhibitors mainly work by physisorption and chemisorption, which create protective films that keep electrolytes from penetrating the system. Scientists have also investigated chitosan, a natural biopolymer made from chitin. Chitosan can make hydrogen-bonded networks with epoxy resins, enhancing film formation and improving corrosion resistance, according to ref. [12–14]. Nanocomposite coatings containing graphene oxide (GO), silicon dioxide (SiO_2), and titanium dioxide (TiO_2) are becoming more popular because they can enhance coating strength, improve thermal stability, and increase corrosion resistance [15]. CVD graphene layers on polymeric coatings made them much more resistant to corrosion, giving them protection for more than 120 days in marine environments [16]. Hybrid siloxane–epoxy coatings have been developed using GO nanosheets functionalised with TEOS. These coatings had higher impedance values and lower corrosion current densities. In addition, hexagonal boron nitride (h-BN) nanosheets are used and to have similar benefits; however, proper dispersion control and functionalisation are important factors [17, 18].

As a passive corrosion prevention technique, superhydrophobic coatings have also been studied, drawing inspiration from the lotus effect. By employing low-surface-energy materials and hierarchical surface roughness, these coatings prevent water and electrolyte contact with the substrate [19]. Using silica–titania nanoparticles, we created superhydrophobic coatings that showed exceptional resistance to salt spray on mild steel. The majority of researchers concur, nevertheless, that superhydrophobic coatings need more work before being widely used because they have durability issues when subjected to mechanical abrasion [20].

Smart coatings, which autonomously respond to damage or environmental stimuli, represent a rapidly evolving field. Redox-mediated protection is provided by the conductive polymers polypyrrole (PPy) and polyaniline (PANI). PANI was added to epoxy matrices, giving them the ability to detect corrosion and heal themselves [21]. Additionally, the use of microencapsulation techniques has made it possible to incorporate healing agents that are released upon coating damage, such as epoxy monomers and di-cyclopentadiene [22]. It also showed that inhibitors can be successfully released in response to pH variations, providing active corrosion protection [23]. These technologies have already started to be adopted in industry. Sol–gel and zirconium-based pre-treatments have been used on a number of aircraft platforms by Airbus and Boeing. Boeing is investigating cerium-based inhibitors and encapsulated healing agents, while Airbus employs trivalent chromium sol–gel pre-treatments to lessen reliance on Cr VI. OEMs in the automotive industry, such as Ford and General Motors, have implemented conversion coatings based on zirconium, which have improved energy efficiency and decreased environmental impact. Green manufacturing techniques are further supported by the use of powder coatings and waterborne electrocoats. Despite these developments, there are still challenges. In terms of long-term durability, many environmentally friendly systems still fall short of chromates, particularly in harsh industrial or marine environments. Research is being conducted on coating-substrate adhesion, polymer compatibility, nanomaterial dispersion, and large-scale process integration. Furthermore, to measure the actual environmental benefits of these green technologies, thorough lifecycle assessments (LCAs) are required. In the future, it is anticipated that the combination of artificial intelligence and machine learning will speed up the process of finding new materials and optimising formulations. The next generation of sustainable protective systems will be defined by smart coatings with multifunctional qualities that combine hydrophobicity, self-healing, conductivity, and inhibition. For protection against corrosion in aerospace and automotive applications to be truly environmentally friendly, it will be essential to integrate recyclability, carbon footprint analysis, and supply chain sustainability.

In the context of aerospace and automotive applications, this chapter provides a comprehensive review of environmentally acceptable corrosion protection solutions by integrating a variety of techniques, from waterborne and sol–gel coatings to smart polymers and biodegradable inhibitors. By connecting scientific advancement with real-world application, it distinguishes itself by emphasising not only the technical capabilities of green coatings but also their compatibility with environmental sustainability objectives. The chapter also highlights new developments like low-VOC formulations, multifunctional materials, and the use of digital technologies like additive manufacturing and artificial intelligence to speed up the creation of new materials. The chapter offers a forward-looking viewpoint that significantly advances sustainable engineering techniques by addressing both the existing constraints and the potential for the future.

13.2 CORROSION CHALLENGES IN AEROSPACE AND AUTOMOTIVE SECTORS

13.2.1 Corrosion Challenges in the Aerospace Sector

In the aircraft industry, corrosion poses serious problems because of the harsh conditions and the need for structural integrity. UV rays, extreme temperature swings, high humidity, salty surroundings, and deicing salts are among a few of the severe circumstances that aircraft materials must withstand. Particularly for high-strength steels, magnesium alloys, and aluminium alloys used in airframes, landing gear, and structural fasteners, these circumstances hasten the corrosion process [24]. Galvanic corrosion, stress-corrosion cracking, intergranular corrosion, and pitting are some of the most frequent deterioration mechanisms seen in aircraft structures, as shown in Figure 13.2(a–e) [25, 26]. Newer lightweight materials are also more vulnerable to chemical and atmospheric deterioration because they frequently lack well-established protective coatings [27].

Additionally, corrosion can result in catastrophic failures if it is not properly controlled, and the aircraft industry requires significantly longer service life [26]. Because of this, aerospace protection systems need to minimise unwanted weight while maintaining reliable, consistent functioning over a long period of time. This is crucial for both aircraft performance and fuel efficiency [28]. Developing efficient corrosion-resistant systems is even more crucial since maintenance and inspection cycles must be kept to a minimum while upholding strict safety standards. Surface coatings and treatments need to be highly durable, work well in a range of operating conditions, be compatible with different substrates, and be resistant to heat cycling. Microbiologically influenced corrosion (MIC) is one of the emerging corrosion

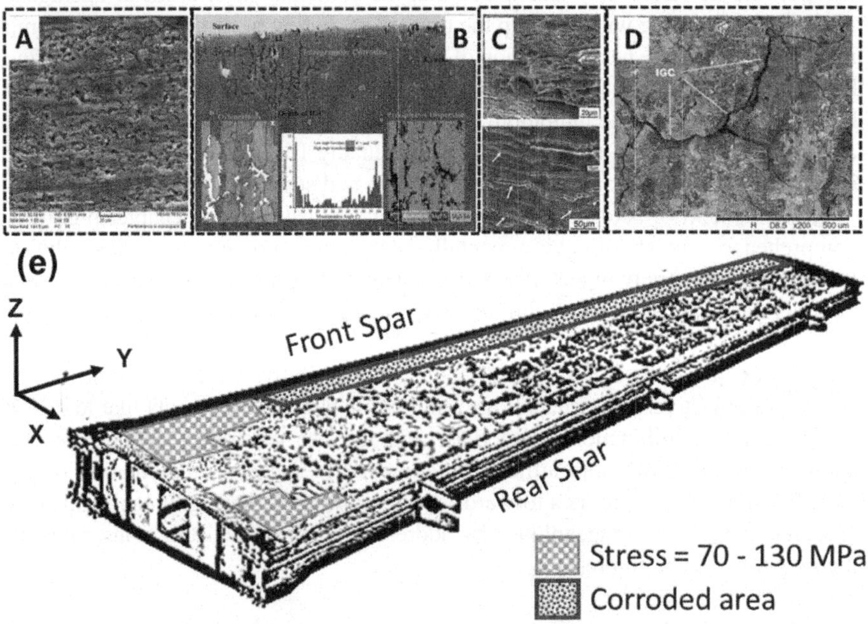

FIGURE 13.2 (a) Typical pit morphology observed on the surface of AA6063 aluminium alloy, (b) intergranular corrosion in 5XXX and 6XXX series aluminium alloys, (c) fracture morphologies and side surface cracks in 2024-T351 aluminium alloy, (d) top-view SEM images of alloy 2060 after 96 hours of exfoliation corrosion testing, and (e) schematic of a horizontal stabiliser indicating corroded regions and areas subjected to principal stresses ranging from 70 to 130 MPa [25].

hazards that has drawn more attention. MIC involves the growth of microorganisms that can produce corrosive by-products, inducing pitting, and degrade materials, especially in fuel tanks and fluid transfer systems [29]. Because MIC has been identified in aviation fuel systems, more research is being done on antimicrobial coatings and surface patterns that prevent the production of biofilms.

13.2.2 Corrosion Challenges in the Automotive Sector

Numerous corrosion problems are prevalent in the automotive sector as a result of exposure to road salts, acid rain, humidity, ultraviolet light, and temperature fluctuations. The vehicle's underbody, brake lines, exhaust systems, battery system, and wheel arches are among the areas where these problems most commonly cause structural and aesthetic deterioration, as shown in Figure 13.3(a and b). Car reliability is still affected by corrosion, even with increased use of protective coatings and galvanisation, particularly in areas experiencing harsh winters. In high-strength steel and aluminium components, road salts speed up localised corrosion, especially at weld zones and dissimilar metal interfaces [30, 31].

In order to comply with international laws like Euro 7 and CAFE standards, automakers are facing mounting pressure to lower the weight of their vehicles in order to enhance fuel efficiency and reduce emissions. As a result, lightweight materials with particular corrosion issues, such as composites, magnesium, and aluminium, have become more popular [32]. Despite its natural passivation, aluminium is vulnerable to pitting in conditions high in chlorides. However, magnesium has low galvanic compatibility and general corrosion resistance. Metal fasteners are frequently used for composites, which results in galvanic cells that accelerate deterioration in wet conditions [33].

The popularity of electric vehicles (EVs) raises additional corrosion-related issues. Special protection is needed for components such as battery enclosures, bus bars, connectors, and power electronics to prevent water intrusion, electrolytic corrosion, and chemical leakage from battery materials [34]. Battery housing is made of lightweight materials, but exposure to road conditions necessitates coatings with the ability to withstand corrosion, electrical insulation, and chemicals. Additionally, EVs remove conventional rust-prone components like exhaust systems and internal combustion engines, but they also create new areas in electronic assemblies that are susceptible to corrosion.

Because hexavalent chromium-based treatments are carcinogenic and subject to stringent regulation under the REACH and RoHS directives, the automobile industry is also under significant regulatory and sustainability pressure to phase them out [35]. However, under demanding service circumstances, conventional technologies such as solvent-based paints, chromate conversion coatings, and cadmium plating still perform better than green alternatives. Finding alternatives that satisfy performance requirements and environmental laws has become a technical problem for manufacturers and researchers.

Vehicle corrosion is also impacted by urban environmental deterioration and climate change. Rising sulphate, chloride, and particulate matter increase the corrosion risk for vehicles in urban or coastal environments, making corrosion problems worse in real-world settings [36].

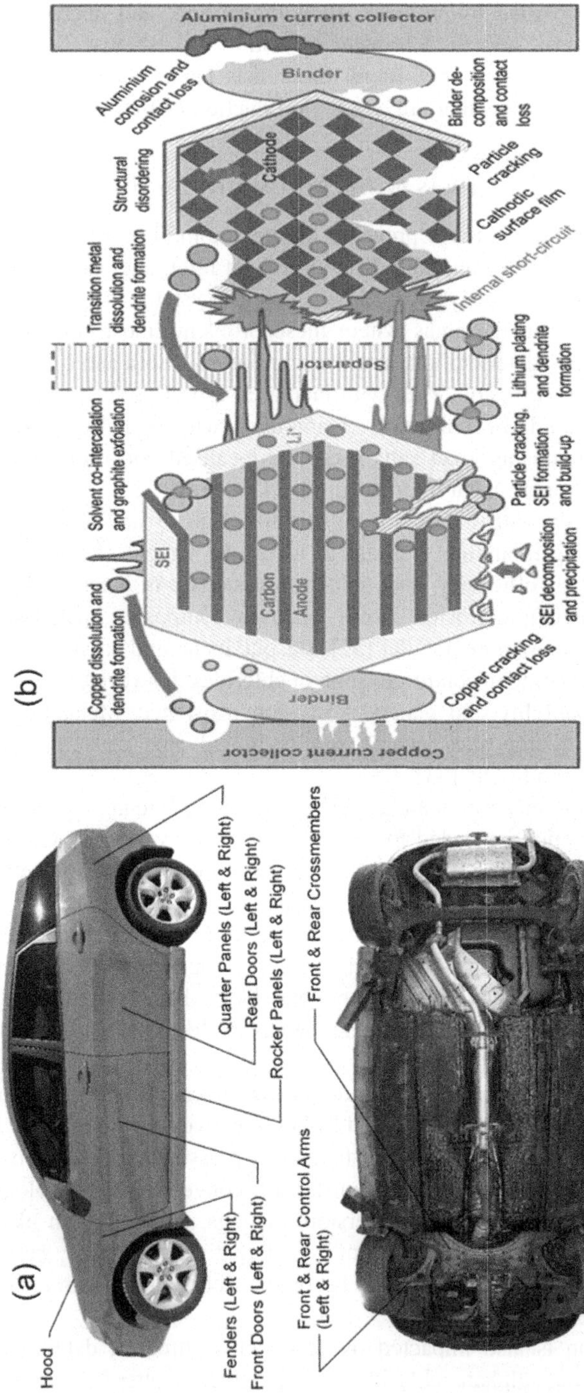

FIGURE 13.3 (a) Corrosion-prone body panels and underbody components of an automobile commonly analysed in automotive corrosion studies, and (b) various internal degradation mechanisms occurring within battery systems [30, 31].

13.3 LIMITATIONS OF CONVENTIONAL PROTECTION METHODS

In the aerospace and automotive industries, traditional corrosion protection techniques like solvent-based paints, chromate conversion coatings, cadmium plating, and metallic sacrificial coatings have long offered strong defences against deterioration. However, there are a number of drawbacks to these approaches, especially when considering sustainability, material compatibility, performance lifespan, and regulatory compliance. Similar environmental and health issues are present with cadmium plating, another popular aerospace technique, particularly regarding bioaccumulation and occupational exposure risks. In addition to restricting their usage, these environmental obligations raise the expense of management and disposal. Newer lightweight materials like magnesium alloys and sophisticated composites are also frequently incompatible with traditional methods. Because their electrochemical characteristics are not compatible, many conventional coatings either cause galvanic corrosion or fail to adhere correctly to these substrates [37]. Although they are used extensively, solvent-based paints greatly increase emissions of VOCs, worsening environmental pollution and posing health hazards to workers during application and drying. As companies shift to low-VOC or waterborne alternatives, the use of these coatings is receiving increasing scrutiny. Performance of numerous conventional coatings deteriorates when exposed to UV light for extended periods of time, heat cycling, salt spray, or mechanical wear—all of which are frequent occurrences in automotive and aerospace settings. This calls for regular upkeep, recoating, or component replacement, all of which have expensive lifecycle costs. Multifunctional qualities that are becoming more and more desirable in contemporary protection systems, such as self-healing, antibacterial activity, or hydrophobicity, are often lacking in conventional barrier coatings. The incapacity of traditional coatings to incorporate intelligent or responsive capabilities is another drawback. Conventional techniques provide very little active defence against corrosion once the barrier is broken; they mostly rely on passive protection. Their long-term dependability is lowered by this restriction, particularly in locations that are challenging to reach for maintenance and inspection. The transition to eco-friendly, multifunctional, and high-performance alternatives is necessary due to the inherent environmental hazards, limited substrate compatibility, regulatory challenges, and performance limitations of conventional protection systems (Figure 13.4(a)), despite their instrumental role in corrosion management.

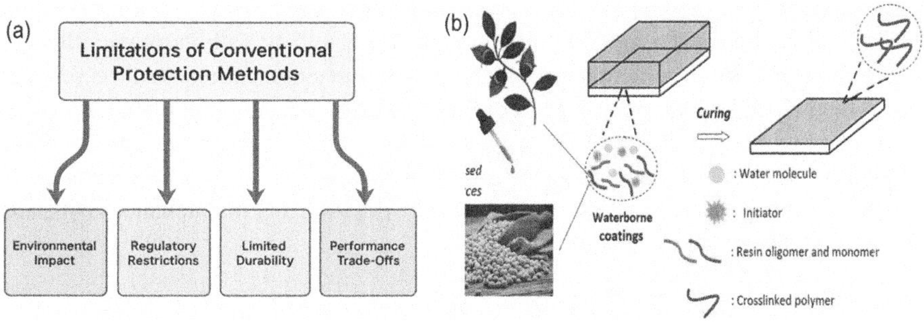

FIGURE 13.4 (a) illustration of the key limitations and challenges in conventional coating methods and (b) Schematic representation of synthesis routes for bio-based coatings in waterborne systems utilising renewable monomers.

13.4 ECO-FRIENDLY CORROSION PROTECTION STRATEGIES

The goal of environmentally friendly corrosion protection techniques is to reduce hazards to human health and the environment while offering an efficient defence against material deterioration. By using sustainable, non-toxic, and biodegradable substitutes that adhere to contemporary environmental requirements like REACH, RoHS, and EPA standards, these initiatives aim to replace dangerous materials like hexavalent chromium and cadmium. Smart multifunctional coatings, sol–gel technology, nanostructured materials, aqueous coatings, and green inhibitors are important strategies. Waterborne coatings (Figure 13.4(b)) have become popular in the automotive and aerospace industries because they minimise or completely eradicate VOCs. These coatings reduce hazardous emissions during application by using water as a solvent and using environmentally benign resins and pigments to preserve adhesion and protective qualities. Similarly, sol–gel-derived coatings have high thermal stability, low environmental effect, and excellent corrosion resistance, particularly those based on siloxane and organically modified silicates. For lightweight metal substrates such as magnesium and aluminium, they are perfect because of their capacity to produce multifunctional, dense, and adherent layers. An efficient and sustainable technique for lowering corrosion rates is the use of green corrosion inhibitors made from plant extracts, amino acids, and biodegradable polymers. These compounds function by adhering to the metal surface and creating a barrier that prevents corrosive agents from reaching the substrate. Among these are extracts from *Lawsonia inermis* (henna), *Azadirachta indica* (neem), and amino acid-based inhibitors such as glutamine and cysteine, which have demonstrated significant effectiveness in settings with high chloride environments. Eco-friendly coating research also heavily relies on nanotechnology. Graphene oxide, nano-TiO_2, and clay platelets are examples of nanoparticles that can be added to organic or hybrid matrices to enhance barrier qualities, decrease permeability, and provide hydrophobic or self-healing behaviour. These nanoparticles successfully postpone the onset of corrosion by forming convoluted pathways for ion diffusion. To provide targeted and long-lasting protection, smart coatings, including pH-sensitive or redox-sensitive nanocontainers, can also release corrosion inhibitors in response to environmental stimuli. Bio-based polymers, including cellulose derivatives, polylactic acid (PLA), and chitosan, are also being researched as coating matrices or binders because of their natural resistance to corrosion, biodegradability, and film-forming capabilities. To improve their protective qualities even more, these materials can be mixed with inorganic fillers or nanostructures. Environmental compliance, materials engineering, surface science, and green chemistry all come together to form environmentally acceptable corrosion protection techniques. In high-performance sectors, these coatings not only meet the urgent need to replace dangerous conventional coatings but also make it possible to integrate intelligent features, increasing service life, lowering maintenance costs, and promoting sustainable development.

13.4.1 Waterborne Coatings

Waterborne coatings are becoming a key component of green corrosion protection technologies because of their reduced flammability, better handling, and low emissions of VOCs. Instead of using conventional hydrocarbons as the principal carrier solvent, these coatings use water, which improves worker safety during the application and curing processes while also lowering environmental pollution [38]. In the automotive and aerospace industries, where extensive coating treatments are frequently required due to complex geometries and huge surface areas, this feature is especially beneficial.

Contemporary waterborne formulations usually include sophisticated resins designed for superior mechanical performance and chemical resistance, such as polyurethanes, acrylics, and epoxies. Aluminium, magnesium, and steel are common substrates in lightweight automotive and aerospace components, and these polymers show excellent adherence to them. Furthermore, curing speeds and performance under challenging conditions have increased due to advancements in crosslinking chemistry, such as self-crosslinkable systems and those activated by temperature or ambient humidity [28]. The effective addition of corrosion inhibitors without compromising dispersion stability has been one of the significant technical turning points in the development of aqueous coatings. Anodic and cathodic protection mechanisms have been effectively provided by the integration of inhibitors like zinc phosphate, calcium molybdate, cerium nitrate, and organic thiols. These elements slow down the rate at which metal dissolves by interfering with electrochemical corrosion reactions. Furthermore, coatings' long-term durability under immersion and salt spray conditions has been further enhanced by the use of organosilanes, which serve as barrier enhancers and adhesion boosters [39]. Metal-rich primers used in the automotive industry are especially well-suited for waterborne epoxy coatings because of their hardness, chemical resistance, and ease of film production. Recent advancements in nanotechnology have made it possible to enhance barrier qualities by using nanoparticles like titanium dioxide, graphene oxide, and nanoclays. By creating tortuous pathways for ions and moisture, these nanoparticles considerably lower the coating's permeability and postpone the onset of corrosion [40]. Bio-based binders made from renewable resources like cellulose, chitosan, and soybean oil can be used to manufacture aqueous coatings from a sustainability perspective. These solutions adhere to the concepts of the circular economy and offer equivalent performance. Furthermore, scientists are looking into incorporating self-healing microcapsules into waterborne matrices that, in response to mechanical damage or pH changes, can release film-forming agents or corrosion inhibitors, prolonging service life without the need for reapplication [41].

The primary obstacles that still exist include obtaining strong corrosion resistance without post-curing treatments and guaranteeing constant performance in extremely changeable environmental circumstances, like extreme cold or humidity. However, hybrid methods that improve mechanical durability and interfacial bonding by combining aqueous organic topcoats with inorganic sol–gel underlayers are showing promise in this area. Waterborne coatings, in summary, are an established, adaptable, and ecologically friendly technology that is becoming more and more essential for contemporary corrosion prevention systems. 3D surface morphology and profile of the PTFE-ATP-SiO_2 composite waterborne hydrophobic coating, along with polarisation curve analysis and electrochemical impedance spectroscopy (EIS) results at different immersion times, highlighting the coating's corrosion resistance

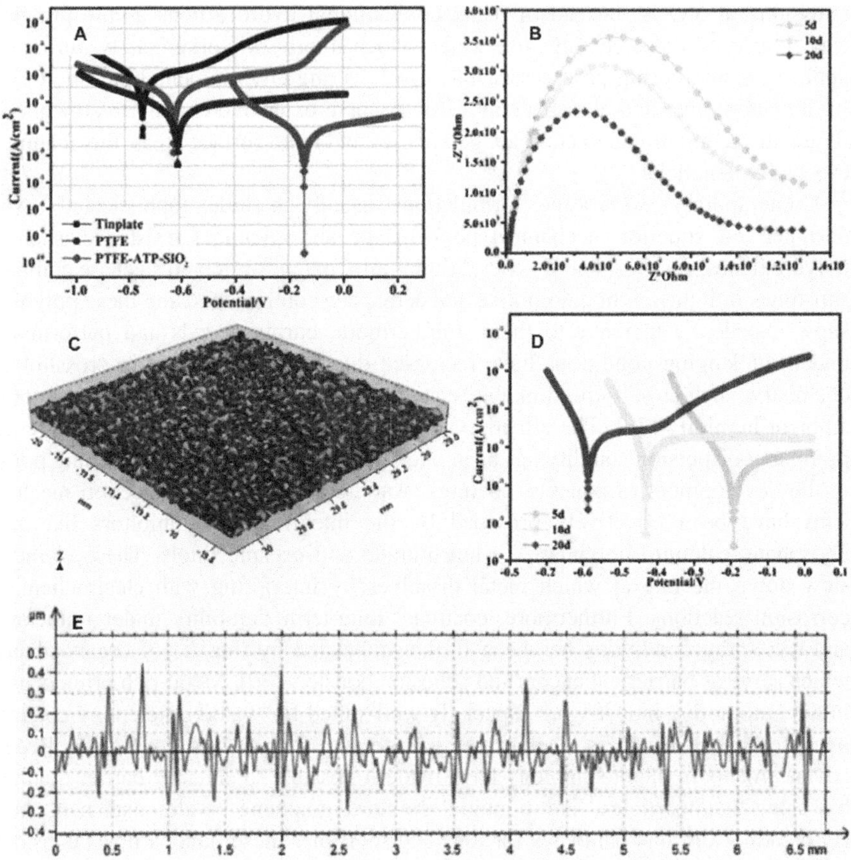

FIGURE 13.5 Morphology of the PTFE-ATP-SiO_2 composite coating: (a) 3D surface morphology of the composite coating, (b and c) Profile of the composite coating, (d) Polarisation curve test, (e) Polarisation curve of the PTFE-ATP-SiO_2 composite coating for different times, and (f) EIS of the PTFE-ATP-SiO_2 composite coating for different times [42].

performance, as shown in Figure 13.5(a–f). Their ability to support bio-based chemistries, smart responses, and functional additives puts them in line with the demands of the automotive and aerospace industries for performance and sustainability in the future [42].

13.4.2 Sol–Gel and Hybrid Coatings

The molecular-level design, high barrier qualities, and environmentally benign synthesis of sol–gel and hybrid coatings make them a revolutionary development in corrosion protection technology (Figure 13.6(h)). In order to create a dense inorganic or organic–inorganic network, metal alkoxides or organosilanes are hydrolysed and condensed to create sol–gel coatings. Because of their exceptional mechanical stability, strong adhesion, and resilience to chemical and thermal stresses,

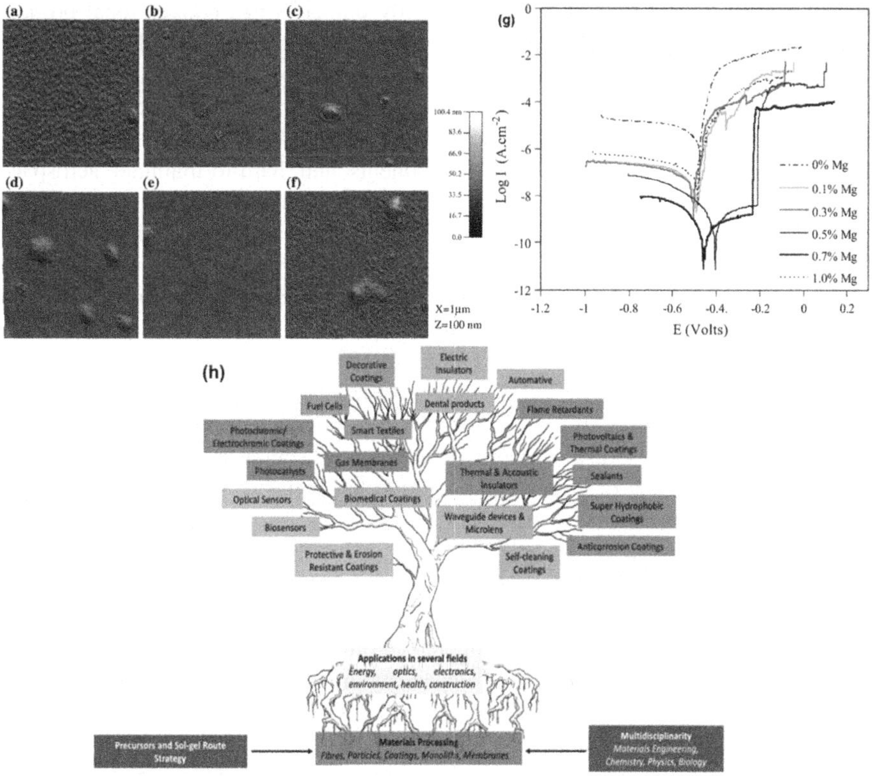

FIGURE 13.6 Atomic force microscopy (AFM) image of sol–gel coating of (a) 0% Mg, (b) 0.1% Mg, (c) 0.3% Mg, (d) 0.5% Mg, (e) 0.7% Mg, (f) 1.0% Mg. (g) Potentiodynamic polarisation curves of sol–gel coating immersed in electrolyte (Harrison's solution); (h) strategy and application of sol–gel in different sectors [43, 44].

these coatings are particularly well-suited for use in automotive and aerospace applications [43]. Sol–gel coatings' capacity to create nanostructured films that serve as efficient physical barriers against moisture, oxygen, and corrosive ions is one of its main benefits, as shown in Figure 13.6(a–g). Functional silanes, including glycidyl oxy-propyl trimethoxy silane (GPTMS) and 3-aminopropyltriethoxysilane (APTES), are added to improve adhesion and crosslinking with metallic substrates [44].

By combining organic polymers (such as epoxy, polyurethane, and acrylic) with inorganic siloxane or silica networks, hybrid sol–gel coatings combine the greatest qualities of both materials: the flexibility and durability of polymers with the thermal stability and barrier performance of ceramics. Automotive components that are subjected to mechanical abrasion, temperature changes, and chemical exposure are especially pertinent to these systems. These hybrid films are resistant to UV deterioration, delamination, and cracking due to the organic-inorganic synergy [45]. Nanomaterials are used in corrosion mitigation, such as silica nanoparticles and nanoclays in sol–gel matrices, to enhance the mechanical and barrier qualities of the substrate.

These nanostructured coatings have frequently outperformed conventional primers and paints in EIS evaluations and salt spray testing [46]. Hybrid sol–gel coatings have also been modified for self-healing uses. These coatings can react to damage and regain their protective function on their own by embedding microcapsules or nanocontainers that are filled with corrosion inhibitors [47]. This method is particularly useful for automobile chassis components and hard-to-maintain aerospace structures that are subjected to environmental stresses and cyclic loading. Crucially, sol–gel coatings are usually applied utilising straightforward and scalable methods like spin, spray, or dip coating. They also cure at very low temperatures, which makes them suitable for sensitive substrates and manufacturing limitations. They are considered a sustainable substitute for chromate-based systems, as they produce little hazardous waste and adhere to laws limiting VOCs and heavy metals [48]. For long-lasting corrosion protection in high-performance settings, sol–gel and hybrid coatings present an appealing, environmentally responsible option. Because of their versatility, multipurpose use, and adherence to green chemistry principles, they are an essential component of protective coating technologies of the future.

13.4.3 Chromate-Free Conversion Coatings

As environmental concerns and hexavalent chromium regulations tighten, CFCs have become essential substitutes for conventional chromate-based systems. Since the aerospace and automotive industries mainly depend on the surface passivation of aluminium and magnesium alloys to guarantee long-term durability and corrosion resistance, these environmentally friendly coatings are especially pertinent to these industries.

By creating an insoluble, sticky, and chemically bound coating on the substrate surface, CFCs prevent electrolyte access and inhibit the electrochemical processes that cause corrosion, as shown in Figure 13.7(a–e) and its polarisation curve in

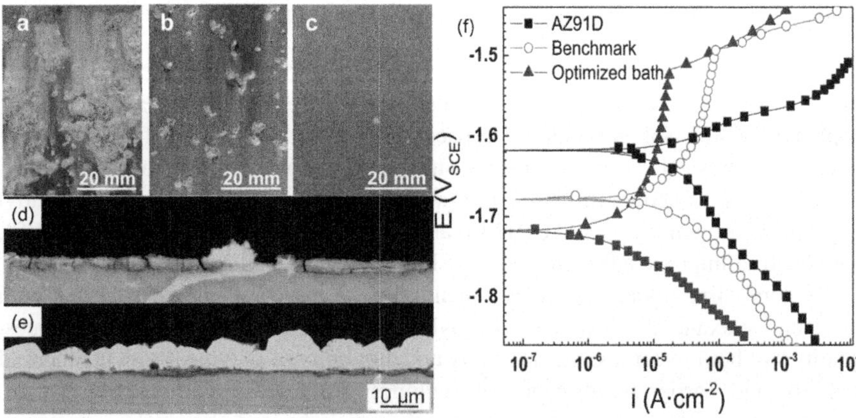

FIGURE 13.7 Macroscopic morphology of the (a) bare and PCCs fabricated from the (b) benchmark and (c) optimised baths after the salt spray test for 120 h. (d) Cross-sectional morphology of the PCCs fabricated from the benchmark and (e) optimised baths. (f) Polarisation curves of the bare alloy and alloys coated with different PCC in 3.5 wt% NaCl solution [49].

Figure 13.7(f). Rare earth metals (including cerium, yttrium, and lanthanum), zirconium, titanium, molybdate, vanadate, and silanes are common foundations for these coatings. Through anodic inhibition or passivation, they provide active protection in addition to barrier qualities. By limiting electron flow and preventing additional corrosion propagation, cerium-based coatings—one of the most researched systems—offer localised self-healing by precipitating cerium hydroxide or oxide into cathodic sites. This behaviour is especially helpful for pitting-prone aerospace-grade aluminium alloys (AA2024-T3). To attain ideal results, pH, precursor concentration, and deposition time can also be changed to modify the shape and thickness of these coatings. Because of their high coverage uniformity, room-temperature functioning, and compatibility with environmentally friendly pre-treatment procedures, conversion coatings based on zirconium and titanium have become more and more common in automobile production lines. These coatings are strong adhesion promoters for subsequent paint layers, forming thin (usually <100 nm), amorphous, and adherent films on steel, aluminium, and magnesium, and galvanised substrates. Zirconium/titanium systems significantly lower sludge production, water use, and hazardous waste disposal as compared to conventional phosphate conversion coatings [49, 50]. Significantly lower environmental and occupational health risks, minimal energy needs, and simplicity of application through spray, immersion, or brush techniques are some of the main benefits of CFCs. These technologies can be easily integrated into industrial processes because they are compatible with the existing painting and surface preparation lines. CFCs are set to become a common solution in the green manufacturing paradigm with further advancements in precursor chemistry, process optimisation, and cooperative pairings with other protective technologies (such as sol–gel and nanoparticle reinforcing).

13.4.4 Natural and Biodegradable Inhibitors

As environmentally friendly and non-toxic substitutes for conventional synthetic substances, natural and biodegradable corrosion inhibitors are becoming more and more popular. These inhibitors, which are mostly derived from plant extracts, animal proteins, or microbial metabolites, offer the dual advantages of reducing corrosion and protecting the environment. Natural chemicals are appealing for use in the environmentally sensitive aerospace and automotive industries because they are biodegradable and frequently break down into innocuous by-products, unlike heavy metal salts or petroleum-derived inhibitors. The schematic of traditional strategies for corrosion mitigation and emerging, more sustainable approaches is shown in Figure 13.8. The presence of alkaloids, tannins, flavonoids, and polyphenols that adsorb onto the metal surface and create protective coatings has led to the substantial anticorrosive capabilities of plant-based inhibitors, particularly those extracted from neem, henna, green tea, garlic, and tobacco leaves. In order to prevent metal from interacting with aggressive ions like chlorides, these chemicals mainly work by chemically or physically adsorbing on active corrosion sites. One example is the successful application of neem (Azadirachta indica) extract on mild steel in acidic environments, which has been shown to reduce corrosion rates [51, 52]. Proteins like casein and gelatin, as well as amino acids, have demonstrated effectiveness in

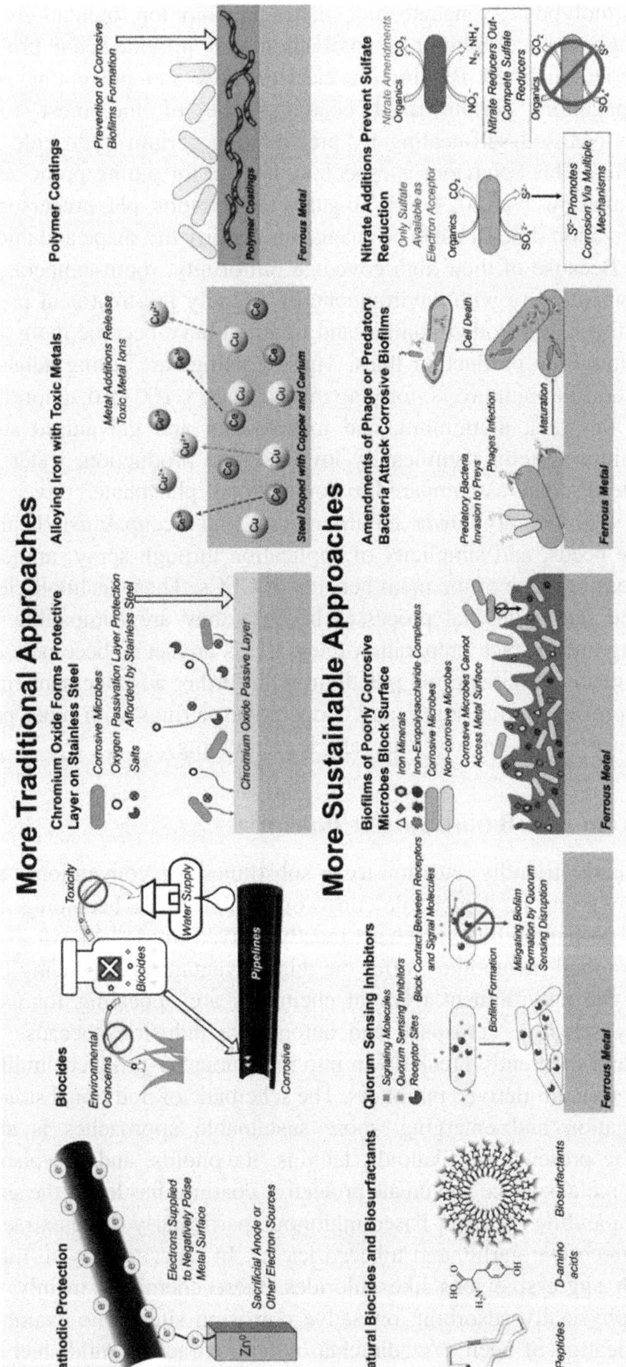

FIGURE 13.8 Traditional strategies for corrosion mitigation and emerging, more sustainable approaches that show promise but require further experimental evaluation [51].

creating complex layers with metallic cations. These biological macromolecules prevent corrosion through pH buffering, film generation, and chelation. In systems where electrolyte compositions vary, like coolant systems in automobile engines or hydraulic lines in aeroplanes, their uses are particularly promising [53]. Recent studies have demonstrated that polysaccharides, including starch, chitosan, and gum Arabic, serve as effective biodegradable corrosion inhibitors. These materials have good water solubility, biodegradability, and film-forming properties. With corrosion inhibition efficiency of 85% in saltwater settings, chitosan, a naturally occurring polymer derived from chitin, has been demonstrated to form protective coatings on steel and aluminium surfaces [54]. Furthermore, the potential of microbial corrosion inhibitors such as biosurfactants and siderophores, which are produced from bacterial or fungal secretions, to chelate metal ions and interfere with microbial corrosion cycles is being investigated. These bio-based substances are becoming more and more popular for preventing MIC in subterranean pipelines, fuel systems, and other automotive parts [55]. Implementing natural inhibitors is primarily difficult due to their varied composition, extract standardisation, and temperature stability. But methods like encapsulating in micro/nano capsules and combining them with other eco-friendly materials (such as sol–gel matrices) are assisting in getting around these restrictions. In order to provide intelligent protection while complying with environmental regulations, these hybrid eco-coatings include biodegradable inhibitors in controlled-release systems. Multi-component blends, customised formulations, and cooperative relationships with other coating technologies are key to the future of natural corrosion inhibitors. As lifetime sustainability and the procurement of bio-based materials gain more attention, natural inhibitors are a revolutionary breakthrough in environmentally friendly corrosion protection.

13.4.5 Super Hydrophobic Coatings

Super hydrophobic coatings have garnered a lot of interest as a potential corrosion prevention strategy because of their ability to repel water and reduce electrolyte contact with metal surfaces. These coatings, which were inspired by natural systems like lotus leaves, have contact angles larger than 150° and sliding angles smaller than 10°. This effectively limits moisture adhesion, which promotes corrosion, and prevents the retention of water droplets. Additionally, the AMCF design demonstrates mechanical and environmental durability, showing changes in oil contact and sliding angles under various treatments such as abrasion, tape peeling, ultrasonic vibration, washing, UV exposure, freezing, and heating. In addition, oil repellence after deformation and the effect of press cycles on the water absorption behaviour of the inner layer are shown in Figure 13.9. A cushion of air is created between the substrate and the aqueous environment in super hydrophobic coatings to prevent the movement of corrosive ions like Cl^- and SO_4^{2-}. This is the mechanism underlying corrosion inhibition. Furthermore, the barrier properties of low surface energy compounds (such as fatty acids, waxes, or fluorinated silanes) are improved when combined with surface roughness [57].

Superhydrophobic coatings can retard the onset and spread of corrosion in the automotive and aerospace sectors, where parts are regularly subjected to harsh

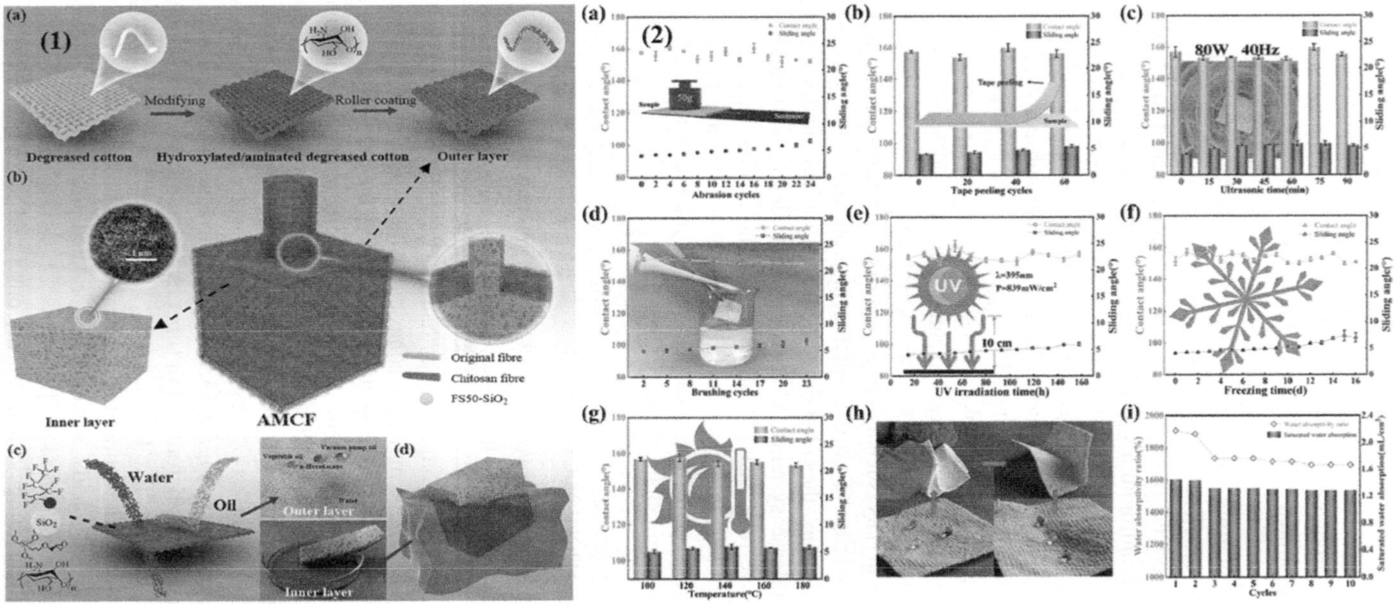

FIGURE 13.9 (1) Schematic of the AMCF design; (2) Evaluation of mechanical and environmental durability. This includes changes in oil contact and sliding angles after (a) mechanical abrasion, (b) tape peeling, (c) ultrasonic treatment, (d) toothbrush washing, (e) UV exposure, (f) freezing, and (g) thermal treatment. (h) Images showing oil repellence of the outer layer after bending, kneading, and twisting, and (i) the effect of press cycles on water absorption capacity and rate of the inner layer [56].

environmental factors such as high humidity and salt-laden atmospheres. Aluminium and magnesium alloys, for instance, have benefited greatly from the effective use of epoxy/silica nanocomposite, titanium nanocomposite films (Figure 13.10(a–g)), and fluorosilane-modified silica nanoparticles, which have been shown to improve corrosion resistance in immersion and salt spray tests [59]. New developments concentrate on creating eco-friendly superhydrophobic coatings with green modifiers such as bio-based silanes, tannic acid, or stearic acid. It has been demonstrated that certain formulations of these coatings maintain their superhydrophobic properties even after extended exposure to UV light and thermal cycling, giving them the benefits of high repellence and decreased environmental effect.

Mechanical stability and durability, however, continue to be the fundamental obstacles. When mechanically damaged, many superhydrophobic surfaces lose their effectiveness due to low abrasion resistance. To counter this, scientists have integrated embedded nanocapsules, dynamic bonding agents (such as Diels–Alder chemistry), or self-healing properties into the coating structure. In order to increase adhesion and endurance in real-world applications, hybrid designs that blend superhydrophobic layers with underlying sol-gel or epoxy primers have also been investigated. Overall, the use of superhydrophobic coatings as environmentally friendly corrosion protection using a ceramic composite is shown in Figure 13.10(h). With further advancements in fabrication methods and materials, these coatings could soon be a standard in high-performance, lightweight automotive and aerospace structures [56, 58].

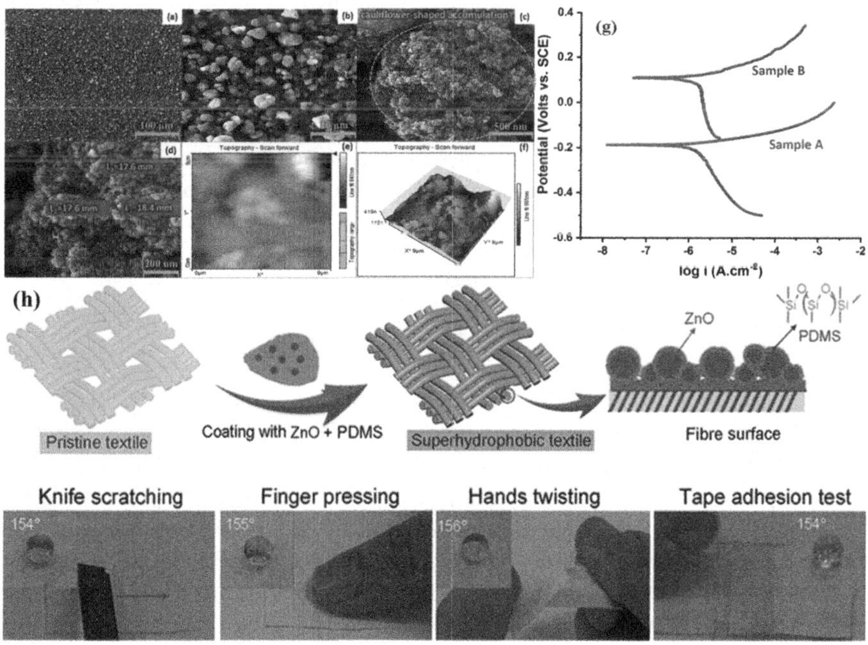

FIGURE 13.10 (a–d) FESEM images of the superhydrophobic nano-TiO_2/5 vol% TMPSi ceramic composite coating in different magnifications, and (e and f) AFM topography of the surface of the coating. (g) Potentiodynamic polarisation curves of the samples in the corrosive medium, 3.5 wt% NaCl solution [58].

13.4.6 Smart and Conductive Polymeric Coatings

An inventive method of environmentally friendly corrosion prevention is represented by smart and conductive polymeric coatings, which incorporate electrical conductivity, responsiveness, and active functionality into the protective layer. In aerospace and automotive applications, where weight reduction, material lifetime, and real-time corrosion monitoring are critical, these coatings are especially beneficial. The word "smart" describes coatings that can detect mechanical damage or changes in the environment and react on their own by repairing or altering their characteristics to reduce corrosion. Due to their capacity to passivate metal surfaces, preserve electrochemical stability, and facilitate charge transport mechanisms that lessen oxidative degradation, conductive polymers like polyaniline (PANI), polypyrrole (PPy), and polythiophene (PTh) have become prominent corrosion-inhibiting substances. These substances inhibit ion exchange and anodic dissolution of the substrate by fostering the development of passive oxide coatings and forming a physical barrier [60]. Maintaining the metal surface in a passivated state even after physical damage is one of the primary mechanisms by which conductive polymeric coatings prevent corrosion. Aluminium, steel, and magnesium alloys have all benefited from the successful application of polyaniline in its emeraldine salt form, which is particularly effective in neutral and acidic settings [61]. According to recent research, under accelerated corrosion testing, PANI-doped hybrid coatings lower corrosion current densities by several orders of magnitude [62]. The synergistic use of conductive polymers with sol–gel matrices, nanoclay, or graphene oxide enhances mechanical integrity and corrosion resistance, while reducing the need for hazardous pigments and solvents. These multifunctional hybrid coatings have been evaluated using EIS, Scanning Vibrating Electrode Technique (SVET), and salt spray tests, showing significant improvement in barrier and self-repairing properties [63]. In summary, smart and conductive polymeric coatings not only reduce corrosion but also provide real-time monitoring, damage sensing, and environmental adaptability, positioning them as next-generation alternatives to conventional protective systems in demanding transportation environments.

13.5 CASE STUDIES AND APPLICATIONS

13.5.1 Aerospace Industry

The aircraft sector has witnessed the practical application of environmentally friendly corrosion prevention techniques. By replacing risky chromate-based methods with sol–gel pretreatments in conjunction with aqueous primers, Airbus has drastically decreased its environmental impact. For aluminium alloys commonly used in fuselage and structural components, these sol–gel coatings—which are frequently based on silane chemistry—offer strong adhesion and superior corrosion protection [61]. In the aerospace sector, environmentally friendly corrosion prevention techniques have been used in practice. These sol–gel coatings, often based on silane chemistry, provide aluminium alloys—commonly employed in fuselage and structural components—with good adhesion and superior corrosion protection. Additionally, magnesium alloys, which are preferred in aircraft due to their potential for weight reduction, have demonstrated improved corrosion resistance and durability when coated with epoxy–silane hybrid coatings. Complex geometrical parts exposed to industrial and maritime environments can benefit from these hybrid systems' improved mechanical qualities under fatigue loading and higher performance in salt spray tests [64]. The use of low-VOC aqueous formulas has also become more common in interior aircraft coatings. In addition to meeting fire resistance requirements, these coatings guarantee lower emissions both during application and curing. For example, cabin interiors have embraced UV-curable coatings based on acrylic dispersions, which offer rapid curing cycles and great durability while still meeting strict toxicity and smoke requirements [65].

13.5.2 Automotive Industry

The application of green corrosion protection technology in the automotive sector has advanced significantly. For car body construction, Ford and General Motors have replaced phosphate-based pretreatments with zirconium-based alternatives. This move reduces the amount of water and energy used in the coating process, as well as the production of sludge, a by-product that needs to be treated as hazardous waste. Furthermore, hybrid coating systems are frequently utilised for EV battery enclosures, aluminium body panels, and underbody components. These systems combine epoxy–siloxane or acrylic resins with nanostructured additives. Even under cyclic thermal and environmental stressors, these coatings provide enhanced corrosion protection, mechanical endurance, and chip resistance [1]. These procedures are part of a larger movement in both sectors to combine sustainability objectives with new material technologies, balancing environmental responsibility, cost-effectiveness, and performance.

13.6 CHALLENGES AND FUTURE DIRECTIONS

Eco-friendly corrosion protection technologies have shown promise, but a number of obstacles still stand in the way of their widespread use in the automotive and aerospace sectors. Achieving performance equivalence with traditional chromate-based systems is one of the biggest obstacles. Many ecologically friendly alternatives perform poorly when exposed to complicated real-world circumstances that include dynamic mechanical stress, UV exposure, and extended salt spray, despite being effective in controlled laboratory settings. Another challenge is their high cost. The production of advanced hybrid and nanocomposite coatings often involves expensive raw materials and complex manufacturing techniques. Process simplification and cost-effective scaling techniques are necessary for broad industrial adoption. Furthermore, there is ongoing research on integrating many functions—like sensing, self-healing, and intelligent inhibitor release—while preserving mechanical robustness. Significant obstacles also exist in the areas of standardisation and regulatory compliance. There aren't many well-recognised methods for assessing these coatings' long-term performance and environmental safety. To create standardised testing procedures and certification standards, industry, academia, and regulatory agencies must work together.

Future studies should concentrate on multipurpose coating systems that combine energy efficiency, recyclability, life-cycle sustainability, and corrosion resistance.

Optimised coating formulations for certain substrates and operating situations can be designed more quickly with the help of AI-driven materials discovery platforms and machine learning. Furthermore, advancements in additive printing may enable the direct deposition of environmentally friendly, functionally graded coatings onto intricate parts. The future of corrosion protection will ultimately depend on integrating surface science, polymer chemistry, materials engineering, and environmental policy to create solutions that satisfy the changing performance, safety, and environmental requirements of next-generation automotive and aerospace systems.

13.7 CONCLUSION

In the automotive and aerospace industries, environmentally friendly corrosion prevention technologies have become an essential part of sustainable materials engineering. Traditional chromate-based coatings and VOC-rich systems are clearly shifting to greener alternatives at an accelerating rate due to strict environmental restrictions, health and safety concerns, and global sustainability obligations. Among them are aqueous coatings, sol–gel and hybrid systems, bio-based and biodegradable inhibitors, superhydrophobic surfaces, and smart polymeric coatings. Their goal is to reduce environmental impact while maintaining or enhancing performance requirements.

The shift from invention at the laboratory scale to industrial implementation is already underway. To increase material durability, lower environmental impact, and guarantee regulatory compliance, automakers like Ford and GM, as well as aerospace behemoths like Airbus, are already incorporating this technology. Achieving cost-effective scalability, guaranteeing durability across a range of operational settings, and harmonising evaluation criteria are three areas where practical deployment still faces difficulties.

In the future, interdisciplinary cooperation, improved materials design supported by artificial intelligence, and the adoption of sophisticated manufacturing technologies like additive manufacturing will be necessary for the successful deployment of next-generation corrosion prevention. These initiatives must come together to create coatings that are recyclable, multipurpose, energy-efficient, and resistant to corrosion.

In conclusion, environmentally friendly corrosion prevention is a proactive approach to innovation and leadership in the materials industry, not merely a reaction to regulatory pressure. The aerospace and automotive sectors may transition to more robust, economical, and ecologically conscious material solutions with sustained research, industry involvement, and policy alignment.

REFERENCES

1. Mamgain, H. P., Pati, P. R., Samanta, K. K., Brajpuriya, R., Gupta, R., Pandey, J. K., . . . & Kanan, M. (2024). A review on bio-inspired corrosion resistant superhydrophobic coating on copper substrate: Recent advances, mechanisms, constraints, and future prospects. *Results in Engineering*, 25, 103868.
2. Mucci, V. L., Aranguren, M. I., Amalvy, J. I., & Hormaiztegui, M. E. (2022). Recent developments in waterborne polyurethanes for coating applications. *Eco-Friendly Waterborne Polyurethanes*, 1, 253–266.

3. Ammar, S., Ramesh, K., Ma, I. A. W., Farah, Z., Vengadaesvaran, B., Ramesh, S., & Arof, A. K. (2017). Studies on SiO2-hybrid polymeric nanocomposite coatings with superior corrosion protection and hydrophobicity. *Surface and Coatings Technology*, 324, 536–545.
4. Rêgo, G. C., Bandeira, R. M., van Drunen, J., Tremiliosi-Filho, G., & Casteletti, L. C. (2023). Multi-layer organic-inorganic hybrid anticorrosive coatings for protection of medium carbon steel. *Materials Chemistry and Physics*, 303, 127841.
5. Alibakhshi, E., Ramezanzadeh, B., & Mahdavian, M. (2021). Self-healing materials in corrosion protection. *Self-Healing Smart Materials and Allied Applications*, 1, 247–296.
6. Antonietti, M., Fechler, N., & Fellinger, T. P. (2014). Carbon aerogels and monoliths: Control of porosity and nanoarchitecture via sol–gel routes. *Chemistry of Materials*, 26(1), 196–210.
7. Zhang, L., Zhang, J., & Liu, R. (2015). Organic–inorganic hybrid sol–gel coatings for corrosion protection of aluminum alloys. *Surface Innovations*, 4(2), 51–69.
8. Chopra, I., Ola, S. K., & Dhayal, V. (2025, January). A comprehensive analysis of advancements in corrosion prevention methods for carbon steel structures: A review. In AIP Conference Proceedings (Vol. 3191, No. 1, p. 020005). AIP Publishing LLC.
9. Yi, A., Du, J., Wang, J., Mu, S., Zhang, G., & Li, W. (2015). Preparation and characterization of colored Ti/Zr conversion coating on AZ91D magnesium alloy. *Surface and Coatings Technology*, 276, 239–247.
10. Zhu, H., & Li, J. (2024). Advancements in corrosion protection for aerospace aluminum alloys through surface treatment. *International Journal of Electrochemical Science*, 19(2), 100487.
11. Al-Amiery, A. A., & Al-Azzawi, W. K. (2023). Organic synthesized inhibitors for corrosion protection of carbon steel: A comprehensive review. *Journal of Bio- and Tribo-Corrosion*, 9(4), 74.
12. Peter, A., & Sharma, S. K. (2021). Plant-based green corrosion inhibition: Need and applications. In Anjali Peter and Sanjay K. Sharma (eds.), *Corrosion Protection of Metals and Alloys Using Graphene and Biopolymer Based Nanocomposites* (pp. 53–63). Boca Raton, FL: CRC Press.
13. Holla, B. R., Mahesh, R., Manjunath, H. R., & Anjanapura, V. R. (2024). Plant extracts as green corrosion inhibitors for different kinds of steel: A review. *Heliyon*, 10(14).
14. Dave, P. N., & Macwan, P. M. (2024). Phytochemicals/plant extracts as corrosion inhibitors for brass in various electrolytes. In Pragnesh N. Dave and Pradip M. Macwan (eds.), *Phytochemistry in Corrosion Science* (pp. 296–328). Boca Raton, FL: CRC Press.
15. Tayel, A., Ramadan, A. R., & El Seoud, O. A. (2018). Titanium dioxide/graphene and titanium dioxide/graphene oxide nanocomposites: Synthesis, characterization and photocatalytic applications for water decontamination. *Catalysts*, 8(11), 491.
16. Lin, W. H., Chen, T. H., Chang, J. K., Taur, J. I., Lo, Y. Y., Lee, W. L., . . . & Wu, C. I. (2014). A direct and polymer-free method for transferring graphene grown by chemical vapor deposition to any substrate. *ACSnano*, 8(2), 1784–1791.
17. Chen, X., Wen, S. F., Liu, Z. Q., & Yue, Z. F. (2021). Hybrid siloxane-epoxy coating reinforced by worm-like graphene oxide with improved mechanical properties and corrosion resistance. *Materials & Design*, 207, 109852.
18. Naclerio, A. E., & Kidambi, P. R. (2023). A review of scalable hexagonal boron nitride (h-BN) synthesis for present and future applications. *Advanced Materials*, 35(6), 2207374.
19. Parvate, S., Dixit, P., & Chattopadhyay, S. (2020). Superhydrophobic surfaces: Insights from theory and experiment. *The Journal of Physical Chemistry B*, 124(8), 1323–1360.
20. Lakshmi, R. V., & Aruna, S. T. (2017). Corrosion protection behaviour of silica–titania hybrid coatings embedded with silica nanoparticles. *Surface Engineering*, 33(6), 467–473.

21. Goswami, S., Nandy, S., Fortunato, E., & Martins, R. (2023). Polyaniline and its composites engineering: A class of multifunctional smart energy materials. *Journal of Solid State Chemistry*, 317, 123679.
22. Zhang, H., Xiao, K., Lin, Z., & Shi, S. (2022). Epoxy microcapsules for high-performance self-healing materials using a novel method via integrating electrospraying and interfacial polymerization. *Journal of Materials Science and Technology*, 112, 59–67.
23. Snihirova, D., Lamaka, S. V., Cardoso, M. M., Condeço, J. A., Ferreira, H. E., & de Fatima Montemor, M. (2014). pH-sensitive polymeric particles with increased inhibitor-loading capacity as smart additives for corrosion protective coatings for AA2024. *Electrochimica Acta*, 145, 123–131.
24. Melhem, G. N. (2018). Aerospace fasteners: Use in structural applications. In George Nadim Melhem (ed.), *Encyclopedia of Aluminum and Its Alloys, Two-Volume Set (Print)* (pp. 30–45). Boca Raton, FL: CRC Press.
25. Kamath, G. M., Mangalgiri, P. D., & Shet, A. (2022). A quantitative assessment of the impact of corrosion on fatigue life of aircraft components. *Engineering Failure Analysis*, 133, 105973.
26. Czaban, M. (2018). *Aircraft Corrosion–Review of Corrosion Processes and Its Effects in Selected Cases*. Fatigue of Aircraft Structures.
27. Andrady, A. L., Pandey, K. K., & Heikkilä, A. M. (2019). Interactive effects of solar UV radiation and climate change on material damage. *Photochemical & Photobiological Sciences*, 18(3), 804–825.
28. Bhat, B. N., Pandey, A. B., Tamirisakandala, S., Nathal, M. V., Ellis, D. L., McGill, P. B., ... & Zhu, D. (2018). *Aerospace Materials Characteristics (No.M17-6171)*. American Institute of Aeronautics and Astronautics.
29. Wang, J., Du, M., Li, G., & Shi, P. (2022). Research progress on microbiological inhibition of corrosion: A review. *Journal of Cleaner Production*, 373, 133658.
30. Sawyer-Beaulieu, S., Tam, E., & Hussein, A. (2022). Measuring corrosion on vehicles, in real-time, using digital imaging and analysis techniques. *Materials*, 15(9), 3053.
31. Kabir, M. M., & Demirocak, D. E. (2017). Degradation mechanisms in Li-ion batteries: A state-of-the-art review. *International Journal of Energy Research*, 41(14), 1963–1986.
32. Hu, H., Nie, X., & Ma, Y. (2014). Corrosion and surface treatment of magnesium alloys. *Magnesium Alloys-Properties in Solid and Liquid States*, 1, 68–83.
33. Montemor, M. F. (2016). Corrosion issues in joining lightweight materials: A review of the latest achievements. *Physical Sciences Reviews*, 1(2), 20150011.
34. Kaliaperumal, M., Dharanendrakumar, M. S., Prasanna, S., Abhishek, K. V., Chidambaram, R. K., Adams, S., ... & Reddy, M. V. (2021). Cause and mitigation of lithium-ion battery failure—A review. *Materials*, 14(19), 5676.
35. Bhatt, H. (2006). Alternatives to hexavalent chromium to comply with European Union's Directives (ELV, RoHS and WEEE). *Plating and Surface Finishing*, 93(10), 20–23.
36. Gupta, V. (2019). Vehicle-generated heavy metal pollution in an urban environment and its distribution into various environmental components. In Vertika Shukla and Narendra Kumar (eds.), *Environmental Concerns and Sustainable Development: Volume 1: Air, Water and Energy Resources* (pp. 113–127). Singapore: Springer Singapore.
37. Ortiz-Ozuna, A., Montoya-Rangel, M., Castaneda, H., Godínez, F. A., & Montoya, R. (2025). Corrosion modeling of Magnesium and its alloys: A short review. *Materials Today Communications*, 1, 112491.
38. Hamidi, N. A. S. M., Kamaruzzaman, W. M. I. W. M., Nasir, N. A. M., Shaifudin, M. S., & Ghazali, M. S. M. (2022). Potential application of plant-based derivatives as green components in functional coatings: A review. *Cleaner Materials*, 4, 100097.
39. Al-Saadi, S., & Singh Raman, R. K. (2022). Silane coatings for corrosion and microbiologically influenced corrosion resistance of mild steel: A review. *Materials*, 15(21), 7809.

40. Kumari, S., Saini, A., & Dhayal, V. (2021). Metal oxide based epoxy coatings for corrosion protection of steel. *Materials Today: Proceedings*, 43, 3105–3109.
41. Cerdan, K., Thys, M., Cornellà, A. C., Demir, F., Norvez, S., Vendamme, R., ... & Brancart, J. (2024). Sustainability of self-healing polymers: A holistic perspective towards circularity in polymer networks. *Progress in Polymer Science*, 152, 101816.
42. Ren, G., Qiao, Z., Hui, Z., Tuo, Y., Zheng, W., Chen, X., & Li, S. (2023). The waterborne superamphiphobic coatings with antifouling, high temperature resistance, and corrosion resistance. *ACS Omega*, 8(15), 13578–13592.
43. Figueira, R. B. (2020). Hybrid sol–gel coatings for corrosion mitigation: A critical review. *Polymers*, 12(3), 689.
44. Varma, P. R., Duffy, B., & Cassidy, J. (2009). Influence of magnesium nitrate on the corrosion performance of sol–gel coated AA2024-T3 aluminium alloy. *Surface and Coatings Technology*, 204(3), 277–284.
45. Huang, C., Liu, J., Zhao, L., Hu, N., & Wei, Q. (2023). Advances in atomic oxygen resistant polyimide composite films. *Composites Part A: Applied Science and Manufacturing*, 168, 107459.
46. Kiosidou, E. D., Karantonis, A., Sakalis, G. N., & Pantelis, D. I. (2018). Electrochemical impedance spectroscopy of scribed coated steel after salt spray testing. *Corrosion Science*, 137, 127–150.
47. Zhang, W., & Xu, J. (2022). Advanced lightweight materials for Automobiles: A review. *Materials & Design*, 221, 110994.
48. Jiménez-López, A. M., & Hincapié-Llanos, G. A. (2022). Identification of factors affecting the reduction of VOC emissions in the paint industry: Systematic literature review-SLR. *Progress in Organic Coating*, 170(106945), 10–1016.
49. Liao, S., Yu, B., Zhang, X., Lu, X., Zhou, P., Zhang, C., ... & Wang, F. (2021). New design principles for the bath towards chromate-and crack-free conversion coatings on magnesium alloys. *Journal of Magnesium and Alloys*, 9(2), 505–519.
50. Jawad, M., Zhao, J., Mubeen, M., Tabish, M., Mahmood, M., Murtaza, H., ... & Fan, B. (2025). Dual-functional sodium 1-dodecanesulfonate-enhanced TiZr conversion coatings on Zn-Al-Mg coated steel: Improved adhesion and filiform corrosion resistance. *Journal of Alloys and Compounds*, 1015, 178856.
51. Wang, D., Zhou, E., Xu, D., & Lovley, D. R. (2023). Burning question: Are there sustainable strategies to prevent microbial metal corrosion? *Microbial Biotechnology*, 16(11), 2026–2035.
52. Rani, B. A., & Basu, B. B. J. (2012). Green inhibitors for corrosion protection of metals and alloys: An overview. *International Journal of Corrosion*, 2012(1), 380217.
53. Kumar, S. B., Hossem, M. S., & Sayed, A. (2024). A comprehensive review of hydraulic systems in aerospace and construction engineering. *Control Systems and Optimization Letters*, 2(3), 266–273.
54. Aguilar-Ruiz, A. A., Dévora-Isiordia, G. E., Sánchez-Duarte, R. G., Villegas-Peralta, Y., Orozco-Carmona, V. M., & Álvarez-Sánchez, J. (2023). Chitosan-based sustainable coatings for corrosion inhibition of aluminum in seawater. *Coatings*, 13(9), 1615.
55. Sharma, P. (2022). Role and significance of biofilm-forming microbes in phytoremediation-a review. *Environmental Technology & Innovation*, 25, 102182.
56. Li, C., Long, C., Cai, Y., Li, Z., Guan, Z., Song, H., ... & Qing, Y. (2023). Design of amphoteric micro/nanofiber-coupling fabric for ultrafast, precise, and robust removal of water droplets from oils. *Chemical Engineering Journal*, 471, 144479.
57. Mohamed, A. M., Abdullah, A. M., & Younan, N. A. (2015). Corrosion behavior of superhydrophobic surfaces: A review. *Arabian Journal of Chemistry*, 8(6), 749–765.
58. Emarati, S. M., & Mozammel, M. (2020). Efficient one-step fabrication of superhydrophobic nano-TiO2/TMPSi ceramic composite coating with enhanced corrosion resistance on 316L. *Ceramics International*, 46(2), 1652–1661.

59. Soliman, H., Zhang, R., Cai, X., Feng, W., Alsarayreh, A. A., Hussain, A. A., & Alsadaie, S. (2025). Multifunctional superhydrophobic coatings for aluminum and magnesium alloys: Applications and performance-review. *Journal of Techniques*, 7(2), 83–100.
60. de Souza, S. (2007). Smart coating based on polyaniline acrylic blend for corrosion protection of different metals. *Surface and Coatings Technology*, 201(16–17), 7574–7581.
61. Snihirova, D., Lamaka, S. V., & Montemor, M. F. (2015) Smart composite coatings for corrosion protection of aluminium alloys in aerospace applications. In M.F. Montemor (ed.), *Smart Composite Coatings and Membranes: Transport, Structural, Environmental and Energy Applications* (pp. 85–121). Cambridge: Elsevier.
62. Ates, M. (2016). A review on conducting polymer coatings for corrosion protection. *Journal of Adhesion Science and Technology*, 30(14), 1510–1536.
63. Zhang, F., Ju, P., Pan, M., Zhang, D., Huang, Y., Li, G., & Li, X. (2018). Self-healing mechanisms in smart protective coatings: A review. *Corrosion Science*, 144, 74–88.
64. Firoozi, A. A., Firoozi, A. A., Oyejobi, D. O., Avudaiappan, S., & Flores, E. S. (2025). Enhanced durability and environmental sustainability in marine infrastructure: Innovations in anti-corrosive coating technologies. *Results in Engineering*, 105144.
65. Rengasamy, S., & Mannari, V. (2014). UV-curable PUDs based on sustainable acrylated polyol: Study of their hydrophobic and oleophobic properties. *Progress in Organic Coatings*, 77(3), 557–567.

14 Microstructure Image-Based Machine Learning for Predicting Fatigue and High-Temperature Properties of Titanium Alloys through Additive Manufacturing

SD Ruksar Begum, Pankaj Kumar, Pulla Sammaiah, and Ande Rohan

14.1 INTRODUCTION

14.1.1 Importance of Titanium Alloys in Aerospace and High-Performance Applications

A lot of research on different materials, such as metals and alloys, composites, and some other advanced materials, has been performed to perfectly meet the needs of the aerospace industry, such as high specific strength, low weight, resistance to higher temperatures, and corrosion resistance [1, 2]. Many materials, such as aluminum, nickel-based alloys, magnesium, titanium, polymer matrix composites, and metal matrix composites, have been developed to meet these requirements. Among all available materials, titanium and its alloys are considered significantly useful owing to their high strength-to-weight ratio, good performance at high temperatures, and better corrosion resistance [1–3].

Despite their excellent mechanical properties in lightweight applications, titanium alloys are less economically feasible in terms of material efficiency during conventional manufacturing because of tool wear during machining, high buy-to-fly ratios, and difficulty in fabricating complex geometries [4]. In search of an effective manufacturing technology, additive manufacturing (AM) for aerospace components is considered one of the best novel processes, which introduces new manufacturing techniques and offers special benefits in terms of design flexibility, shortened lead

DOI: 10.1201/9781003653226-14

times, and cost-effectiveness [4]. AM is an innovative technology that creates objects with complex geometries layer-by-layer and has rapidly prospered in the manufacture of metallic parts for structural and functional applications [5]. Titanium alloys, along with AM, are excellent for working with metal AM processes.

The materials selected for aerospace components must withstand high temperatures, impact loads, and harsh environmental conditions while remaining lightweight. Titanium alloys are attractive in the aerospace industry owing to their high specific strength (i.e., strength-to-weight ratio), fracture toughness, fatigue resistance, and excellent corrosion resistance [6].

14.1.2 Challenges in Fatigue and High-Temperature Property Prediction

Generally, aerospace components are subjected to various temperature and pressure variations, vibrations, fatigue loads, and harsh environmental conditions, such as corrosive and erosive atmospheres. Hence, such parts should be properly designed to withstand failure, and properties such as fatigue life should be predicted well in advance for better performance. However, there are many challenges in property prediction, such as in additively manufactured materials, where fatigue properties are influenced by numerous parameters, such as processing conditions, microstructure, and post-treatments, which necessitate comprehensive evaluation methods that are often inefficient and costly.

14.1.3 Microstructure–Property Relationships and Variability in Additive Manufacturing

AM processes, such as selective laser melting (SLM) and electron beam melting (EBM), build parts layer by layer, creating unique thermal histories and microstructures. These microstructures are defined by the grain size, phase distribution, porosity, and residual stress, which strongly influence properties such as strength, ductility, and fatigue resistance. Rapid cooling and directional solidification cause anisotropy, whereas thermal cycling affects the dimensional accuracy [7]. AM is highly sensitive to parameters such as laser power and scan speed, causing variability in properties [8]. Researchers have used techniques such as electron backscatter diffraction (EBSD) and X-ray diffraction (XRD) to study these relationships, aiming to standardize parameters and apply in situ monitoring for consistent, high-quality AM component production. Figure 14.1 summarizes the critical parameters affecting microstructure and properties in AM processes.

Pribe et al. developed a physics-based process–structure–property simulation framework for additively manufactured Ti-6Al-4V parts, which helped generate process–structure simulations. A Monte-Carlo–based model and a fast Fourier transform–based crystal plasticity solver were used for porosity development and to predict the fatigue indicator parameter, respectively. The authors concluded that the keyhole porosity severely affects the fatigue life and that the pore-microstructure interaction raises the uncertainty in the extreme values of the fatigue indicator parameter [9].

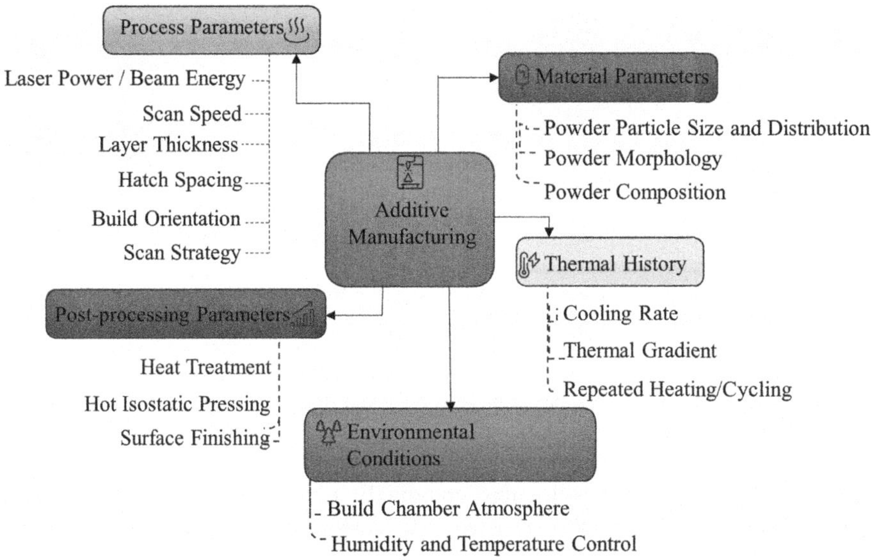

FIGURE 14.1 Key parameters influencing microstructure–property relationships in additive manufacturing.

14.1.4 Emerging Role of Machine Learning and Image-Based Analysis

In addition to prototypes, additively manufactured parts are increasingly being used as functional parts in multiple domains. Owing to multiple interdependent factors in AM processes, achieving error-free parts for industrial use is a concern [10]. The integration of machine learning (ML) with image-based characterization is rapidly gaining prominence in AM research, offering a robust and non-destructive means to predict mechanical performance in real time. This combined approach is particularly valuable for assessing the fatigue behavior and high-temperature properties.

14.2 LITERATURE REVIEW

14.2.1 Overview of Titanium Alloy Microstructures and Fatigue Mechanisms

At normal pressure, Ti has two stable forms. At low temperatures, it has an alpha form with a hexagonal structure. At high temperatures, it has a beta form with a cubic structure [6, 11]. When pure titanium cools, it changes from the beta to the alpha form. The alpha form is more tightly packed and has a different shape than the beta form [3]. Titanium alloys are grouped according to the amount of alpha and beta stabilizers they contain: alpha, alpha–beta, and beta alloys. The main alpha stabilizers are Al, Sn, Zr, O, N, and C. The main beta-stabilizers are Mo, V, Fe, Cr, Co, Mn, Nb, Ni, W, and Ta [6, 12].

Some of the titanium alloys used in the aerospace application include Ti–5Al–5Mo–5V–3Cr-0.5Fe (Ti-5553), Ti–5Al–5Mo–5V–3Cr–1Zr (Ti-55,531), Ti–5Al–4Mo–4V–4Cr–3Zr (Ti-1300), Ti–5Al–3Mo–3V–2Cr–2Zr–1Nb–1Fe (Ti-5321), and

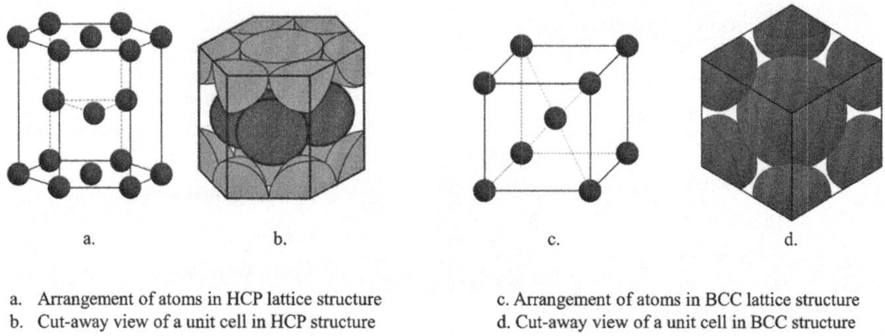

FIGURE 14.2 Atomic arrangement and unit cell representations of α and β phases in titanium alloys. (a–b) Show the hexagonal close-packed (HCP) structure of the α-phase, while (c–d) illustrates the body-centered cubic (BCC) structure of the β-phase.

Ti-6Al-4 V [4, 13]. Ti64 is the most commonly used titanium alloy because of its high strength-to-weight ratio, resistance to corrosion, and temperature stability. It consists of 90% Ti, 6% Al, and 4% V [4]. In general, AM microstructures exhibit microvoids and low fatigue resistance owing to the extremely high cooling rate of the molten pool [14]. Ti64 parts must be carefully evaluated for their fatigue properties because they are subjected to cyclic loading in service, specifically for structural applications [15]. Studies have found that the fatigue performance can be enhanced by microstructural additions and hot isostatic pressing (HIP) treatment [14, 15]. The α-phase in titanium alloys exhibits a hexagonal close-packed (HCP) structure, whereas the β-phase follows a body-centered cubic (BCC) arrangement, as illustrated in Figure 14.2.

14.2.2 High-Temperature Degradation Phenomena in AM Titanium Alloys

Titanium alloys, which are most commonly used in the aerospace industry, often experience corrosion and high-temperature oxidation, resulting in electrochemical and thermal-induced degradation, respectively, which critically impact the service life of the part [16].

With the addition of various elements like aluminum (Al) and molybdenum (Mo), the high-temperature resistance of Ti alloys can be improved by adding various elements, such as Al and Mo. Among them, Al is the most essential thermally stable strengthening element, and Mo refines the crystal structure and improves the resistance to high temperatures [17].

14.2.3 Traditional versus AI-Based Property Prediction Approaches

Property prediction for structural components is largely needed and is critical, as AM involves layer-by-layer deposition, making it difficult to predict the final part properties because of the involvement of a variety of process parameters. In contrast, material properties are mostly predetermined in traditional manufacturing processes.

Traditional property prediction depends mostly on empirical testing, extensive experimentation, and physics-based modeling, whereas recent advances in AI/ML have opened new possibilities for property prediction and are capable of handling the involved parameter complexity of AM. A modern approach using AI, especially ML, has emerged as a successful technology for the real-time monitoring of 3D printed parts. This approach has the potential to optimize process parameters, forecast costs, predict properties, and other considerations by training the respective model [18]. Figure 14.3 illustrates the categorization of property prediction methods into traditional and AI-based approaches, each leveraging different modeling strategies and data types for predicting material behavior.

Zhan and Li used artificial neural networks (ANN) and random forest (RF) methods to predict how long materials can last before they get tired. They compared these results with traditional methods called continuum damage mechanics (CDM) theory [19]. Zhao et al. suggested a new method for predicting the fatigue life of lamellar titanium alloys. They used algorithms with different fatigue indicators, defect shapes, and lamellar patterns to study the performance of these materials. They used ANN, random forest regression (RFR), support vector regression (SVR), and convolutional neural networks (CNN) to predict the fatigue life [20].

14.2.4 Machine Learning Applications in Microstructural Image Analysis

AM of alloys, such as titanium, results in complex thermal cycles, which lead to residual stresses, porosity, and grain anisotropy, and affect the fatigue performance and mechanical reliability. Promising solutions for such problems are offered by recent advancements in ML and image-based microstructural analysis, which help in quick and automated microstructure throughput evaluation, whereas conventional characterization methods are laborious and time-consuming. For critical fatigue parts, such as aerospace structures, ML-based microstructure analysis helps in the early identification of defects and grain anomalies and even real-time quality monitoring.

Bordekar et al. presented a novel explainable artificial intelligence (AI) methodology for defect characterization and detection. They used a supervised ML algorithm called the support vector machine (SVM) in Ti6Al4V specimens to differentiate the pixel data into pores and inclusions from CT scan data [21]. In a review by Deshmankar et al., it was concluded that significant work is still required in microstructural predictions and the integration of diverse data types [22].

14.3 MATERIALS AND METHODS

14.3.1 Additive Manufacturing Process—Powder Bed Fusion or Directed Energy Deposition for Ti-6Al-4V

AM is becoming increasingly popular because it can create custom parts, complex shapes, and items from tough materials. There are seven types of AM: vat polymerization, material extrusion, binder jetting, material jetting, powder bed fusion, directed energy deposition, and sheet lamination. Powder bed fusion and directed energy deposition are the two main methods for working with metals and alloys. In powder bed fusion, heat from a laser or electron beam melts the powder materials.

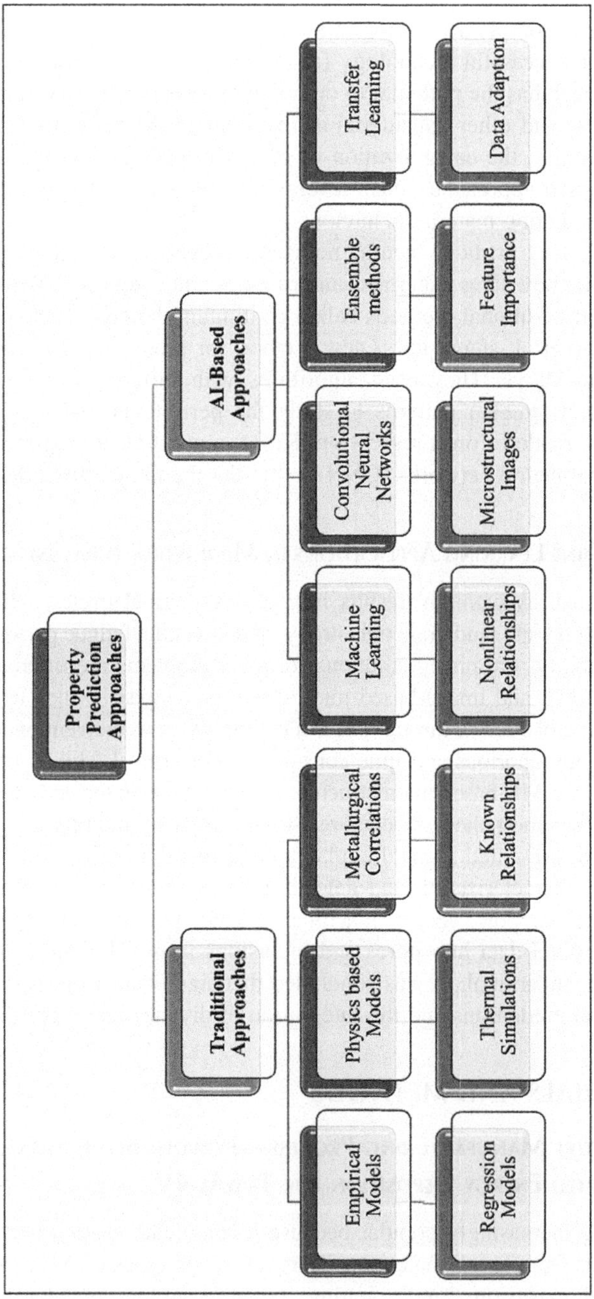

FIGURE 14.3 Classification of property prediction approaches used in materials science.

This method is applicable to powdered polymers, metals, and alloys. Depending on the heat source, powder bed fusion is divided into selective laser sintering (SLS), SLM/laser powder bed fusion (L-PBF), EBM, and direct metal laser sintering (DMLS). The working mechanism of the powder bed fusion process is shown in Figure 14.4, highlighting the coordinated movement of the powder delivery system, heat source, and build platform to produce additively manufactured parts (Table 14.1).

Directed energy deposition (DED) is an exceptional AM process that is suitable not only for the fabrication of new parts but also for repairing and adding features to existing parts. In this process, focused thermal energy from a laser, electron beam, or plasma arc is used to melt the material in either powder or wire form. The working principle of the DED process is illustrated in Figure 14.5, highlighting the simultaneous material feeding and melting during AM (Table 14.2).

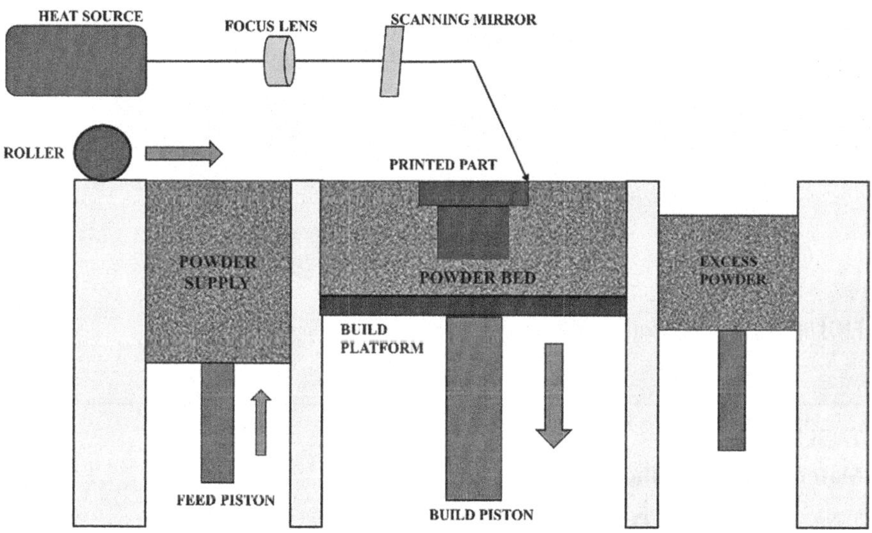

FIGURE 14.4 Schematic diagram of the Powder Bed Fusion (PBF) process illustrating key components.

TABLE 14.1
Types of Powder-Bed Fusion Processes

S. No.	Process	Material Type	Energy Source
1	Selective Laser Sintering (SLS)	Polymers	Laser
2	Selective Laser Melting (SLM)/Laser Powder Bed Fusion (L-PBF)	Metals	Laser
3	Electron Beam Melting (EBM)	Metals	Electron beam
4	Direct Metal Laser Sintering (DMLS)	Metals	Laser

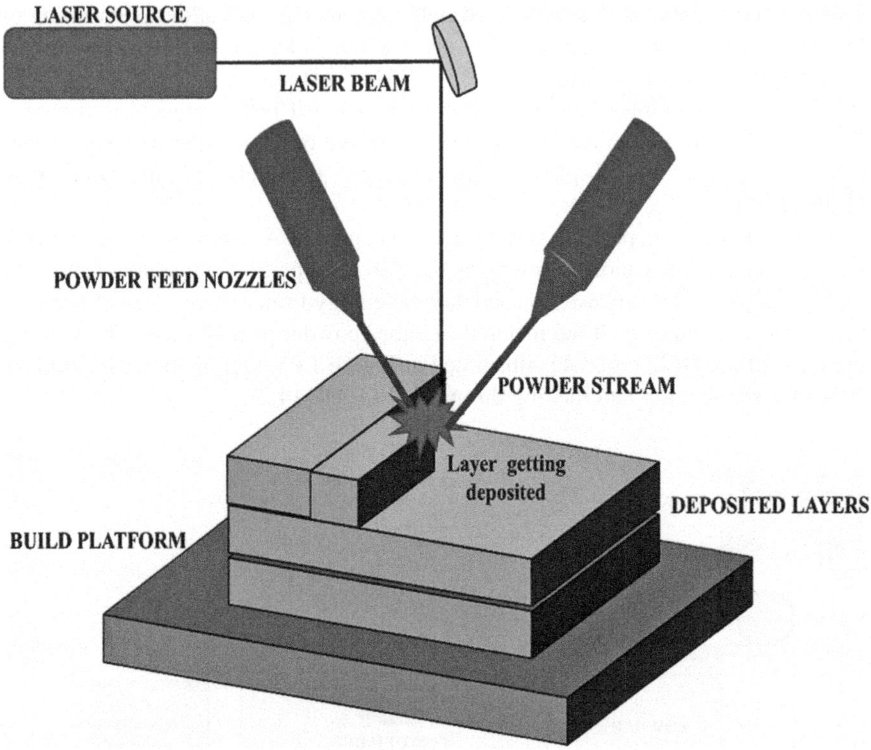

FIGURE 14.5 Schematic of the directed energy deposition (DED) process.

TABLE 14.2
Materials Used in the Processes

S. No.	Process	Material
1	SLS (Selective Laser Sintering)	Nylon (Polyamide (PA)) PA11, PA12, glass-filled nylon (PA-GF), carbon fiber-filled nylon (PA-FR), aluminum-filled polyamide (alumide), polypropylene (PP), polyetherketoneketone (PEKK) or polyetheretherketone (PEEK), and thermoplastic polyurethane.
2	SLM (Selective Laser Melting) DMLS (Direct Metal Laser Sintering) EBM (Electron Beam Melting)	Metals: Stainless steels (316L, 17-4PH), titanium and its alloys, aluminum and its alloys, cobalt chrome alloys, nickel alloys (inconel and hastelloy), tool steels, precious metals (gold and silver), and copper and its alloys.
3	DED	In addition to above-mentioned metals, DED is capable of processing refractory metals (tungsten and niobium) and ceramics (zirconia, alumina, and silicon nitride) composites.

14.3.2 Machine Learning Algorithms for Property Prediction

Many ML algorithms are available to predict the mechanical properties of products manufactured using powder bed fusion and DED. Regression-based ML models were employed to develop a relationship between the LPBF process parameters and mechanical properties. To predict the properties, such as the hardness, tensile strength, and density of the parts made by LPBF, models such as K-nearest neighbors (KNN), SVMs, and ANNs are used. In DED, ML models are employed to predict the mechanical properties based on the thermal history and process parameters. The CNN, XGBoost (eXtreme Gradient Boosting), and RF models showed excellent predictions of mechanical properties based on the thermal history data and optimization of process parameters, respectively.

14.3.3 Sample Preparation and Heat Treatments—Processing Conditions to Introduce Microstructural Variation

To investigate the influence of processing conditions on microstructural evolution, various AM techniques were employed to fabricate metal alloys, followed by comparative analyses. Wang et al. [23] fabricated Cu-15Ni-8Sn alloy blanks using Laser DED (L-DED) and compared them with those produced via L-PBF. The study revealed that L-DED samples exhibited coarser grains, dendritic structures, and segregated phases due to the slower cooling rates inherent to the process. These features contributed to stress concentration and reduced yield strength as a result of work hardening mechanisms.

In contrast, Ti alloys produced via L-PBF generally demonstrate superior mechanical properties—specifically higher strength and hardness—compared to those fabricated using DED or electron beam powder bed fusion (EB-PBF) methods [24].

It is reported that commercially pure titanium (CP–Ti) fabricated using L-PBF contains a mixture of platelet-shaped α and acicular α' phases. However, CP–Ti produced via DED predominantly exhibits a plate-like α phase. Due to the higher energy input required in DED for achieving high-density components, the process generates a deeper melt pool and subjects the material to multiple thermal cycles. This increased thermal exposure enhances heat transfer to the deposited layers, promoting the dissolution and transformation of microstructural phases—particularly the α phase into a more homogenized α structure. A comparison of L-DED and L-PBF processes showing differences in grain structure, strength, cooling rate, energy input, and heat cycles is presented in Figure 14.6.

14.3.3.1 STEP-1: Sectioning and Mounting

To preserve the microstructural integrity, the Ti6Al4V samples were carefully sectioned and prepared based on the specific analysis requirements. Sectioning was performed orthogonal to the build direction to study the layer-by-layer microstructural evolution and interlayer bonding characteristics, parallel to the build direction to assess columnar grain structure and texture development along the build height, and at a 45° orientation to evaluate intermediate orientations and potential anisotropic effects. After sectioning, the samples were mounted in epoxy resin for further examination.

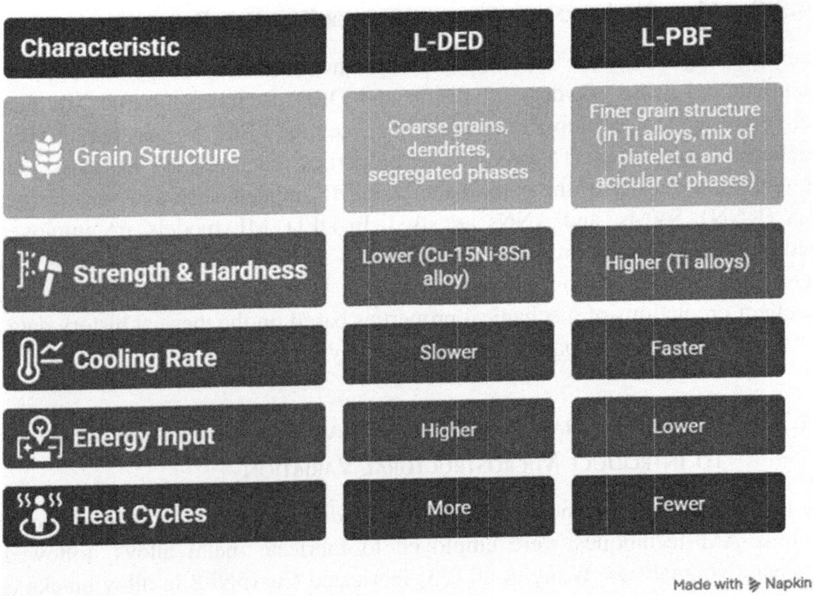

FIGURE 14.6 Comparison of L-DED and L-PBF processes showing differences in grain structure, strength, cooling rate, energy input, and heat cycles.

14.3.3.2 STEP-2: Sample Preparation

Once the specimen is ready after sectioning, it should be followed by sequential grinding and polishing for better analysis.

Grinding: samples were ground with a minimal force of application to prevent work hardening and mechanical twinning on progressive SiC paper grits (220, 400, 800, and 1200).

Polishing: This is performed for advanced characterization. Diamond paste progression (6 μm, 3 μm, and 1 μm) was followed by a final polish using colloidal silica (0.05 μm) to achieve scratch-free surfaces.

Vibratory Polishing: If the sample has to be analyzed under EBSD, then extended vibratory polishing for about 4–8 hours is carried out with colloidal silica to eliminate residual subsurface deformation and gain superior surface quality.

For high-resolution microstructural analysis, electropolishing or ion beam milling may be employed:

Electropolishing: This process is carried out for transmission electron microscopy specimens to obtain an excellent surface finish using perchloric acid-based electrolytes (60% methanol, 35% butanol, and 5% perchloric acid) at −30°C.

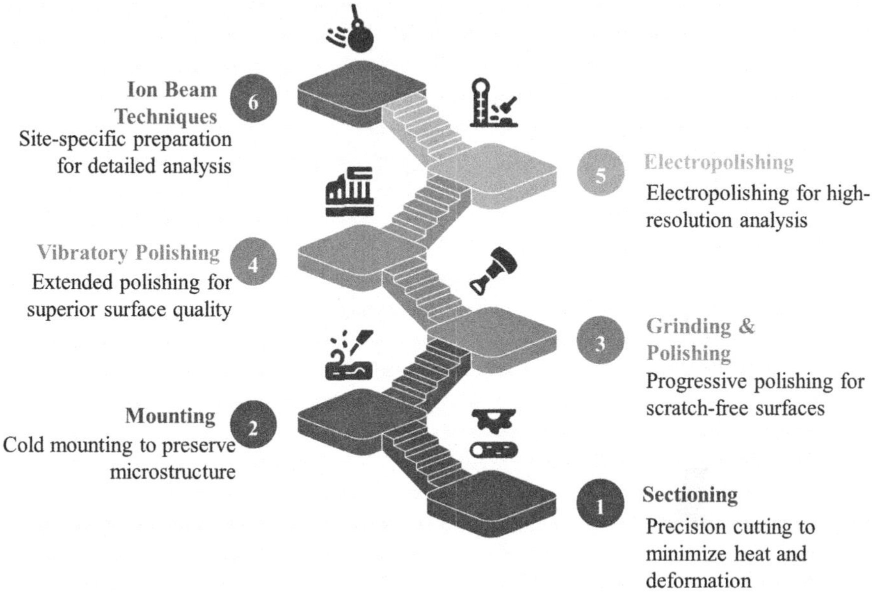

FIGURE 14.7 Stepwise schematic of optimal sample preparation for microstructural analysis.

Ion beam milling: Site-specific specimens with artifact-free surfaces can be prepared using either focused ion beam (FIB) or broad ion beam (BIB) techniques, which can be used for detailed phase identification and defect analysis. Figure 14.7 outlines the essential steps involved in achieving optimal sample preparation for advanced microstructural characterization.

14.3.4 Heat Treatment

Additively manufactured parts are subjected to constrained thermal expansion and rapid thermal cycling, and they contain significant thermal residual stresses. Several heat treatment methods were also employed on the specimens to relieve residual stresses.

Various heat treatment strategies were employed to tailor the microstructure of Ti6Al4V components. Low-temperature stress relief at 650°C for 2–4 hours under vacuum or inert gas followed by furnace cooling primarily relieved residual stresses with minimal α' decomposition. Intermediate stress relief at 704°C for 2–4 hours in a vacuum or argon atmosphere, followed by air cooling, resulted in partial α' decomposition and the formation of a fine $\alpha + \beta$ structure. Duplex heat treatments were applied to precisely control phase balance and morphology, involving solution treatment above the β-transus, followed by aging below it. A typical solution treatment at 1050°C for 1 hour with air cooling, followed by aging at 720°C for 2 hours and air cooling, produced a lamellar $\alpha + \beta$ structure with controlled α-plate thickness. A modified duplex treatment comprising β-annealing at 1020°C for

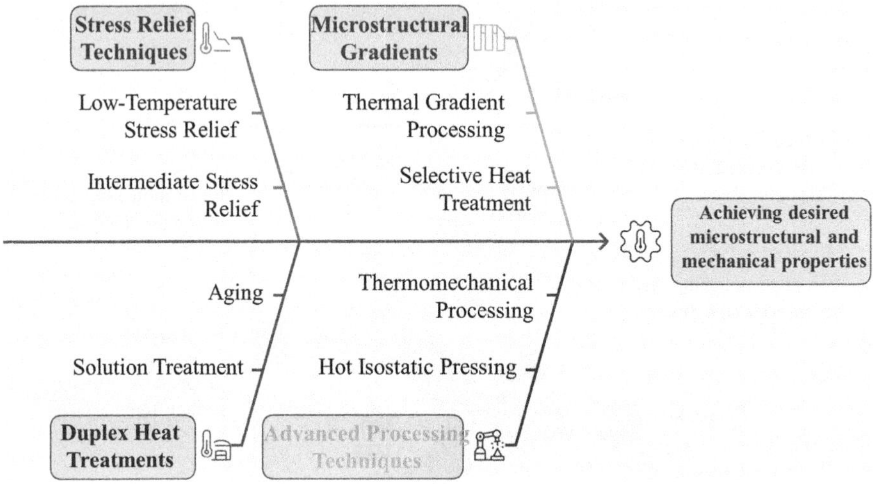

FIGURE 14.8 Heat treatment strategies for Ti-6Al-4V components categorized into stress relief techniques.

2 hours with furnace cooling and α + β annealing at 843°C for 2 hours with air cooling resulted in a bimodal α + β structure characterized by primary α and transformed β regions. Figure 14.8 illustrates various heat treatment pathways employed to optimize the microstructure and mechanical performance of Ti-6Al-4V alloy components.

The microstructural features of the parts fabricated by either of the two processes (PBF and DED) vary in many ways. In PBF, each layer experiences repeated thermal cycles, which leads to tempering effects. The scan strategies used also affect the final microstructure owing to local thermal gradients and cooling rates, and sometimes even result in a fine microstructure if the layer thickness is small. Hence, it can be concluded that the thermal cycling effects, scanning strategy, and layer thickness influence the microstructure of the part. In DED, the microstructure control parameters are heat input variability, substrate interaction, and multi-material interaction.

14.3.5 Microstructure Characterization—SEM, EBSD, and Optical Microscopy Image Acquisition

Generally, microstructure analysis is performed to understand and reveal information about the phases, grain size, porosity, defects, and crystallography of a specimen. Many advanced microscopy techniques are available and are highly productive compared to traditional techniques [25].

Optical microscopy uses visible light and lenses to magnify surface features, which form the foundation of microstructural analysis, whereas electron microscopy techniques, such as scanning electron microscopy (SEM) and transmission electron microscopy, provide high-resolution and atomic-level resolution imaging using a beam of electrons. Both techniques can be coupled with X-ray spectroscopy (EDS/

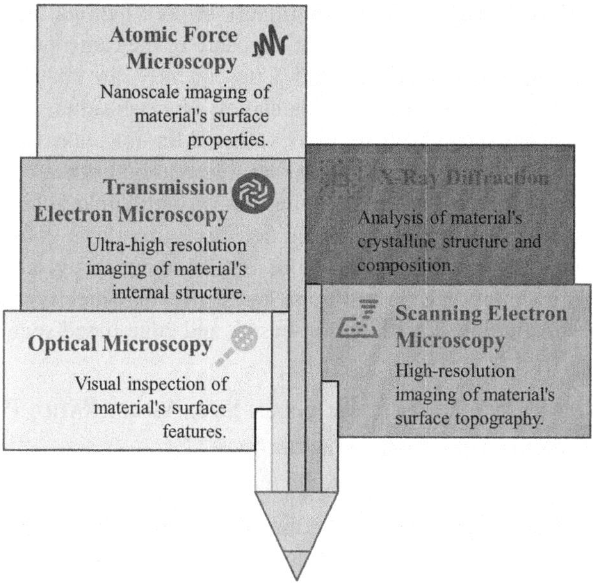

FIGURE 14.9 Overview of key material characterization techniques used in microstructural analysis.

EDX) for compositional analysis, called energy dispersive EDS/EDX and EBSD. Figure 14.9 presents the primary characterization tools used to analyze materials at various resolution levels, from surface features to internal crystalline structures.

14.3.6 Image Processing and Feature Extraction

It is known that the parts fabricated through the AM process, specifically PBF and DED, require thorough process monitoring in order to control the quality of the build and defect detection. This can be achieved by equipping high-resolution in-situ cameras to monitor the layer-wise process, CT scans, and optical microscopy, thereby dealing with a variety of images. These images can then be used to detect defects by preprocessing and feature extraction.

14.3.7 Preprocessing Techniques: Filtering, Normalization, and Segmentation

Preprocessing involves filtering, normalization, and segmentation. Filtering is used to improve image quality by emphasizing important features, such as edges and defects, by comparing pixel values. Gaussian, median, and edge filters are examples of spatial filters. The filtering process enhances the visibility of defects, detects cracks, lack-of-fusion regions, and even melt pool irregularities, and prepares the images for further analysis.

There may be a chance of varying brightness or poor lighting on a few layers in certain directions, either due to the lack of exposure of the camera on a few layers or due to the laser intensity during the build. Consequently, the image quality may be poor and may show changes in pixel intensities. Normalization was performed to standardize the image intensities and pixel values. Min–max normalization, Z-score normalization, and adaptive normalization are a few to be listed.

Segmentation is the process of grouping pixels with similar characteristics into meaningful regions, such as distinguishing defective areas from good material. This process is generally used in images from optical microscopy and CT scans to segment porosity, count defects, and track crack growth. Three types of segmentation techniques include threshold, region-based, and edge-based segmentation.

14.3.8 Morphological Features: Grain Size, Aspect Ratio, Porosity, and Alpha-Beta Phase Distribution

In AM, morphology refers to the size, shape, texture, and structure of the particles of the raw material, especially powder, and the features of the final printed component. It must be monitored and optimized because morphology directly affects the mechanical properties, such as tensile strength, creep resistance, and fatigue life. To understand the part performance and material properties, some critical morphological features require special focus, such as grain size, aspect ratio (AR), phase distribution, and porosity.

Grain size is defined as the area of individual grains in the microstructure or the average diameter of the crystallites. Grain size has a direct influence on the mechanical properties of the part; the finer the grain size, the higher the strength and fatigue resistance, whereas coarser grains lead to lower strength but improve creep resistance. Grain size refinement can be achieved, as per the requirement, by optimizing the laser power, scan speed, and hatch spacing. It can be measured using optical microscopy, EBSD, and image analysis software.

In simple terms, the AR is the ratio of the length to the width of the grains. In LPBF, owing to directional heat flow, columnar grain growth can be observed, and with a high AR, elongated grains are visible, leading to the anisotropy of the mechanical behavior.

Porosity in AM can be quantified as the number of pores, voids, or gaps within the material. The pores can be due to gas entrapment or a lack of fusion. It must be reduced as much as possible, as it is considered a defect that lowers strength and fatigue life and leads to stress concentration zones. It can be measured by image analysis using X-ray computed tomography (CT) scanning and optical microscopy. In titanium alloys, such as Ti6Al4V, a spatial phase distribution in the α and β phases is observed. This phase distribution is formed owing to the rapid cooling during the printing process. The phase distribution also influences mechanical properties. It can be observed via SEM, EBSD, XRD, and optical microscopy. The morphological features of additively manufactured alloys—such as grain size, phase distribution, AR, and porosity—play a critical role in determining mechanical and functional performance (Figure 14.10).

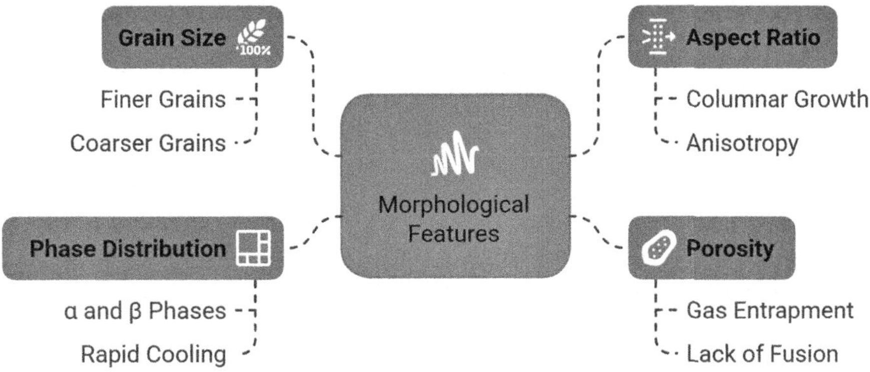

FIGURE 14.10 Key morphological features in additively manufactured alloys.

14.4 MACHINE LEARNING FRAMEWORK

The (shape) complexity of titanium alloys in the aerospace field using AM poses a wide range of problems because of the variety in the design concepts, material properties, and issues in quality control. Evaluating this large number of process parameters is increasingly difficult using traditional statistical techniques. To overcome these problems, various studies have reported the use of ML techniques, which help in process parameter optimization, model adaptation, and quality control. ML improves product quality and predicts data-driven costs, mechanical properties, printability, and anomaly detection [26]. ML is a subset of AI that uses algorithms to train models using a dataset to achieve specific requirements, such as fault and property prediction. The ML approach is categorized into three groups: supervised learning (SL), unsupervised learning (UL), and reinforcement learning (RL) [27]. Figure 14.11 illustrates the three primary categories of ML based on data supervision and learning mechanisms.

SL algorithms use a dataset comprising both input and output pairs (called labeled datasets) to train the model [28]. As a result, these algorithms, after thorough training and testing, can predict the unknown output for a given new input data. These algorithms can be of the regression or classification type. ML algorithms such as ANNs, SVMs, decision trees (DT), and Naïve Bayes are commonly used SL algorithms. UL algorithms use statistical methods from a dataset and are trained to represent input patterns without using explicit target outputs for each input. Two types of UL algorithms are currently in practice: clustering and dimensionality reduction algorithms. Examples include K-means clustering and principal component analysis. The third category of ML is the RL algorithm. It is a closed-loop method in which a specific objective of a problem can be solved with interaction between the challenge and an environment [29]. Q-learning falls under RL. Other ML algorithms include Gaussian process regression (GPR), CNNs, KNN, and RF.

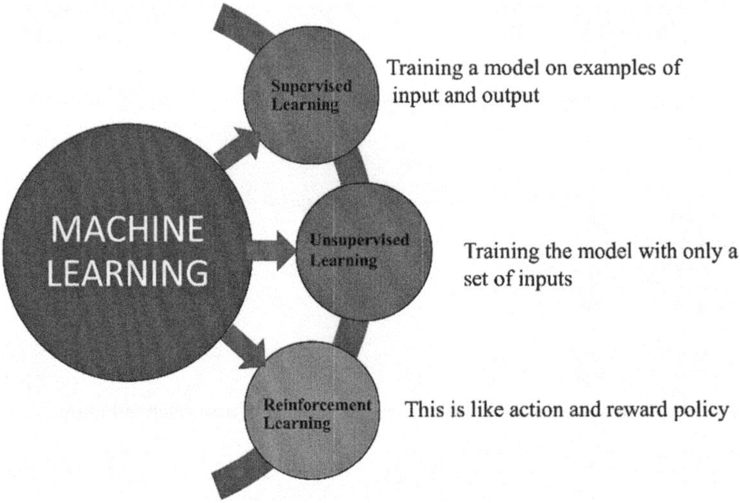

FIGURE 14.11 Fundamental types of machine learning approaches—supervised learning, unsupervised learning, and reinforcement learning.

14.4.1 Supervised Learning Models: SVM, Random Forest, and Gradient Boosting

Gobert et al. [29] have used SVM-Linear, a binary classification technique, in identifying the defects in a component printed by LPBF, either as a flaw or a normal build condition, and even cross-validated by CT scan data. Their work consisted of an in-situ defect detection strategy using supervised ML. During the building process, the authors collected multiple images at each build layer using a high-resolution digital single-lens reflex (DSLR) camera. These images were stacked layer-wise, and the features in multiple dimensions were extracted and used as binary classifiers [30].

Zhan and Li [30] have identified that the theoretical methods for fatigue analysis in AM parts cannot directly consider the effect of process parameters and are even time-consuming. Hence, they developed a novel approach using three ML models, namely, SVM, RF, and ANN, to predict the fatigue life results. The training dataset for these three models was produced by a data-driven analysis called CDM-based fatigue life prediction applied to additively manufactured stainless steel 316L, which considered the effects of the process parameters, namely, laser power, scan speed, hatch space, and powder layer thickness [31].

14.4.2 CNN Models for Image Classification/Regression

Huu et al. [28], in their review, have proposed that the integration of SVM and extreme learning machine (ELM) algorithms in the CNN architecture can enhance the ML models for defect detection in the AM process. The hybrid model aims to overcome the high costs associated with data collection, data unavailability, difficulties in data labeling, and even overfitting and underfitting issues [30].

A data-informed knowledge discovery framework was developed by Alfred and Amiri [31] using various AM fatigue datasets from the literature. This framework extracted meaningful characteristics from the datasets after analysis, which aided in ML model training for solving the classification and regression problems, keeping the fatigue ratio as the key property. KNN and SVM methods were used to classify the process variables into different fatigue ratio categories, and the GPR model, along with the Shapley Additive explanations (SHAP) method, was used to predict the fatigue ratio and explain the predictions of the GPR model. The framework was validated using datasets from three alloys: Ti–6Al–4V, IN718, and AlSi10Mg [32].

14.4.3 Labeling Data with Fatigue Life, Creep Rupture Strength, and Thermal Resistance Metrics

Bao et al. [32] adopted an ML method to explore the effect of the position, size, and morphology of defects on the fatigue life of an SLM Ti64 alloy. The data related to the geometric features of the critical defects and the actual fatigue life were obtained using high-resolution synchrotron X-ray microtomography and high-cycle fatigue (HCF) testing. These data were used to train the SVR, a regression version of SVM, to predict the fatigue life of the alloy from the defect parameters. A grid search approach with cross-validation was used to optimize the model parameters. This study validated that SVM outperformed in predicting the fatigue life, maintaining the coefficient of determination between the predicted and actual fatigue life up to 0.99 [32].

Sanchez et al. [33] studied the creep behavior of additively manufactured nickel alloys (Alloy 718) printed using L-PBF. They have applied ML techniques such as regularized linear regression, RF, gradient boosted tree (GBT), SVR, and deep neural network (DNN) to determine the influence of process parameters on the creep rate of the alloy and to predict the creep rate. AM process parameters, such as build orientation, number of lasers, scan strategies, and geometrical material descriptors, such as part density and number of pores from optical microscope images, were used as input data for image analysis techniques. The authors concluded that RF and GBT produced accurate predictions, while SVR achieved an accuracy of approximately 98.6%. The effects of parameters were also evaluated, and the number of lasers had an insignificant effect, whereas the density, number of pores, and build orientation significantly affected the creep rate [33].

14.4.4 Cross-Validation, Hyperparameter Tuning, and Evaluation Metrics

Parameter tuning, also called hyperparameter optimization, is considered one of the important processes in ML, as it helps in selecting the best values for the hyperparameters for a model so that model performance can be improved. The data or settings that were set manually and not included in the training set are called hyperparameters, and the behavior of the learning algorithm is controlled by them. Grid search, random search, Bayesian optimization, and cross-validation are a few techniques

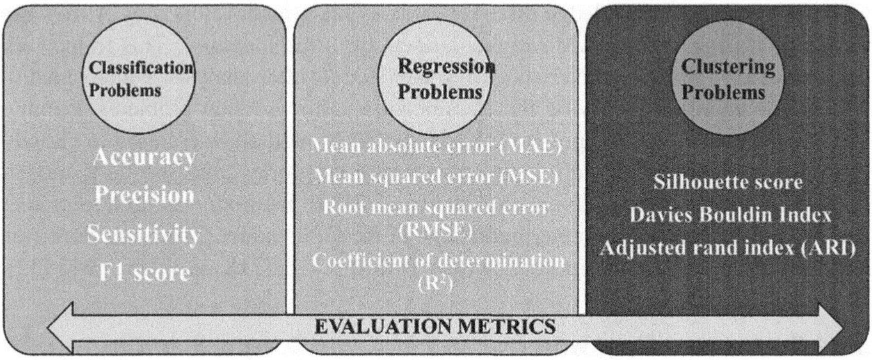

FIGURE 14.12 Overview of evaluation metrics used in machine learning for different problem types.

used for parameter tuning. Bao et al. adopted a grid search with cross-validation techniques in their work [32]. The grid search finds the best combination of parameters for a given model, and cross-validation helps identify the superior parameters from the grid search [32].

The quantitative measures used to assess the performance of an ML model on a given task are called evaluation metrics. Various evaluation metrics are available and are chosen depending on the type of problem being solved using the model. Figure 14.12 categorizes common evaluation metrics based on classification, regression, and clustering problem domains in ML.

In general, to predict the fatigue or creep rate, MSE, RMSE, or R2 are used as evaluation metrics because they are continuous values, whereas for classifying the fatigue life, accuracy or F1 scores are required. The MSE for the SVR model was used as the evaluation metric for predicting the fatigue life in the work carried out by Bao et al. [32]. Sanchez et al. [33] adopted various evaluation metrics to predict the creep rate, namely the Median Absolute Deviation (MAD), Coefficient of Determination (R2), Mean Absolute Error (MAE), Root Mean Square Error (RMSE), and Mean Absolute Percentage Error (% error) [33].

14.5 CONCLUSIONS

This chapter demonstrates the potential of integrating microstructure image analysis with ML techniques to predict fatigue and high-temperature properties of additively manufactured titanium alloys. By leveraging high-resolution imaging and advanced ML models, critical microstructural features can be quantitatively linked to mechanical performance. The proposed framework enables non-destructive, rapid, and accurate predictions, offering a powerful tool for material design, quality assurance, and process optimization. This approach not only enhances our understanding of structure–property relationships in AM titanium alloys but also lays the groundwork for intelligent materials engineering driven by data and computation.

REFERENCES

1. R. Soni, R. Verma, R. Kumar Garg, and V. Sharma, "A critical review of recent advances in the aerospace materials," *Mater Today Proc*, vol. 113, pp. 180–184, 2023, doi: 10.1016/j.matpr.2023.08.108
2. P. Pushp, S. M. Dasharath, and C. Arati, "Classification and applications of titanium and its alloys," *Mater Today Proc*, vol. 54, pp. 537–542, 2022, doi: 10.1016/j.matpr.2022.01.008
3. M. Najafizadeh, "Classification and applications of titanium and its alloys: A review," *Alloy. Compd. Commun.*, vol. 3, no. June, p. 100019, 2024, doi: 10.1016/j.jacomc.2024.100019
4. M. Srivastava *et al.*, "Additive manufacturing of Titanium alloy for aerospace applications: Insights into the process, microstructure, and mechanical properties," *Appl. Mater. Today*, vol. 41, p. 102481, 2024, doi: 10.1016/j.apmt.2024.102481
5. T. Zhang and C. T. Liu, "Design of titanium alloys by additive manufacturing: A critical review," *Adv. Powder Mater.*, vol. 1, no. 1, p. 100014, 2022, doi: 10.1016/j.apmate.2021.11.001
6. Z. Liu, B. He, T. Lyu, and Y. Zou, "A review on additive manufacturing of titanium alloys for aerospace applications: Directed energy deposition and beyond Ti-6Al-4V," *JOM*, vol. 73, no. 6, pp. 1804–1818, 2021, doi: 10.1007/s11837-021-04670-6
7. N. Zohdi, R. (Chunhui) Yang, S. Torres-Giner, and M. Vargas, "Material anisotropy in additively manufactured polymers and polymer composites: A review," *Polymers*, vol. 13, no. 19, p. 3368, 2021, doi: 10.3390/POLYM13193368
8. Kumar Ramavajjala, T. R. Dandekar, R. K. Khatirkar, C. Joshi, R. N. Chouhan, and A. Agnihotri, "A review on the correlation between microstructure, heat treatment and mechanical properties of additively manufactured AlSi10Mg by LPBF," *Crit. Rev. Solid State Mater. Sci.*, vol. 50, no. 3, pp. 239–274, 2025, doi: 10.1080/10408436.2024.2414012
9. J. D. Pribe *et al.*, "A process-structure-property simulation framework for quantifying uncertainty in additive manufacturing: Application to fatigue in Ti-6Al-4V," *Integr. Mater. Manuf. Innov.*, vol. 12, pp. 231–250, 2023, doi: 10.1007/s40192-023-00303-9
10. S. Deshpande, V. Venugopal, M. Kumar, and S. Anand, "Deep learning-based image segmentation for defect detection in additive manufacturing: An overview," *Int. J. Adv. Manuf. Technol.*, vol. 134, no. 5, pp. 2081–2105, 2024, doi: 10.1007/S00170-024-14191-6
11. D. Banerjee and J. C. Williams, "Perspectives on titanium science and technology," *Acta Mater.*, vol. 61, no. 3, pp. 844–879, 2013, doi: 10.1016/J.ACTAMAT.2012.10.043
12. J. Mantione, M. Garcia-Avila, M. Arnold, D. Bryan, and J. Foltz, "Properties of novel high temperature titanium alloys for aerospace applications," *MATEC Web Conf.*, vol. 321, p. 04006, 2020, doi: 10.1051/matecconf/202032104006
13. T. S. Tshephe, S. O. Akinwamide, E. Olevsky, and P. A. Olubambi, "Additive manufacturing of titanium-based alloys- A review of methods, properties, challenges, and prospects," *Heliyon*, vol. 8, no. 3, p. e09041, 2022, doi: 10.1016/j.heliyon.2022.e09041
14. Z. Qu *et al.*, "High fatigue resistance in a titanium alloy via near-void-free 3D printing," *Nature*, vol. 626, no. 8001, pp. 999–1004, 2024, doi: 10.1038/s41586-024-07048-1
15. X. Shui, K. Yamanaka, M. Mori, Y. Nagata, K. Kurita, and A. Chiba, "Effects of post-processing on cyclic fatigue response of a titanium alloy additively manufactured by electron beam melting," *Mater. Sci. Eng. A*, vol. 680, pp. 239–248, 2017, doi: 10.1016/J.MSEA.2016.10.059
16. S. Mazumder, M. V. Pantawane, S. S. Joshi, and N. B. Dahotre, "Electrochemical and thermal-induced degradation of additively manufactured titanium alloys: A review,"

Crit. Rev. Solid State Mater. Sci., vol. 47, no. 6, pp. 915–954, 2022, doi: 10.1080/10408436.2021.1989664

17. J. Sun, H. Lu, H. Zhang, K. Luo, and J. Lu, "Effect of thermal exposure on microstructure and mechanical properties of Ti65 high-temperature titanium alloy deposited by laser direct energy deposition," *Mater. Sci. Eng. A*, vol. 908, p. 146757, 2024, doi: 10.1016/J.MSEA.2024.146757
18. M. Rastak, S. Vanaei, S. Vanaei, and M. Moezzibadi, "Machine learning in 3D printing," *Ind. Strateg. Solut. 3D Print. Appl. Optim.*, pp. 273–294, 2024, doi: 10.1002/9781394150335.CH14
19. Z. Zhan and H. Li, "A novel approach based on the elastoplastic fatigue damage and machine learning models for life prediction of aerospace alloy parts fabricated by additive manufacturing," *Int. J. Fatigue*, vol. 145, p. 106089, 2021, doi: 10.1016/J.IJFATIGUE.2020.106089
20. Y. Zhao, Y. Xiang, and K. Tang, "Machine learning-based fatigue life prediction of lamellar titanium alloys: A microstructural perspective," *Eng. Fract. Mech.*, vol. 303, p. 110106, 2024, doi: 10.1016/J.ENGFRACMECH.2024.110106
21. H. Bordekar, N. Cersullo, M. Brysch, J. Philipp, and C. Hühne, "eXplainable artificial intelligence for automatic defect detection in additively manufactured parts using CT scan analysis," *J. Intell. Manuf.*, vol. 36, no. 2, pp. 957–974, 2025, doi: 10.1007/s10845-023-02272-4
22. A. P. Deshmankar, J. S. Challa, A. R. Singh, and S. P. Regalla, "A review of the applications of machine learning for prediction and analysis of mechanical properties and microstructures in additive manufacturing," *J. Comput. Inf. Sci. Eng.*, vol. 24, no. 12, pp. 1–14, 2024, doi: 10.1115/1.4066575
23. J. Wang, X. Zhou, J. Li, J. Zhu, and M. Zhang, "A comparative study of Cu–15Ni–8Sn alloy prepared by L-DED and L-PBF: Microstructure and properties," *Mater. Sci. Eng. A*, vol. 840, p. 142934, 2022, doi: 10.1016/J.MSEA.2022.142934
24. H. Y. Ma *et al.*, "Advances in additively manufactured titanium alloys by powder bed fusion and directed energy deposition: Microstructure, defects, and mechanical behavior," *J. Mater. Sci. Technol.*, vol. 183, pp. 32–62, 2024, doi: 10.1016/J.JMST.2023.11.003
25. S. Doddapaneni *et al.*, "Advancements in EBSD techniques: A comprehensive review on characterization of composites and metals, sample preparation, and operational parameters," *J. Compos. Sci.*, vol. 9, no. 3, 2025, doi: 10.3390/jcs9030132
26. U. M. R. Paturi, S. T. Palakurthy, S. Cheruku, B. Vidhya Darshini, and N. S. Reddy, "Role of machine learning in additive manufacturing of titanium alloys—A review," *Arch. Comput. Method Eng.*, vol. 30, no. 8, pp. 5053–5069, 2023, doi: 10.1007/s11831-023-09969-y
27. M. Trovato, L. Belluomo, M. Bici, M. Prist, F. Campana, and P. Cicconi, "Machine learning in design for additive manufacturing: A state-of-the-art discussion for a support tool in product design lifecycle," *Int. J. Adv. Manuf. Technol.*, vol. 137, no. 5, pp. 2157–2180, 2025, doi: 10.1007/s00170-025-15273-9
28. P. N. Huu *et al.*, "Review: Enhancing additive digital manufacturing with supervised classification machine learning algorithms," *Int. J. Adv. Manuf. Technol.*, vol. 133, no. 3–4, pp. 1027–1043, 2024, doi: 10.1007/s00170-024-13781-8
29. C. Gobert, E. W. Reutzel, J. Petrich, A. R. Nassar, and S. Phoha, "Application of supervised machine learning for defect detection during metallic powder bed fusion additive manufacturing using high resolution imaging," *Addit. Manuf.*, vol. 21, pp. 517–528, 2018, doi: 10.1016/J.ADDMA.2018.04.005
30. Z. Zhan and H. Li, "Machine learning based fatigue life prediction with effects of additive manufacturing process parameters for printed SS 316L," *Int. J. Fatigue*, vol. 142, p. 105941, 2021, doi: 10.1016/J.IJFATIGUE.2020.105941

31. S. O. Alfred and M. Amiri, “A data-informed knowledge discovery framework to predict fatigue properties of additively manufactured Ti–6Al–4V, IN718 and AlSi10Mg alloys using fatigue databases,” *Prog. Addit. Manuf.*, vol. 10, no. 9, pp. 6183–6210, 2025, doi: 10.1007/s40964-025-00965-1
32. H. Bao, S. Wu, Z. Wu, G. Kang, X. Peng, and P. J. Withers, “A machine-learning fatigue life prediction approach of additively manufactured metals,” *Eng. Fract. Mech.*, vol. 242, p. 107508, 2021, doi: 10.1016/J.ENGFRACMECH.2020.107508
33. S. Sanchez, D. Rengasamy, C. J. Hyde, G. P. Figueredo, and B. Rothwell, “Machine learning to determine the main factors affecting creep rates in laser powder bed fusion,” *J. Intell. Manuf.*, vol. 32, no. 8, pp. 2353–2373, 2021, doi: 10.1007/s10845-021-01785-0

15 Next-Generation Surface Engineering

Trends in Coating and Cladding Technologies

Bappa Das, Shakti Kumar, Bishub Choudhury, Sumit Kumar Sinha, Arka Ghosh, and Royal Madan

15.1 INTRODUCTION

Coating and cladding technologies have become essential in modern materials engineering, providing solutions to improve surface properties such as corrosion resistance, wear resistance, thermal insulation, and aesthetic appeal. In contrast to cladding, which applies a larger layer to surfaces, usually by bonding, welding, or mechanical joining, coatings are usually thin layers applied to surfaces using a variety of techniques, such as spraying, deposition, or chemical processes [1, 2]. As industrial demands have increased, the importance of these technologies has likewise grown, particularly in fields where component longevity and operational dependability are critical, such as biomedical engineering, power generation, automotive, and aerospace [3].

The evolution of coating and cladding processes mirrors larger technological advances in materials science and manufacturing. Traditional processes like electroplating, galvanizing, thermal spraying, and roll bonding lay the groundwork for protective surface treatments. However, these approaches are frequently hampered by difficulties such as poor adhesion, porosity, and limited control over microstructure. With the advent of precision engineering, new processes such as chemical vapor deposition (CVD), physical vapor deposition (PVD), sol–gel processing, and cold spray have been developed, giving higher coating density, compositional control, and tailored performance [4–6]. Similarly, recent cladding techniques like laser cladding, explosive cladding, and friction stir cladding have permitted metallurgical bonding with less dilution and improved mechanical characteristics [7, 8].

Recent research has focused on the development of smart and functional coatings. Smart coatings can respond to environmental stimuli, including temperature, light, and pH, enabling self-healing, anti-corrosion, and thermal regulation [9, 10]. Self-healing polymeric coatings packed with microcapsules or nanocontainers can release

 DOI: 10.1201/9781003653226-15

healing agents when cracks form, restoring protective properties [11–13]. Nanostructured coatings are used in high-performance industries such as aerospace and energy to improve heat barrier characteristics, reduce friction, and increase hardness [14, 15].

Advanced cladding techniques are rapidly gaining traction, notably in additive manufacturing (AM) and laser-based systems. Laser cladding allows for precision deposition of high-performance alloys, ceramics, or composites onto metal substrates, resulting in dense, well-bonded layers with customized features [16]. When combined with directed energy deposition (DED) in metal 3D printing, cladding makes it easier to repair, refurbish, and hybrid manufacture essential components [17]. Recent cladding innovations, including the utilization of laser and cold metal transfer (CMT) technologies, are improving precision, performance, and sustainability in a variety of industrial applications [18]. These advances not only improve component functionality but also promote sustainability by extending service life and reducing material waste.

The purpose of this chapter is to provide a complete review of the most recent developments and innovations in advanced coating and cladding technologies. It will start with core concepts and classifications, followed by a critical comparison between established and innovative techniques. Advanced materials, such as nanocomposites, functionally graded materials (FGMs), and high-entropy alloys (HEAs), will be examined in relation to next-generation coatings and cladding. Case studies from areas such as maritime, energy, and biomedicine will demonstrate real-world applications. In addition, the chapter will discuss present constraints, environmental and economic factors, and future research areas, such as the integration of artificial intelligence (AI), robotics, and sustainable materials. This chapter hopes to serve as a useful resource for researchers, engineers, and industry practitioners interested in surface engineering and materials development.

15.2 FUNDAMENTALS OF COATING AND CLADDING TECHNOLOGIES

Coating and cladding are two different, but frequently complementary, surface engineering techniques used to improve the qualities and performance of materials by altering their surface features. While both approaches preserve substrates and improve service life, they have different principles, thickness scales, and bonding mechanisms. Coating is the application of a relatively thin layer—ranging from a few nanometers to several hundred micrometers—on the surface of a component via physical, chemical, or mechanical methods [4]. These coatings can be functional, decorative, or protective and are chosen based on specific operational requirements, such as temperature resistance, corrosion resistance, or reduced friction [2].

Cladding, on the other hand, is the deposition or attachment of a thicker layer, typically 1 mm or greater, via metallurgical or mechanical bonding to improve load-bearing capacity, wear resistance, or structural integrity [1]. Cladding techniques are very beneficial for repairing worn parts, increasing mechanical strength, and combining dissimilar materials to offer multifunctional features. Traditional cladding technologies include roll bonding, explosion welding, and overlay welding, while newer methods include laser cladding, friction stir cladding, and CMT [7, 18].

A vast range of coating methods exists, each adapted to specific material systems and performance needs. PVD and CVD are popular methods for thin-film applications in electronics, optics, and tool coatings. These techniques provide superior adhesion, purity, and microstructural control [19]. Thermal spray coatings—such as plasma spray, flame spray, and high-velocity oxy-fuel (HVOF)—are popular in industrial applications because they can coat vast areas with thick, wear-resistant layers [15]. Furthermore, modern technologies like cold spray and aerosol deposition enable coating without melting the feedstock, allowing for use on heat-sensitive substrates [6].

Laser and CMT cladding—a subset of DED—melts a metal powder or wire feedstock onto a substrate using a high-power laser beam, resulting in a dense, metallurgically bonded layer with precise geometry and dilution control [16]. This technology allows for precision coating of complex geometries and is increasingly used in aerospace and tool repair applications. Friction stir cladding, on the other hand, is a solid-state method that generates heat by friction between a rotating tool and the materials, resulting in effective bonding without melting. This method is especially appealing for materials like aluminum and magnesium alloys, which are prone to oxidation during fusion-based procedures [20].

Coatings can be manufactured out of metals, ceramics, polymers, or composites, depending on the application. Ceramic coatings, such as alumina or zirconia, are commonly employed as thermal barriers and insulation due to their low thermal conductivity and high temperature stability [21]. Metal-based coatings, such as chromium, titanium, or nickel alloys, are used to prevent corrosion and wear in mechanical components [22]. Polymer-based coatings are more chemically resistant and flexible, but their temperature sensitivity limits their use. Cladding materials, on the other hand, are predominantly metallic, particularly stainless steels, nickel-based alloys, and cobalt-based superalloys, due to their better mechanical qualities and compatibility with substrate metals [18].

The bonding process is an important distinction between coating and cladding. Coatings can adhere using mechanical interlocking, Van der Waals forces, or chemical bonding, whereas cladding typically relies on metallurgical bonding via heat and pressure or fusion [9]. Surface preparation—such as grit blasting, machining, or chemical treatment—is essential for assuring bond strength, reducing flaws, and increasing service life. Temperature, feed rate, and ambient environment are all important process characteristics that influence the end product's quality and performance [23].

15.3 TRADITIONAL VERSUS ADVANCED TECHNOLOGIES

Surface engineering has advanced substantially over the last few decades, from relatively simple protective approaches to sophisticated processes designed for high-performance and multifunctional use. The distinction between classic and sophisticated coating and cladding methods is critical for understanding the evolution of current manufacturing and maintenance approaches.

Electroplating, hot-dip galvanizing, and painting have long been used to protect industrial surfaces. Electroplating is still popular because of its ability to produce smooth, corrosion-resistant surfaces from metals such as chromium, zinc, and nickel. Despite its efficacy, electroplating frequently uses hazardous electrolytes, posing environmental and occupational health concerns [1]. Similarly, hot-dip galvanizing, in which steel is immersed in molten zinc, provides excellent corrosion resistance and is frequently employed in building and infrastructure. However, it has little microstructural control and may produce coatings that are uneven or excessively thick for precise applications.

In the field of cladding, traditional methods such as manual welding, roll bonding, and explosive welding have been used to apply thick protective coatings or bond different materials. These methods are generally affordable and accessible; however, they are limited by precision, material compatibility, and process control. For example, explosive welding produces strong metallurgical bonding but is difficult to apply to small or complex components, and the use of explosives raises safety issues [1].

The development of modern technology has substantially increased the capabilities of surface modification. Laser cladding, for example, provides fine control over heat input, resulting in metallurgically bonded layers with little dilution and heat-affected zones. This allows for the application of high-performance materials, such as nickel-based alloys or ceramic-metal composites, to complex geometries with outstanding wear and corrosion resistance [7]. When compared to traditional welding, laser cladding provides more dimensional accuracy and material efficiency, making it excellent for aerospace, machining, and power-generating applications.

Another important breakthrough is cold spray coating, a solid-state method that accelerates powder particles without melting them. Unlike thermal spray technologies that use high temperatures, cold spray preserves the feedstock material's intrinsic qualities while preventing thermal degradation, oxidation, and substrate distortion. This method is especially useful for repairing delicate components in the aerospace and defense industries, where preserving the original microstructure is crucial [6].

In addition, PVD and CVD procedures enable atomic-level control of coating thickness, content, and structure. These vacuum-based methods are routinely employed to create thin, hard coatings for cutting instruments, semiconductors, and optics. They provide benefits such as high purity, great adhesion, and the capacity to create coatings with nanometer-scale precision [23]. However, they are extremely expensive and need controlled settings, thus limiting their use in low-cost applications.

Advanced technologies have various benefits over traditional methods, including improved performance, material efficiency, and environmental sustainability. Many current procedures aim to reduce hazardous waste and energy consumption while increasing coating quality. Laser cladding and cold spray, for example, generate less waste and require less post-processing, making them compatible with green manufacturing aims [6, 7].

Despite these benefits, traditional technologies continue to play an important role, especially in large-scale, low-cost, and noncritical applications. However, as sectors want longer-lasting components, greater reliability, and specialized functions, sophisticated techniques are projected to become more common. The growing integration of automation, AI, and digital monitoring improves the precision and repeatability of these sophisticated processes, establishing them as critical instruments for future-ready production systems [1, 23, 24].

15.4 EMERGING COATING TECHNOLOGIES

Coating technologies have advanced far beyond their traditional function of surface protection, embracing multifunctionality, environmental sustainability, and digital technology integration. The increasing demands of contemporary engineering applications in the fields of electronics, biomedicine, energy, aerospace, and automotive are driving this change. Surface engineering is changing as a result of the demand for surfaces that have specific optical, thermal, electrical, or self-healing qualities in addition to resistance to corrosion and wear. As a result, nanoscale materials, intelligent response systems, sustainable formulations, hybrid manufacturing, and intelligent control mechanisms are becoming more and more prominent in current coating development trends.

15.4.1 Nanostructured and Multifunctional Coatings

The development of nanostructured coatings, which outperform traditional coatings in terms of mechanical, thermal, and chemical properties, is one of the most important developments. These coatings are created by modifying materials at the nanoscale, giving them special qualities, including more surface area, better adherence, and more effective barriers. Nanocomposite coatings, for instance, which are composed of metallic, ceramic, or polymeric nanoparticles scattered throughout a matrix, provide exceptional hardness, low friction, and resistance to wear and oxidation [25]. Through the use of processes like PVD and atomic layer deposition (ALD), multilayered nanostructured coatings allow for the creation of interfaces that regulate the spread of cracks, improving toughness and fatigue life [23].

These coatings are now widely used in cutting tools, turbine blades, medicinal implants, and electronic gadgets. For example, in machining processes, TiAlN and CrN multilayers with nanometric thicknesses have demonstrated enhanced oxidation resistance and thermal stability. Additionally, fluorinated silica or carbon-based hydrophobic and superhydrophobic nanocoatings are being investigated for surfaces that are self-cleaning, anti-fouling, and anti-icing [24]. These features are essential for solar panel technologies, aircraft, and maritime situations, where surface contamination significantly affects performance.

15.4.1.1 Smart and Responsive Coatings

The creation of smart coatings, sometimes known as functional or responsive coatings, represents another emerging area. These materials are designed to respond when exposed to environmental stimuli like light, stress, pH, humidity, temperature, or electrical fields. One extensively researched class consists of self-healing coatings, which use networks of polymers or microcapsules that release healing chemicals in response to cracks or other flaws. After being released, these agents fill the gaps and polymerize, increasing service life and restoring barrier integrity [26]. Other smart coatings include conductive or piezoelectric coatings, which transform mechanical deformation into electrical impulses and are useful for sensors and energy harvesting devices, as well as thermochromic and photochromic systems, which change color in response to temperature or light [27]. Drug-eluting coatings, which release therapeutic compounds in response to biological signals, are becoming more and more popular in the biomedical area, particularly in wound dressings and implantable devices. Real-time monitoring of stress, temperature, or corrosion in structural materials is now possible because of embedded sensing layers created by the convergence of microelectronics and coatings. These advancements support transportation and critical infrastructure systems by enabling structural health monitoring and predictive maintenance.

15.4.1.2 Sustainable and Environmentally Friendly Coatings

The transition to green and sustainable coatings is being accelerated by rising environmental consciousness and stricter laws governing hazardous compounds. Significant volumes of volatile organic compounds (VOCs) are released by conventional solvent-based coatings, endangering human health and the environment. Due to their lower toxicity and reduced carbon footprint, aqueous coatings, powder coatings, and UV-curable formulations are becoming more and more popular [21]. Waterborne coatings, which employ water as a solvent, maintain good adhesion, gloss, and durability while drastically reducing VOCs. Powder coatings completely eliminate the need for solvents and enable high material utilization rates with little waste by electrostatically spraying dry powder and then thermally curing it. UV-curable coatings are energy-efficient and appropriate for substrates that are sensitive to heat since they cure quickly at room temperature.

Bio-based coatings are another novel class that are made from renewable feedstocks such as cellulose, chitosan, and lignin. Antimicrobial, corrosion-resistant, and even self-healing qualities can be incorporated into these materials. The biodegradability and antimicrobial properties of chitosan-based coatings have demonstrated potential for use in food packaging and biomedical implants [28].

15.4.2 Additive Manufacturing and Hybrid Deposition

Advanced surface functionalization is becoming possible thanks to the convergence of coating technology and AM. Specifically, methods like laser surface engineering, cold spray AM, and DED are being incorporated into AM platforms to create components with integrated functionality or surface protection [29].

These hybrid technologies enable compositionally customized surfaces, gradient coatings, and site-specific reinforcement—all of which are especially useful in the medicinal and aerospace sectors. For example, in a single build process, a turbine blade might be 3D printed with a corrosion-resistant metallic alloy at the base and a wear-resistant ceramic-metal layer at the leading edge. Additionally, additive-enabled coatings are suitable for contemporary digital manufacturing methods because they facilitate quick prototyping and minimize material waste.

Furthermore, the creation of dense, metallurgically bonded layers with few heat-affected zones is made possible by the exceptional control over microstructure and phase composition provided by laser-based surface modification techniques, including laser remelting, laser alloying, and laser cladding [7]. These procedures are increasingly being employed for customization, life extension, and component repair.

15.4.3 Integration of AI, Sensors, and Process Automation

With the introduction of Industry 4.0, coating processes are becoming more digitalized by integrating AI, machine learning (ML), and real-time monitoring. The development, application, and maintenance of coatings are being revolutionized by these tools. Now, coating performance may be predicted by AI algorithms based on formulation factors and environmental variables, greatly cutting down on experimental cycles and expenses [17].

To keep an eye on crucial factors, including layer thickness, curing status, adhesion, and surface temperature during deposition, coating equipment is incorporating in-situ sensing technologies, such as acoustic emission sensors, laser interferometry, and infrared thermography. This enables closed-loop control systems to automatically modify processing variables in order to guarantee constant quality and reduce errors.

Additionally, coating processes are being simulated and optimized before implementation using digital twins, which are virtual representations of actual systems. By enabling the testing of various materials, techniques, and environmental factors, these virtual copies lower risk and increase efficiency. It is anticipated that surface engineering will increasingly include AI and IoT as processing power and sensor technologies improve.

15.5 EMERGING CLADDING TECHNOLOGIES

Enhancing surface qualities, including resistance to corrosion, wear, thermal stability, and aesthetic appeal, has long been the goal of cladding technologies. Nonetheless, the contemporary cladding technique prioritizes multifunctionality, accuracy, sustainability, and tolerance to harsh environments, much like coatings do. In sectors where structural components are subjected to demanding working circumstances, such as aerospace, nuclear, oil and gas, defense, and power production, cladding has become particularly important. Utilizing cutting-edge materials, digital tools, hybrid methods, and bespoke microstructures to create surface layers with unique performance characteristics characterizes the current innovation wave.

15.5.1 Advanced Materials and Alloy Design

In order to endure severe mechanical, thermal, and chemical stresses, modern cladding methods are increasingly depending on metal matrix composites and high-performance alloys. These include titanium alloys, cobalt–chromium alloys, nickel-based superalloys, and HEAs, which are becoming more and more well-known because of their remarkable mechanical strength and resistance to corrosion [30]. Due to their exceptional hardness, wear resistance, and thermal stability, HEAs—which are composed of five or more primary elements in nearly equal proportions—are appropriate for high-temperature and nuclear settings.

Furthermore, chromium carbide–cobalt and tungsten carbide–nickel ceramic-metal (cermet) claddings provide exceptional wear resistance and are frequently used in material handling, drilling, and mining applications. In addition to improving microstructural stability and toughness, cladding layers enhanced with nanoparticles or fibers provide improved fatigue resistance and longer service life [31].

15.5.2 Laser and Electron Beam Cladding

The increasing use of high-energy beam techniques, especially laser and electron beam cladding, is one of the most significant developments in cladding. These methods provide dense, metallurgically bonded layers with great surface integrity and less deformation by offering outstanding control over heat input and dilution [7].

With laser cladding, cladding material—which can be supplied in wire or powder form—is fused onto a substrate using a concentrated laser beam. Thin, flawless overlays with a personalized composition are made possible by the high precision and localized heating. Applications include mold restoration, biomedical implants, and turbine blade repair. In contrast, electron beam cladding offers greater penetration and is perfect for thick, dense layers on structural elements. These beam-based procedures now frequently incorporate closed-loop feedback control, ensuring consistent metallurgical quality and uniform layer shape.

Additionally, hybrid systems are being explored that combine laser cladding with other methods to reduce porosity and refine grain structure, such as mechanical stirring, magnetic fields, or ultrasonic vibration [32]. These techniques provide improved bonding with the substrate and homogeneous distribution of reinforcing particles.

15.5.3 Cold Spray and Additive-Based Cladding

The use of solid-state cladding has greatly increased with the development of cold spray technology. Cold spray, which works at very low temperatures, propels powdered materials at supersonic speeds to generate dense, oxide-free coatings upon impact, in contrast to thermal spray or fusion-based techniques [6]. This method is especially appealing for reactive metals like magnesium, titanium, and aluminum because it maintains the inherent qualities of the substrate and cladding materials.

Cold spray is particularly well-suited for defense equipment, aerospace parts, and corrosion-prone structures where minimizing thermal distortion and metallurgical changes is crucial. Additionally, the method facilitates gradient layers and functional transitions between incompatible materials by supporting multi-material deposition [6].

DED and wire arc additive manufacturing (WAAM) techniques, capable of both building components and applying cladding layers simultaneously, are also outcomes of the combination of cladding and AM platforms [33, 34]. Figure 15.1 shows the CMT cladding process of a steel-based alloy on an Al substrate. These hybrid solutions reduce production time and material waste by enabling repair, customization, and functional grading all in one step [29].

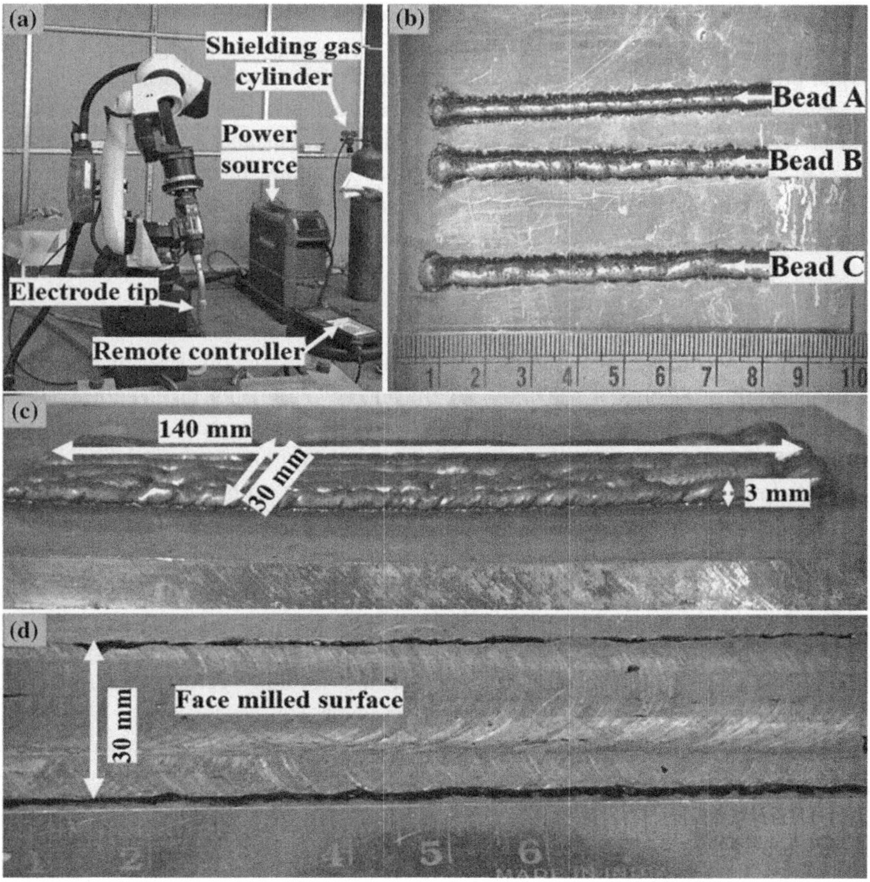

FIGURE 15.1 Cladding layer produced by cold metal transfer technology: (a) CMT setup, (b) three types of beads, (c) as-deposited bead Type C, and (d) machined bead Type C [33].

15.5.4 Functionally Graded and Multi-layered Clads

The use of FGMs, whose composition or structure progressively varies from one surface to another, is a significant advancement in advanced cladding. This gradient lessens the chance of delamination and cracking under operating stress by minimizing sudden changes in thermal expansion, hardness, or chemical characteristics [35]. FGMs are being extensively investigated for use in biomedical implants, engine cylinders, and turbine blades.

Similar to this, materials having several purposes, such as an exterior layer that resists corrosion, an intermediate layer that resists wear, and a ductile base layer, can be stacked in multi-layered clad systems. These tailored designs make it possible to create surfaces that are best suited for particular uses, including impact resistance in armor systems or wear and corrosion in slurry conditions [36]. Figure 15.2 shows the possible classifications of production methods for functionally graded coatings.

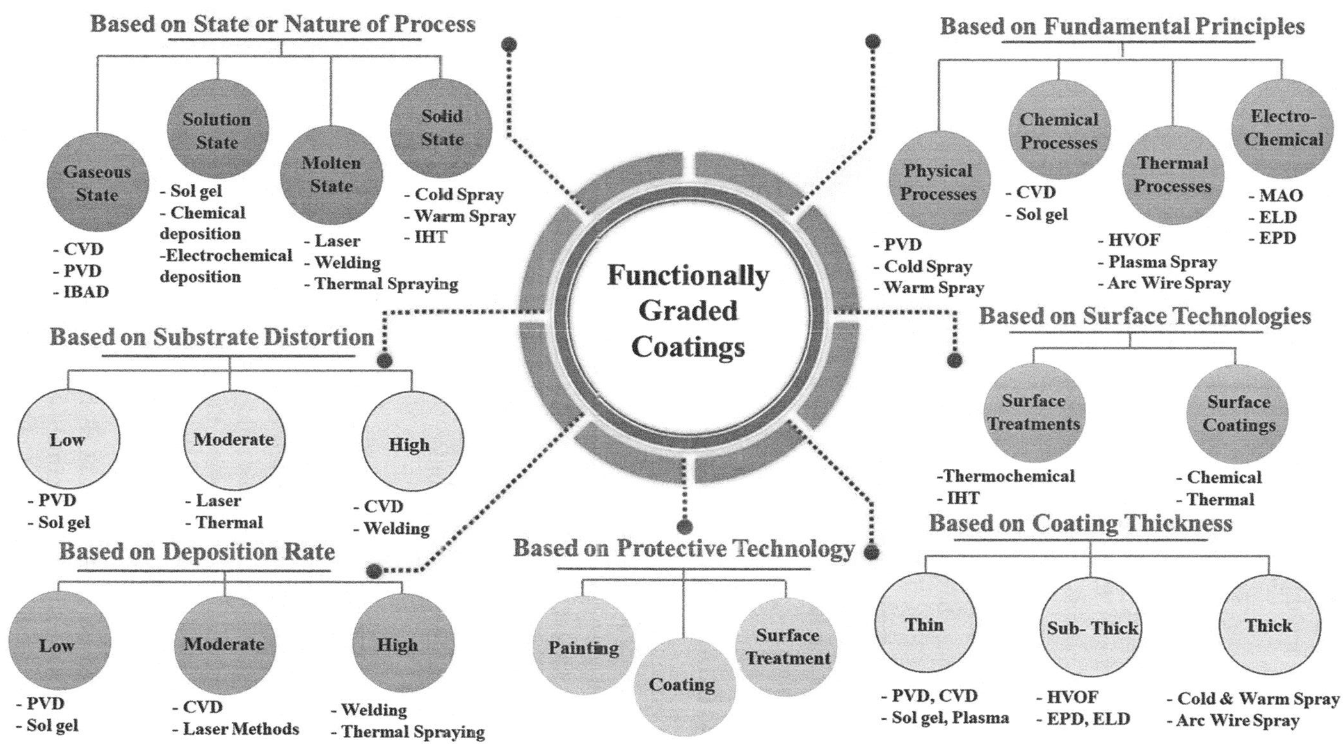

FIGURE 15.2 Possible classifications of production methods for functionally graded coatings [36]. **Reproduced with permission.**

Advanced control techniques, which coordinate feedstock delivery, heat input, and deposition path, are frequently used to produce these intricate systems. To guarantee homogeneity and reproducibility, recent advancements include the utilization of robotic tool paths, multi-powder feeders, and real-time heat mapping.

15.5.5 Digitalization, Simulation, and AI Integration

Cladding technologies are being equally impacted by the digital transformation of manufacturing. Digital twins, ML, and AI are currently being used to optimize cladding operations in terms of quality control, material utilization, and energy efficiency. Rapid process development and adaptive control are made possible by these systems' ability to forecast bead geometry, residual stress, and thermal history based on process parameters [38].

Engineers can observe phase change, powder flow, and melt pool dynamics using simulation techniques based on Finite Element Analysis (FEA) and Computational Fluid Dynamics (CFD). This degree of process awareness is essential for reducing trial-and-error and accelerating scale-up. Additionally, real-time temperature, velocity, and surface roughness input can be obtained from sensor-integrated cladding heads, allowing for intelligent control systems and reliable performance across production batches.

15.6 MATERIAL INNOVATIONS

Advances in material science are inextricably connected to the creation of sophisticated coating and cladding systems. The need for materials with excellent qualities is constant as industries push the limits of performance in harsh situations, like high temperatures, corrosive media, and abrasive conditions. HEAs, amorphous metals, nanocomposites, ceramics, and bio-derived materials are now among the new materials for coatings and claddings, going beyond conventional alloys. In addition to improved protection, these new material classes are intended to provide multifunctionality, sustainability, and adaptability.

15.6.1 High-Entropy Alloys (HEAs)

The design of HEAs is a radical departure from traditional alloys. The presence of five or more principal elements in nearly equal ratios in HEAs results in high mixing entropy and the production of simple solid-solution phases with exceptional properties. Their exceptional wear resistance, thermal stability, and corrosion resistance have led to a significant increase in their use in coatings and claddings, particularly in harsh service environments such as nuclear reactors, power plants, and aerospace engines [38].

Coatings based on HEA, like CoCrFeNiAl and CoCrFeMnNi, provide exceptional hardness and resilience to thermal fatigue. Figure 15.3 depicts the morphologies and elemental distribution of CoCrFeNiAl coatings. Applying them using thermal spray or laser cladding techniques produces thick, sticky coatings that are resistant to oxidation at high temperatures and cyclic loading [39]. Additionally,

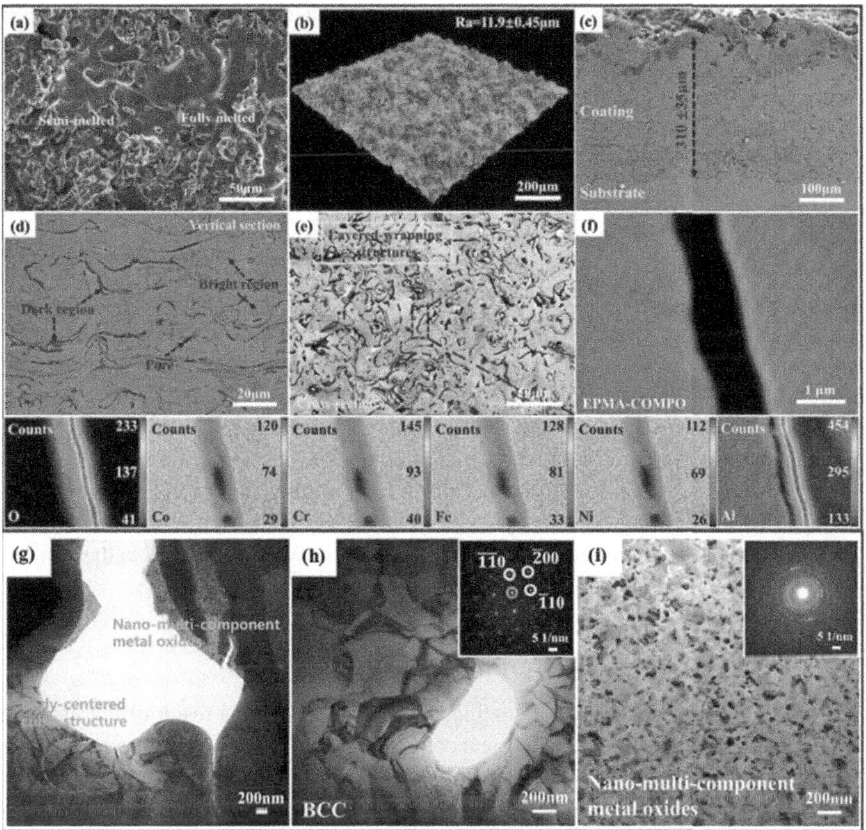

FIGURE 15.3 Morphologies and elemental distribution of CoCrFeNiAl coatings: original surface (a); three-dimensional surface (b); vertical section and cross-section (d–f); TEM images and SAED patterns of matrix and nano multi-component oxides (g–i) [39]. **Reproduced with permission.**

HEAs are tunable, allowing property optimization for specific applications by varying elemental composition.

15.6.2 Amorphous and Metallic Glass Coatings

Using amorphous metals, sometimes referred to as metallic glasses, represents another emerging material breakthrough gaining traction. These materials' exceptional strength, superior corrosion resistance, and minimal friction are all results of their disordered atomic structure. When corrosion and wear resistance are critical, amorphous alloy coatings are especially helpful in oil and gas, marine, and biomedical applications [40].

These coatings are usually deposited utilizing quick solidification techniques like HVOF spraying or laser cladding because of their metastable nature. In corrosive and erosive conditions, alloys like Fe-based or Zr-based metallic glasses can function better than stainless steel or chrome coatings, providing long-term performance with lower maintenance costs [41].

15.6.3 Ceramic and Cermet Composites

Coatings made of ceramic have long been prized for their chemical inertness and thermal stability. Recent advancements in materials science, however, have produced sophisticated ceramic composites and cermets that overcome earlier shortcomings, including brittleness and low toughness. For instance, zirconia-based ceramics stabilized with magnesia or yttria are currently frequently utilized in thermal barrier coatings (TBCs) due to their great phase stability at high temperatures and low thermal conductivity [42].

The hardness of ceramics and the toughness of metals are combined in cermets, which are composites of ceramic particles embedded in a metallic matrix (such as WC–Co or TiC–Ni). In the drilling, cutting, and mining industries, they are extensively used in cladding layers to provide protection against wear. New research has focused on nano-structured and gradient cermets, which enhance adhesion and reduce residual stresses [43].

15.6.4 Nanocomposites and Reinforced Polymers

A novel class of nanocomposite coatings has been created as a result of the incorporation of nanoparticles into conventional coatings. These materials improve mechanical, barrier, and functional properties by utilizing the large surface area and reactivity of nanoscale reinforcements, such as carbon nanotubes, graphene, SiC, or TiO_2. In addition to decreasing permeability and surface roughness, nanocomposite coatings can dramatically increase hardness, scratch resistance, and thermal conductivity [44].

Nanofillers improve flame retardancy, UV stability, and antibacterial activity in polymeric coatings, which is especially helpful in applications related to electronics, packaging, and healthcare. Furthermore, hybrid organic–inorganic systems, particularly in protective coatings and flexible electronics, provide a compromise between performance and flexibility.

15.6.5 Bio-based and Sustainable Materials

Bio-derived and sustainable coating materials are increasingly becoming popular as environmental rules tighten and the circular economy movement gains momentum. The development of biodegradable coatings that are non-toxic and renewable involves innovations such as the usage of proteins, chitosan, lignin, and cellulose nanofibers [45]. These materials show great promise in low-impact construction, medical gadgets, and food packaging.

For instance, starch-based polymers and polyurethanes produced from lignin are being investigated as binders in anticorrosive coatings. Additionally, chromates and other hazardous chemicals that have historically been utilized in metal coatings are being replaced by green corrosion inhibitors made from plant extracts and amino acids [46]. Although still in their early stages, these materials offer surface engineers with a strategic pathway toward sustainability.

15.6.6 Material Tailoring for Smart Coatings

Finally, the creation of responsive materials that can respond to environmental stimuli like temperature, pH, moisture, or mechanical damage has become necessary due to the rise of smart coatings. Metal-organic frameworks (MOFs) that offer properties like self-healing, sensing, or energy storage; conductive polymers; and shape-memory alloys are examples of material breakthroughs in this area [47].

Coatings having phase-change materials (PCMs) incorporated in them are used to regulate temperature in the marine and aerospace industries. Similar to this, materials that can mend themselves using reversible polymer networks or encapsulated healing agents are being actively developed for long-term defense and infrastructure uses.

15.7 APPLICATIONS AND CASE STUDIES

The widespread use of sophisticated coating and cladding technologies in important industrial areas underscores their practical usefulness. These technologies are now essential for improving the lifespan, performance, and durability of vital parts exposed to challenging conditions. Applications in the actual world, such as biomedical implants and aeronautical engines, demonstrate the worth and efficiency of contemporary surface engineering methods.

15.7.1 Aerospace and Defense Applications

Protective coatings are crucial in the aerospace industry because aircraft components are subjected to high mechanical and thermal loads. To lessen heat transfer and oxidation, TBCs, especially those based on yttria-stabilized zirconia (YSZ), are applied to combustion chambers and turbine blades. Electron-beam PVD (EB-PVD), which creates a columnar microstructure perfect for enduring thermal cycling, is one sophisticated use. In military jet engines, where component longevity and efficiency are crucial, this method is frequently applied to turbine airfoils [48].

Additionally, laser cladding has become more and more common for restoring aeronautical components like landing gear and compressor blades. Hard-facing materials, such as alloys based on cobalt or nickel, allow engineers to prolong component life without having to replace the entire part. In order to provide ballistic resistance without appreciably increasing weight, armored vehicles and aircraft panels employ composite cladding materials, such as ceramics and titanium alloys.

15.7.2 Energy and Power Generation

Technologies for coating and cladding are essential for safeguarding equipment in energy systems, where erosion, corrosion, and high temperatures are ongoing dangers. Ceramic coatings are used in gas turbines to prevent thermal wear on superalloy blades. In the meantime, fossil fuel facilities are increasingly using laser or cold spray cladding to strengthen the surfaces of heat exchangers, boiler tubes, and valves.

A real-world example is a thermal power plant in India where boiler tubes were coated with Cr_3C_2–NiCr using HVOF spraying. As a result, the frequency of maintenance was greatly reduced, with wear being reduced by 70% and corrosion resistance improved by over 60% [49]. By applying corrosion-resistant coatings to reactor components, laser cladding helps nuclear plants better tolerate thermal stress and radiation damage.

15.7.3 Oil, Gas, and Petrochemical Industries

Coatings are essential for both performance and safety in the oil and gas industry, which frequently encounters aggressive corrosive media and abrasive slurries. Cladding layers made of high-nickel alloys, such as Inconel 625, are frequently used to safeguard pipelines, valves, and subsea components. Because it enables precise deposition with little heat input, maintaining the integrity of the underlying material, laser cladding has proven very helpful in this industry.

In one well-known instance, a large offshore drilling company used automated laser cladding to fix worn valve seats made of alloys based on nickel. By doing this, the component's functionality was restored, and its operating life was increased by more than three years. Cermet coatings, especially those containing tungsten carbide, are frequently used to protect components like sand screens and mud motor rotors. These coatings provide exceptional erosion resistance in challenging downhole environments [39].

15.7.4 Automotive and Heavy Machinery

The automotive industry uses advanced coatings extensively to increase part durability, lower emissions, and improve fuel efficiency. To better control heat, engine pistons and cylinder heads are coated with thermal coatings that are plasma-sprayed. Diamond-like carbon (DLC) coatings are applied to camshafts and valve lifters in high-performance engines to reduce wear and friction. In the heavy machinery and mining industries, cladding plays a crucial role in extending the service life of equipment exposed to abrasive environments. Hardfacing using chromium carbide overlays is commonly applied to excavator teeth, crusher liners, and drill rods.

15.7.5 Biomedical and Healthcare Devices

In the biomedical field, material biocompatibility, sterility, and resistance to bodily fluids are essential. Coatings such as titanium nitride (TiN) and hydroxyapatite are used extensively on implants, including hip joints, dental prosthetics, and spinal fusion devices. These coatings enhance bone integration, reduce inflammation, and prevent bacterial colonization.

One case involved spinal implants coated with TiO_2 via PVD, which showed significantly lower ion release and improved healing outcomes compared to uncoated versions. Similarly, in surgical tools and catheters, silver nanoparticle-embedded coatings have been adopted to provide antimicrobial functionality and reduce hospital-acquired infections [38].

15.7.6 Marine and Coastal Infrastructure

Biofouling and rust are recurring problems in marine environments. Ship hulls, offshore platforms, and undersea pipelines are coated with protective materials, usually based on epoxy or polyurethane systems with incorporated zinc or aluminum particles. In salty settings, these coatings greatly increase the service life of steel buildings by acting as sacrificial barriers.

One prominent instance was the use of nickel–aluminum bronze that had been laser-cladded on naval ship propeller shafts. Compared to conventional techniques, this coating system fared far better against cavitation erosion and saltwater corrosion, which led to a threefold increase in maintenance intervals and a reduction in lifecycle costs [39]. Aligned with sustainability objectives, the adoption of foul-release coatings in marine vessels has also decreased the demand for biocidal paints, which are hazardous to the environment.

15.8 POST-PROCESSING IN COATING AND CLADDING TECHNOLOGIES

In advanced coating and cladding workflows, post-processing is an essential step that is frequently needed to optimize surface properties, fix irregularities in deposition, or boost the finished product's functional performance. Although protective and functional layers can be produced by deposition methods, including thermal spraying, laser cladding, or PVD, they may also contribute surface roughness, residual stress, porosity, or adhesion issues that call for further care.

15.8.1 Surface Finishing and Polishing

Surface finishing, which involves lapping, polishing, or grinding to attain the required surface roughness and dimensional precision, is one of the most popular post-processing procedures. Extremely smooth and flawless surfaces are necessary for applications such as biomedical implants or aeronautical engine components in order to minimize friction, improve wear resistance, and guarantee biocompatibility. For example, thermal spray coatings usually result in rough surfaces that need to be polished or machined to fulfill tolerances [2].

For intricate geometries or high-precision components, like microelectronics or surgical instruments, electropolishing or chemical-mechanical polishing (CMP) is utilized in addition to mechanical finishing. By removing burrs and tiny surface flaws, these procedures lessen the possibility of fatigue failure and stress concentrations when in use.

15.8.2 Heat Treatment and Stress Relief

Rapid heating and cooling cycles in coatings and clads deposited by high-energy methods, such as arc welding, CMT cladding, or laser cladding, frequently result in residual thermal strains. Annealing or heat treatment after deposition is frequently employed to reduce these pressures and enhance the coated layer's microstructural stability. In multi-layer clads, where thermal mismatch between layers can result in warping or delamination if improperly addressed, stress-relieving heat treatments are particularly crucial [49].

Phase change is also required in certain coatings to activate their final properties. For instance, controlled heat treatment may enhance the hardness or corrosion resistance of nanocrystalline formations or amorphous metallic coatings. In other situations, post-processing ensures metallurgical integration between the coating and the substrate by facilitating diffusion bonding [50, 51].

15.8.3 Sealing and Impregnation

Sealing procedures are essential for closing open porosity and improving corrosion resistance in many porous coatings, particularly those made using flame spray or arc spray processes. When sealants—which can range from organic polymers to ceramic slurries—are impregnated into the coating matrix and allowed to cure, they form a barrier that stops the entry of moisture, chemicals, or gases. In chemical processing or marine conditions, where unsealed coatings are vulnerable to underfilm corrosion, this is especially important.

Sealed TBCs in aircraft have shown noticeably better resistance to gas erosion and high-temperature oxidation. Sealing enhances biocompatibility and lowers inflammatory reactions in biomedical applications by halting the leaching of metallic ions.

15.8.4 Inspection and Non-destructive Testing (NDT)

Quality assurance procedures, which usually involve non-destructive testing (NDT) to confirm coating integrity, thickness, and adherence, are part of post-processing. X-ray tomography, dye penetrant analysis, eddy current inspection, and ultrasonic testing are examples of common NDT techniques. Without causing any harm to the component, these methods identify subsurface cracks, voids, or delamination.

Furthermore, adhesion testing—such as pull-off or scratch tests—is frequently carried out after post-processing to make sure that coatings fulfill industry requirements prior to deployment. A developing approach that makes predictive maintenance and long-term performance monitoring possible is the use of digital twin models and AI-assisted inspection.

15.8.5 Challenges in Post-processing

Although post-processing improves performance, it also makes the manufacturing cycle more complex, expensive, and time-consuming. Each additional step must be supported by the requirements of the application. Excessive polishing, for instance, can weaken engineered surface textures or decrease coating thickness. Furthermore, if heat treatment is not carefully managed, it may cause undesired phase shifts or residual stress accumulation.

Consequently, process integration—the co-optimization of deposition and post-processing steps—is emerging as a major area of study. To increase production efficiency and reproducibility, hybrid manufacturing systems that integrate deposition, finishing, and inspection onto a single platform are also being developed.

15.9 CHALLENGES AND LIMITATIONS

Advanced coating and cladding technologies have many benefits, yet there are barriers to their widespread use. Their extensive use across industries may be impacted by technical difficulties, financial factors, and scalability constraints. For future advancement and to close the gap between laboratory innovation and industry implementation, it is imperative to comprehend these difficulties.

15.9.1 Technical and Material Constraints

The choice and compatibility of materials are one of the main obstacles. The integrity of coatings can be jeopardized by problems, including delamination, residual stress accumulation, and temperature mismatch, and not all coating materials adhere well to all substrates. For example, variations in thermal expansion coefficients can cause peeling or cracking during operation or thermal cycling when dissimilar metals are bonded by thermal spraying or laser cladding [42]. Furthermore, phase transformation or sintering can cause some advanced coatings, including ceramic or oxide layers, to lose their stability over time in high-temperature applications.

Another technical challenge is controlling the microstructure and homogeneity of coatings, particularly in procedures like cold spraying and plasma spraying, where the final layer quality is significantly impacted by temperature, particle velocity, and environmental factors. Achieving consistent results over complex geometries and large surface areas remains a difficult task, even with the help of robotics and real-time monitoring systems.

15.9.2 Process Complexity and Equipment Costs

Numerous cutting-edge coating techniques, including laser cladding, electron-beam EB-PVD, and PVD, call for specialized tools and extremely regulated conditions. Due to their high cost and need for skilled workers, these installations are less feasible for small and medium-sized businesses. Particularly when the coated components have little added value, the initial outlay and ongoing expenses may be unaffordable.

Furthermore, high-energy procedures (such as those based on lasers or electron beams) require a lot of power and process optimization, which might not be acceptable for disposable or low-end parts. The adoption of these technologies is slowed by this economic hurdle in sectors like consumer goods production that have narrow profit margins.

15.9.3 Scalability and Throughput Limitations

Scaling up to industrial numbers is still difficult, even though several coating and cladding methods show great results in research and pilot-scale operations. It can be challenging to maintain consistent coating thickness, repeatability, and process speed when moving from small-scale components to large quantities or large-surface items like structural parts or pipelines.

Additionally, certain methods—like additive laser cladding—are slower than others and better suited for repair applications rather than high-throughput manufacturing. Downtime for coating application can also be a logistical problem in large infrastructure projects, especially in remote or offshore areas with limited on-site equipment and skilled labor.

15.9.4 Environmental and Regulatory Concerns

The selection of materials and techniques for coating and cladding is increasingly influenced by environmental requirements. For example, some high-performance coatings include hazardous heavy metals or VOCs, posing a risk to human health and the environment. Even if eco-friendly coatings are being developed, they are still in their infancy when compared to well-established industrial formulations, even though sustainability is becoming a major design factor.

Waste management is further complicated by the recycling of cladded components and the disposal of leftover coating ingredients. Even coatings that appear to be inert might hinder the recyclability of base metals, requiring extra processing steps that raise the complexity and total expense of the overall lifecycle.

15.9.5 Knowledge and Skill Gaps

Finally, the absence of skilled staff and regulated processes is a major drawback. Proficiency in materials science, process control, and surface engineering is necessary for advanced coating systems. The lack of widespread access to this knowledge hinders the capacity of local industries to successfully implement or sustain such technology in many areas, particularly in emerging economies.

Poor adherence or a coating's quick failure in use might result from incorrect process settings or insufficient surface preparation, even in cases where equipment is available. Investments in education, training, and cooperative industry-academia projects will be necessary to close this talent gap.

15.9.5.1 Future Outlook and Research Directions

The combination of sophisticated material science, sustainable practices, and digital tools is expected to propel the revolutionary expansion of coating and cladding technologies in the future. The next generation of surface engineering solutions will prioritize predictability, sustainability, and efficiency in addition to performance enhancement as industry demands change and environmental laws become more stringent. A number of important topics are emerging as priorities for future studies and industry implementation.

Integration with AI/ML for Predictive Coating Behavior: AI and ML are expected to play a major role in coating behavior prediction and process parameter optimization. Historically, the primary techniques for creating novel coatings have relied on empirical testing and trial-and-error, which are costly and time-consuming. AI algorithms trained on large datasets of material properties and performance factors have made it possible to forecast and simulate the durability, adhesion strength, and thermal resistance of coatings under a variety of service conditions. For example, by establishing a correlation between coating microstructure and performance and feed rate, spray distance, and temperature, ML models can be utilized to optimize spray parameters in thermal spray operations. Additionally, real-time quality control is made possible by predictive analytics, which lowers process variability and material waste. As more data becomes available, AI systems will become central to automated coating design, making surface engineering more intelligent, efficient, and application-specific.

Green and Sustainable Coating Materials: The creation of sustainable and environmentally friendly coatings is accelerating due to the increased emphasis on environmental responsibility on a worldwide scale. Traditional coatings frequently use energy-intensive production techniques, VOCs, or hazardous heavy metals. Future studies are concentrating on environmentally friendly nanomaterials, aqueous systems, and bio-based polymers that match or surpass the performance of conventional formulations. Recent developments in graphene-based coatings and sol–gel technology present viable low-impact substitutes for wear prevention and corrosion resistance. Additionally, scientists are investigating temporary uses for recyclable or biodegradable coatings, especially in consumer electronics, packaging, and biomedical equipment. In order to fulfill future legal and societal demands, it will be essential to incorporate sustainability factors into coating lifecycle analysis, from the sourcing of raw materials to the disposal of end-of-life materials [40].

15.9.6 Automation and Robotics in Application Processes

The application of coatings in industrial settings is changing due to automation. Because they provide accuracy, reliability, and safety, robotic arms and CNC-controlled spray systems are already commonplace in automotive and aerospace production lines. Robotic applicators will be directed by AI-based decision-making to adjust in real-time to surface geometry, environmental conditions, and flaw identification in the next wave of development, which will concentrate on fully integrated smart systems.

Process monitoring and feedback loops will be further improved by developments in sensor technology and machine vision, enabling self-adjusting operations with little assistance from humans. This lowers exposure to dangerous compounds while simultaneously increasing quality and reproducibility. Future robotic platforms will likely include mobile units for large-scale infrastructure, such as ships or bridges, enabling in-situ coating and inspection, even in remote or underwater environments.

15.9.7 Hybrid and Multifunctional Coating Technologies

The upcoming generation of coating strategies is anticipated to be dominated by hybrid technologies, which integrate various deposition techniques or material systems. For instance, functionally graded coatings that seamlessly transition from ductile to hard layers can be created by combining laser cladding with plasma spraying, providing specialized performance across the material's depth.

Research is also being done on multifunctional coatings that combine electrical conductivity, corrosion prevention, heat resistance, and self-healing properties. These coatings are especially useful in fields where weight, space, and dependability are crucial, such as biomedicine, electronics, and aircraft. In order to increase system endurance and safety, future advancements will also concentrate on smart coatings that can detect changes in the environment or structural damage and react on their own.

15.10 CONCLUSION

The traditional protective functions of advanced coating and cladding technologies have rapidly evolved into multipurpose solutions that improve performance, efficiency, and durability in vital industries. This chapter examined how the field is changing as a result of advances, including smart coatings, nanostructured materials, AI-driven process optimization, and sustainable practices. Although these technologies offer many benefits, such as increased corrosion control, wear resistance, and thermal protection, they also present certain limitations. Scalability and broad adoption may be hampered by problems, including exorbitant processing expenses, residual strains, and the requirement for post-processing procedures like heat treatment or polishing. Furthermore, meticulous process control and quality assurance are still necessary when incorporating sophisticated coatings into intricate manufacturing environments. In the future, considerable potential lies in combining intelligent systems, eco-friendly materials, and hybrid deposition techniques. Future coatings will be more flexible, effective, and sustainable with continued research and interdisciplinary cooperation. These developments will not only satisfy performance requirements but also support more general objectives of resource efficiency and environmental responsibility.

REFERENCES

1. Davis, J. R. (Ed.). (2001). *Surface Engineering for Corrosion and Wear Resistance*. ASM International.
2. Pawlowski, L. (2008). *The Science and Engineering of Thermal Spray Coatings*. John Wiley & Sons.
3. Burakowski, T., & Wierzchon, T. (1998). *Surface Engineering of Metals: Principles, Equipment, Technologies*. CRC Press.
4. Leyland, A., & Matthews, A. (2006). Optimization of nanostructured tribological coatings. In Albano Cavaleiro and Jeff Th. M. Hosson (eds.), *Nanostructured Coatings* (pp. 511–538). New York: Springer New York.
5. Beake, B. D. (2022). The influence of the H/E ratio on wear resistance of coating systems–Insights from small-scale testing. *Surface and Coatings Technology, 442*, 128272.
6. Assadi, H., Kreye, H., Gärtner, F., & Klassen, T. J. A. M. (2016). Cold spraying–A materials perspective. *Acta Materialia, 116*, 382–407.
7. Toyserkani, E., Khajepour, A., & Corbin, S. F. (2004). *Laser Cladding*. CRC Press.
8. Lekala, M. B., Van Der Merwe, J. W., & Pityana, S. L. (2012). Laser surface alloying of 316L stainless steel with Ru and Ni mixtures. *International Journal of Corrosion, 2012*(1), 162425.

9. Montemor, M. F. (2014). Functional and smart coatings for corrosion protection: A review of recent advances. *Surface and Coatings Technology*, *258*, 17–37.
10. Baghdachi, J. (2009). *Smart Coatings*. American Chemical Society.
11. White, S. R., Sottos, N. R., Geubelle, P. H., Moore, J. S., Kessler, M. R., Sriram, S. R., . . . & Viswanathan, S. (2001). Autonomic healing of polymer composites. *Nature*, *409* (6822), 794–797.
12. Yang, Y., Wang, Y., Li, M. X., Wang, T., Wang, D., Wang, C., . . . & Wang, H. Y. (2023). Recent progress in self-repairing coatings for corrosion protection on magnesium alloys and perspective of porous solids as novel carrier and barrier. *Journal of Magnesium and Alloys*, *11*(10), 3585–3608.
13. Zhang, F., Ju, P., Pan, M., Zhang, D., Huang, Y., Li, G., & Li, X. (2018). Self-healing mechanisms in smart protective coatings: A review. *Corrosion Science*, *144*, 74–88.
14. Voevodin, N. N., Balbyshev, V. N., Khobaib, M., & Donley, M. S. (2003). Nanostructured coatings approach for corrosion protection. *Progress in Organic Coatings*, *47*(3–4), 416–423.
15. Mondal, K., Nuñez III, L., Downey, C. M., & Van Rooyen, I. J. (2021). Recent advances in the thermal barrier coatings for extreme environments. *Materials Science for Energy Technologies*, *4*, 208–210.
16. Dutta Majumdar, J., & Manna, I. (2011). Laser material processing. *International Materials Reviews*, *56*(5–6), 341–388.
17. Heralić, A., Christiansson, A. K., & Lennartson, B. (2012). Height control of laser metal-wire deposition based on iterative learning control and 3D scanning. *Optics and Lasers in Engineering*, *50*(9), 1230–1241.
18. Das, B., Panda, B. N., Sharma, F., & Dixit, U. S. (2024). Recent developments in cladding and coating using cold metal transfer technology. *Journal of Materials Engineering and Performance*, *33*(7), 3130–3147.
19. Anders, A. (Ed.). (2000). *Handbook of Plasma Immersion Ion Implantation and Deposition* (Vol. 8). New York: Wiley.
20. Mishra, R. S., & Ma, D. Z. (2005). Friction stir welding and processing. *Materials Science and Engineering: R: Reports*, *50*(1–2), 1–78.
21. Chen, H. F., Zhang, C., Liu, Y. C., Song, P., Li, W. X., Yang, G., & Liu, B. (2020). Recent progress in thermal/environmental barrier coatings and their corrosion resistance. *Rare Metals*, *39*, 498–512.
22. Dennis, R. V., Patil, V., Andrews, J. L., Aldinger, J. P., Yadav, G. D., & Banerjee, S. (2015). Hybrid nanostructured coatings for corrosion protection of base metals: A sustainability perspective. *Materials Research Express*, *2*(3), 032001.
23. Mattox, D. M. (2010). *Handbook of Physical Vapor Deposition (PVD) Processing*. William Andrew.
24. Farooq, S. A., Raina, A., Mohan, S., Arvind Singh, R., Jayalakshmi, S., & Irfan Ul Haq, (2022). Nanostructured coatings: Review on processing techniques, corrosion behaviour and tribological performance. *Nanomaterials*, *12*(8), 1323.
25. Bai, Y., Zhang, H., Shao, Y., Zhang, H., & Zhu, J. (2021). Recent progresses of superhydrophobic coatings in different application fields: An overview. *Coatings*, *11*(2), 116.
26. Yang, Y., & Urban, M. W. (2013). Self-healing polymeric materials. *Chemical Society Reviews*, *42*(17), 7446–7467.
27. Baidya, K., Roy, A., & Das, K. (2024). A review of polymer-matrix piezoelectric composite coatings for energy harvesting and smart sensors. *Journal of Coatings Technology and Research*, *21*(1), 55–85.
28. Budiarso, I. J., Rini, N. D., Tsalsabila, A., Birowosuto, M. D., & Wibowo, A. (2023). Chitosan-based smart biomaterials for biomedical applications: Progress and perspectives. *ACS Biomaterials Science & Engineering*, *9*(6), 3084–3115.

29. Gu, D. D., Meiners, W., Wissenbach, K., & Poprawe, R. (2012). Laser additive manufacturing of metallic components: Materials, processes and mechanisms. *International Materials Reviews*, *57*(3), 133–164.
30. Cui, H., & Jiang, D. (2021). Research progress of high-entropy alloy coatings. *Acta Metallurgica Sinica*, *58*(1), 17–27.
31. Yakovlev, A., Bertrand, P., & Smurov, I. (2004). Laser cladding of wear resistant metal matrix composite coatings. *Thin Solid Films*, *453*, 133–138.
32. Meng, L., Zhao, W., Hou, K., Kou, D., Yuan, Z., Zhang, X., ... & Zeng, X. (2019). A comparison of microstructure and mechanical properties of laser cladding and laser-induction hybrid cladding coatings on full-scale rail. *Materials Science and Engineering A*, *748*, 1–15.
33. Das, B., Panda, B. N., & Dixit, U. S. (2022). Microstructure and mechanical properties of ER70S-6 alloy cladding on aluminum using a cold metal transfer process. *Journal of Materials Engineering and Performance*, *31*(11), 9385–9398.
34. Das, B., Panda, B. N., & Dixit, U. S. (2024). Effect of layer thickness in cold metal transfer cladding of Fe-based ER70S-6 alloy on AA 6061-T6 aluminum alloy. *Journal of Materials Engineering and Performance*, *34*(8), 6440–6462.
35. Bandyopadhyay, A., Dash, A., & Bose, S. Additive manufacturing of multi-material structures. In Bandyopadhyay, A., Dash, A., and Bose, S. (eds.), *Additive Manufacturing* (pp. 244–266). CRC Press.
36. Fathi, R., Wei, H., Saleh, B., Radhika, N., Jiang, J., Ma, A., ... & Ostrikov, K. K. (2022). Past and present of functionally graded coatings: Advancements and future challenges. *Applied Materials Today*, *26*, 101373.
37. Chen, X., Yao, M., Kong, F., Fu, Y., Wu, J., & Zhang, H. (2023). In-situ quality monitoring of laser hot wire cladding process based on multi-sensing diagnosis and machine learning model. *Journal of Manufacturing Processes*, *87*, 183–198.
38. Zhang, Y., Zuo, T. T., Tang, Z., Gao, M. C., Dahmen, K. A., Liaw, P. K., & Lu, Z. P. (2014). Microstructures and properties of high-entropy alloys. *Progress in Materials Science*, *61*, 1–93.
39. Mu, Y. K., Jia, Y. D., Xu, L., Jia, Y. F., Tan, X. H., Yi, J., ... & Liaw, P. K. (2019). Nano oxides reinforced high-entropy alloy coatings synthesized by atmospheric plasma spraying. *Materials Research Letters*, *7*(8), 312–319.
40. Inoue, A., & Takeuchi, A. (2011). Recent development and application products of bulk glassy alloys. *Acta Materialia*, *59*(6), 2243–2267.
41. Zhang, Z., Yuan, J., Jing, Z., Cheng, Y., & Liang, X. (2022). Research progress and development trend of amorphous and nanocrystalline composite coatings: A review. *JOM*, *74*(12), 4597–4611.
42. Padture, N. P., Gell, M., & Jordan, E. H. (2002). Thermal barrier coatings for gas-turbine engine applications. *Science*, *296*(5566), 280–284.
43. Medvedovski, E. (2024). *Advanced Ceramics and Coatings for Wear and Corrosion Related Applications in Modern High-Efficient Coal Production and Processing: A Technical Review*. Ceramics International.
44. Nguyen-Tri, P., Nguyen, T. A., Carriere, P., & Ngo Xuan, C. (2018). Nanocomposite coatings: Preparation, characterization, properties, and applications. *International Journal of Corrosion*, *2018*(1), 4749501.
45. Iqbal, Y., Ahmed, I., Irfan, M. F., Chatha, S. A. S., Zubair, M., & Ullah, A. (2023). Recent advances in chitosan-based materials; The synthesis, modifications and biomedical applications. *Carbohydrate Polymers*, *321*, 121318.
46. Ahmed, M. A., Amin, S., & Mohamed, A. A. (2024). Current and emerging trends of inorganic, organic and eco-friendly corrosion inhibitors. *RSC Advances*, *14*(43), 31877–31920.

47. Nagappan, S., Moorthy, M. S., Rao, K. M., & Ha, C. S. (2016). Stimuli-responsive smart polymeric coatings: An overview. *Industrial Applications for Intelligent Polymers and Coatings*, *1*, 27–49.
48. Seitov, B., Kurbanbekov, S., Baltabayeva, D., Kakimzhanov, D., Katpayeva, K., Temirbekov, A., . . . & Mussakhan, N. (2025). Review of physical and mechanical properties, morphology, and phase structure in Cr3C2-NiCr composite coatings sprayed by HVOF method. *Coatings*, *15*(4), 479.
49. Das, B., Panda, B. N., & Dixit, U. S. (2024). Effects of heat treatment on the mechanical properties of Fe-based ER70S-6 cladding on aluminum substrate using cold metal transfer process. *Journal of Materials Engineering and Performance*, *33*(1), 173–193.
50. Jing, Z., Xu, P., Liu, Q., & Yu, C. (2023). Residual stress release during laser cladding process: A review. *Journal of Laser Applications*, *35*(3), 031201.
51. Zhou, T., Song, Z., & Sundmacher, K. (2019). Big data creates new opportunities for materials research: A review on methods and applications of machine learning for materials design. *Engineering*, *5*(6), 1017–1026.

16 Dielectric Behavior of $BiVO_4$

Potential Impacts and Applications in Mechanical Engineering Systems

Anima Ghosh and Surja Kiran Ghorui

16.1 INTRODUCTION

Dielectric materials, which generally store and release electric energy, are increasingly in demand in numerous applications of mechanical engineering domains, including sensors, capacitors, actuators, and energy storage devices. The dielectric properties of any compound are a critical aspect in the era of material science and engineering, particularly in storage device applications. One of the emerging compounds, bismuth vanadate ($BiVO_4$), has received significant attention from the scientific community because of its intriguing dielectric characteristics. It is a semiconducting oxide that has been extensively explored for various energy ranges for application in photocatalysis and thermoelectric energy conversion [1–3]. Further, $BiVO_4$-based sensor devices find applications in the aerospace and automotive industries because real-time monitoring of mechanical systems is curtailed for safety and optimal performance. The polarization ability of $BiVO_4$ under an electric field makes it a more suitable candidate for actuator technologies. Here, the compound can convert electrical energy into mechanical energy. A detailed understanding of the dielectric properties of $BiVO_4$ and its potential in mechanical engineering systems will open new avenues for the development of efficient and innovative technologies. The dielectric properties are identified by checking their ability to polarize in the presence of an electric field, which leads to electrical energy storage that can be released when the electric field is removed. The key parameters used to evaluate the dielectric properties of materials are breakdown strength, dielectric constant, and loss factor. $BiVO_4$ with a favorable combination of electronic, structural, and optical responses suits as a potential candidate for high-performance dielectric applications. A recent scientific interest has focused on understanding the impact of various factors such as grain size, temperature, and doping on the dielectric behavior of $BiVO_4$, in order to realize the impact in use in mechanical systems.

DOI: 10.1201/9781003653226-16

$BiVO_4$ is a wide-bandgap semiconductor with a tetragonal crystal structure, composed of bismuth (Bi) and vanadium (V) in a 1:1 stoichiometric ratio. The dielectric properties of $BiVO_4$ have been extensively studied in recent years, primarily owing to their applications in photocatalysis and other electronic devices [2, 3]. The material's dielectric behavior is mainly influenced by its crystal structure, electronic configuration, and the presence of impurities or dopants. The dielectric constant and loss factor of $BiVO_4$ were investigated as functions of temperature and frequency. The results showed that the dielectric constant decreased with increasing frequency, which is typical for most dielectric materials. However, $BiVO_4$ exhibited high dielectric constants at low frequencies, which could be beneficial for applications that require a high capacitive performance. The temperature dependence of the dielectric constant suggested that $BiVO_4$ could operate efficiently in a wide range of temperatures, making it suitable for thermally stable applications [1].

In addition, the dielectric characteristics of $BiVO_4$ with various dopants, such as transition metals (Fe, Co), rare-earth elements (La, Ce), and non-metallic ions (S, N), have been studied to improve their dielectric activity [4–6]. In fact, the addition of dopants helps to change the electronic structure of materials and significantly influences the polarization behavior of compounds. Therefore, it affects the dielectric constant and dielectric loss. Further, rare-earth element doping on $BiVO_4$ results in a significant improvement in the dielectric constant at low frequencies. The direct effect of doping on the electronic structure of $BiVO_4$ results in polarization effects [6]. $BiVO_4$ with doping exhibits enhanced energy storage capabilities and reduced energy loss, making it a demanding material for mechanical engineering applications. Furthermore, the $BiVO_4$ doped with phosphorus and nitrogen enhances the conductivity of the modified compound and increases its dielectric performance. This has opened up new eras for use in applications of sensor and actuator technologies. Additionally, the presence of oxygen vacancies and structural defects in $BiVO_4$ results in a significant enhancement of its dielectric properties. Basically, the defects create localized states in the electronic structure, which can trap more charges and significantly improve the overall dielectric behavior of the compound. The $BiVO_4$ with controlled oxygen vacancy percentage exhibited improved dielectric properties by a higher dielectric constant and a lower loss factor, making it more suitable in storage applications [7].

The Projected Augmented Wave (PAW) method was employed by Walsh et al. [8] to investigate the electronic structure of $BiVO_4$ and reported the electronic structure of its band-edge composition. A comparison study of the electronic structure and photocatalytic activity analyzed with other components of $InMO_4$ (M = V, Nb, Ta) showed that the improved photocatalytic behavior enhanced the mobility of photogenerated holes [9]. In addition, DFT and time-domain density functional theory (TDDFT) were used to determine the minimized recombination from the controlled oxygen vacancy distributions. The localized oxygen vacancies act as charge recombination centers within the band edge, enhancing the nonadiabatic coupling of the valence band maximum and conduction band minimum. This enhances the fast charge and energy losses [10]. Se doping and O-vacancy were explored in a monoclinic $BiVO_4$ structure with a DFT study, and it was reported that Sc doping in vacancy sites creates a mid-gap level and changes the dielectric constant values significantly [11]. A heterojunction study with Bi_2O_4 and $BiVO_4$ reported by Xiaofei et al. [12] showed favorable band alignment for enhancing the hole transfer and enhanced visible light utilization. This report proves that interface engineering with oxides can enhance the charge separation in pristine $biVO_4$ hetero-structures. Other heterojunction and hybrid structures, such as Au/$BiVO_4$, $BiVO_4$/TiO_2, $BiVO_4$/grapheme, $BiVO_4$@C/Ag nanotube, $BiVO_4$–$CaTiO_3$ composite matrix, have also been studied [13–19]. The literature results and reports highlight that $BiVO_4$ exhibits decent electronic properties due to its direct band gap values (~2.3–2.4 eV) and prominent 6s-Bi lone pair hybridization with a tunable band edge. This makes them suitable for photoactive applications.

In this work, we studied the dielectric properties of $BiVO_4$ via Density Functional Theory (DFT) as implemented in the Wien2k code. A deep understanding of the dielectric properties of $BiVO_4$ could drive advances in the development of more efficient and functional materials for a wide range of applications in the mechanical engineering domain.

16.2 COMPUTATIONAL DETAILS

To study the dielectric properties of $BiVO_4$, we have studied DFT as implemented in the Wien2k code [20]. This was framed using the full-potential and linearized augmented plane-wave method. This method is suitable for calculating the electronic structures and optical properties of crystalline materials with high accuracy. Here, the generalized gradient approximation (GGA) along with the Perdew, Burke, and Ernzerhof (PBE) functional was used for the exchange-correlation interactions [21]. To address the typical underestimation of the bandgap in GGA and to achieve better accuracy in optical spectra, additional calculations were performed using the Tran-Blaha modified Becke–Johnson (TB-mBJ) potential [22]. The starting structure of monoclinic $BiVO_4$ (space group I2/b) was obtained from experimental crystallographic data, and complete structural optimization was performed until the forces on all atoms were below 1 mRy/a.u, ensuring a relaxed and stable ground-state geometry. The muffin-tin radii were set to 2.5 a.u. for Bi, 1.9 a.u. for V, and 1.6 a.u. for O, with the plane-wave cutoff defined by $R_{MT} \times K_{max} = 7.0$, providing a converged basis set. The Brillouin zone [23] was framed using a 7 × 7 × 1Monkhorst-Pack k-point mesh, which yielded reliable results for both the electronic and dielectric properties.

The real and imaginary contributions of the complex dielectric function, considering the frequency dependency $\varepsilon(\omega) = \varepsilon_1(\omega) + i\varepsilon_2(\omega)$, can be calculated using WIEN2k's OPTIC and joint modules. The imaginary part, $\varepsilon_2(\omega)$, was derived from the momentum matrix elements between the occupied and unoccupied states, whereas the real contribution, $\varepsilon_1(\omega)$, was obtained through the Kramers–Kronig transformation. The optical properties were calculated for both the parallel and perpendicular components of the dielectric tensor for the crystallographic axes over a photon energy range of 0–15 eV. A Gaussian energy broadening of 0.1 eV was applied to simulate the experimental spectral resolution. This computational approach offers a detailed insight into the dielectric response of $BiVO_4$, which is crucial for assessing its potential in optoelectronic and energy-related applications.

16.3 RESULTS AND DISCUSSION

The fundamentals of the optical properties can be analyzed from the dielectric constant measurements of the materials. The complex form of the dielectric function $\varepsilon(\omega) = \varepsilon_1(\omega) + i\varepsilon_2(\omega)$ characterizes the response of the material to electromagnetic radiation. Here, we study both the real and imaginary components of the dielectric function. The real value of the dielectric function, $\varepsilon_1(\omega)$, provides the information about the capability of materials to store electric energy; on the other hand, the imaginary part, $\varepsilon_2(\omega)$, is related to the energy absorption based on electronic transitions. These components are linked to a complex form of the refractive index. Furthermore, $\varepsilon_2(\omega)$ can be directly calculated from the density of states and the matrix of momentum between the unoccupied and occupied levels of the electronic states. Once the imaginary part $\varepsilon_2(\omega)$ is evaluated, the real form of the dielectric function is formulated using the Kramers–Kronig transformation [24]. The values of $\varepsilon_1(\omega)$ and $\varepsilon_2(\omega)$ are computed using the standard optical equations.

The formula used for calculating $\varepsilon_2(\omega)$ is,

$$\varepsilon_2(\omega) = \left(\frac{4\pi^2 e^2}{m^2\omega^2}\right)\sum_{i,j}\int \langle i|M|j\rangle^2 f_i(1-f_i)\delta(E_f - E_i - \omega)d^3k \tag{16.1}$$

$\varepsilon_1(\omega)$ was extracted using the Kramers–Kronig relation [25]:

$$\varepsilon_1(\omega) = 1 + \frac{2}{\pi}p\int_0^\infty \frac{\omega'\varepsilon_2(\omega')d\omega'}{\omega'^2 - \omega^2} \tag{16.2}$$

where ω represents the frequency, ω' represents the integration variable, and p represents the principal value of the integration.

Figures 16.1 and 16.2 show the real and imaginary contributions of the dielectric function for a wide range of energies. Here, we plot the energy variation up to 10 eV.

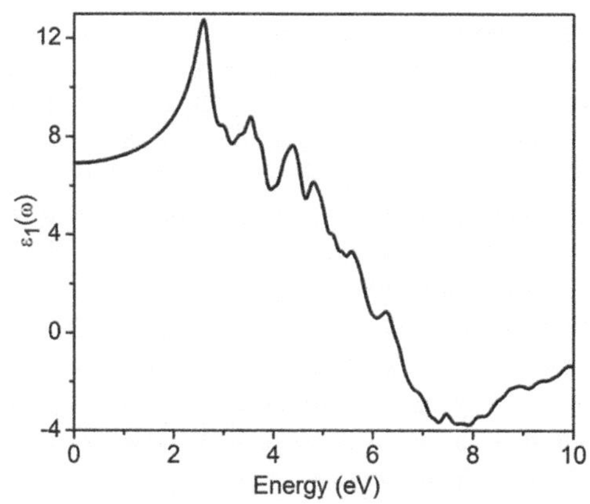

FIGURE 16.1 The real part of the complex dielectric function.

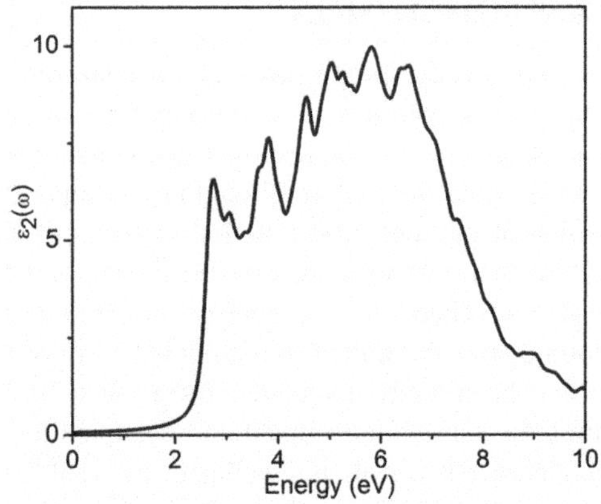

FIGURE 16.2 The imaginary part of the complex dielectric function.

The static dielectric constant $\varepsilon_1(0)$ of $BiVO_4$ is 7.3, which closely matches with the earlier reported value [26]. We can observe from Figure 16.1 that, in the low energy range of 0–2.5 eV, the ε_1 value increases with energy. Further, when the incident energy is higher than 2.5 eV, the ε_1 values begin to show noticeable variation. This reflects the complex interband electronic transitions of these materials. The main peaks are located at 2.5 eV. The imaginary part of the dielectric function $\varepsilon_2(\omega)$ is fully associated with the optical absorption spectra owing to electronic transitions. We can estimate the optical band gap of $BiVO_4$ from the $\varepsilon_2(\omega)$ spectra using the following formula [27] of the coefficient of absorption $\alpha(\omega)$:

$$\alpha(\omega) = \sqrt{2}\omega\left[\sqrt{\varepsilon_1^2(\omega) + \varepsilon_2^2(\omega)} - \varepsilon_1(\omega)\right]^{1/2} \tag{16.3}$$

16.4 CONCLUSIONS

The WIEN2k code, based on the Full-Potential Linearized Augmented Plane Wave (FP-LAPW) method, was utilized to understand the dielectric properties of $BiVO_4$. Our results for both the real (ε_1) and imaginary (ε_2) components of the dielectric function are in strong agreement with previously reported experimental and theoretical data. A detailed analysis of the real part of the dielectric function, $\varepsilon_1(\omega)$, reveals a distinct energy-dependent behavior. We found that in the lower energy range of 0–2.5 eV, the dielectric function ε_1 increases steadily with energy, which can be attributed to the strong polarization response in this spectral region. Above 2.5 eV, the dielectric function ε_1 exhibits the non-linear variations, which basically correspond to interband transitions as well as indicate the critical points in the band structure. The fluctuations in the dielectric function reflect the complex nature of the electronic excitation in a material. The imaginary part of the dielectric function $\varepsilon_2(\omega)$, which is associated with optical absorption due to electronic transitions, provides further insight. From the onset of $\varepsilon_2(\omega)$, one can estimate the optical band gap of $BiVO_4$. In our study, this method yielded a band gap value consistent with known experimental values, confirming the suitability of the computational parameters and the accuracy of the electronic structure. In conclusion, the calculated complex dielectric properties of $BiVO_4$ not only support the validity of our approach but also offer crucial insights into its optoelectronic behavior, reinforcing its potential for applications in photovoltaics, photocatalysis, and optoelectronic devices.

REFERENCES

1. Wu, F.–F., Zhou, D., Du, C., Xu, D., Shi, Z.-Q., Darwish, M. A., Zhou, T., Sun, S.-K., Yao, X.-G., Lin, H.-X., Zhang, M.-R., Liang, Q.-X., Chen, Y.-W., Chen, G.-H., Jantunen, H. (2024). A comprehensive study on crystal structure, phase compositions of the $BiVO_4$–$LaVO_4$ binary dielectric ceramic system and a typical design of dielectric resonator antenna for C-band applications. *Applied Materials Today, 38*, 102222.
2. Alijani, H., Iravani, S., Varma, R. S. (2023). Bismuth vanadate ($BiVO_4$) nanostructures: Eco-friendly synthesis and their photocatalytic applications. *Catalysts, 13* (1), 59.
3. Abdel Maksoud, M.I.A., Fahim, R.A., Kassem, S.M., Awed, A.S. (2023). Gamma irradiation-induced changes in structural, linear/nonlinear optical, and optoelectrical properties of PVB/$BiVO_4$ nanocomposite for organic electronic devices. *Optical and Quantum Electronics 55*, 1165.
4. Tian, K., Wu, L., Han, T., Gao, L., Wang, P., Chai, H., Jin, J. (2022). Dual modification of $BiVO_4$ photoanode by rare earth element neodymium doping and further $NiFe_2O_4$ co-catalyst deposition for efficient photoelectrochemical water oxidation. *Journal of Alloys and Compounds, 923*, 166352.
5. Tian, K., Wu, L., Yang, B., Chai, H., Gao, L., Wang, M., Jin, J. (2023). Anchored lithium-rich manganese nanoparticles boosting Nd-$BiVO_4$ photoanode for efficient solar-driven water splitting. *Colloids and Surfaces A: Physicochemical and Engineering Aspects, 2023*, 662, 130976.
6. Lee, S., An, M., Park, J., Lee, J., Yoo, D., Cho, E. (2024). Zffect of Fe and Co co-doping on the electronic and dielectric properties of $BiVO_4$ – A combined experimental and theoretical study. *Journal of Materials Chemistry A, 12*, 325–338.
7. Wang, W., Strohbeen, P. J., Lee, D., Zhou, C., Kawasaki, J. K., Choi, K.-S., Liu, M., Galli, G. (2020). The role of surface oxygen vacancies in $BiVO_4$. *Chemistry of Materials, 32*(2), 1278–1288.
8. Walsh, A., Yan, Y., Huda, M. N., Mowafak, M.A-J. (2009). Band edge electronic structure of $BiVO_4$: Elucidating the role of the Bi s and V d orbitals. *Chemistry of Materials, 21*, 547–551.
9. Oshikiri, M., Boero, M., Jinhua, Y., Zhigang, Z., Giyuu, K. (2002). Electronic structures of promising photocatalysts InMO4 MÄV, Nb, Ta ··· and $BiVO_4$ for water decomposition in the visible wavelength region. *Journal of Chemical Physics, 117* (15), 7313.
10. Shuyi, L., Xiaodan, Y., Jinlu, H. (2023). Enhancing the photoelectrochemical activity of monoclinic $BiVO_4$ by discretizing oxygen vacancies: insights from DFT calculations. *Physical Chemistry Chemical Physics,25*, 18866–18873.
11. Huang, Y., Zhang, X., Wang, J., Zhang, J., Wei, X., Xu, Y. (2020). The effects of Scandium doping and oxygen vacancies on the electronic states and optical properties of monoclinic $BiVO_4$: A DFT + U study. *Canadian Journal of Physics, 98*(5), 509–517.
12. Xiaofei, C., Xin, Z., Jun, H., Zhong, C. (2020). First-principles investigation of the electronic properties of the Bi_2O_4(101)/$BiVO_4$(010) heterojunction towards more efficient solar water splitting. *Physical Chemistry Chemical Physics*, 22, 2449–2456

13. Hu, Y., Li, D.Z., Zheng, Y., Chen, W., He, Y.H., Shao, Y., Fu, X.Z., Xiao, G.C. (2011). $BiVO_4/TiO_2$ nanocrystalline heterostructure: A wide spectrum responsive photocatalyst towards the highly efficient decomposition of gaseous benzene. *Applied Catalysis B: Environmental*, 104(1), 30–36
14. Zhao, S., Chen, C., Ding, J., Yang, S., Zang, Y., Ren, N. (2022). One-pot hydrothermal fabrication of $BiVO_4/Fe_3O_4$/rGO composite photocatalyst for the simulated solar light-driven degradation of Rhodamine B. *Frontiers of Environmental Science & Engineering*, *16*, 36.
15. Fu, Y.S., Sun, X.Q., Wang, X. (2011). $BiVO_4$–graphene catalyst and its high photocatalytic performance under visible light irradiation. *Materials Chemistry and Physics* 131(1), 325–330.
16. Oluwole, A. O., Yusuf, T. L., Tichapondwa, S. M., Daramola, M. O., Iwarere, S. A. (2024). Fabrication of $BiVO_4/Ag_2CrO_4$ heterojunction composites modified with graphene oxide for enhanced photoelectrochemical and photocatalytic performance. *RSC Advances 14*, 38044–38058.
17. Lee, G.-J., Lee, X.-Y., Lyu, C., Liu, N., Andandan, S., Wu, J. J. (2020). Sonochemical synthesis of copper-doped $BiVO_4$/g-C_3N_4 nanocomposite materials for photocatalytic degradation of bisphenol A under simulated sunlight irradiation. *Nanomaterials 10* (3), 498.
18. Rodríguez, M. E. M., Rehman, A., Hashim, A. (2018). Dielectrical and structural studies of composite matrix $BiVO_4$–$CaTiO_3$ and temperature effects by impedance spectroscopy. *Journal of Materials Science: Materials in Electronics*, *29*, 16248–16258.
19. Dabodiya, T.S., George, T., Murugan, A. V. (2023). Enhanced solar-photo (electro) catalysis of microwave-hydrothermal synthesized $BiVO_4$, ms-$BiVO_4$/RGO/Ag_3PO_4 nanohybrid: Crystalline phase-dependent interfacial charge carrier dynamics. *Energy Technology*, *11* (3), 2201099.
20. Blaha, P., Schwarz, K., Madsen, G. K. H., Kvasnicka, D., Luitz, J., Laskowski, R., Tran, F., Marks, L.D. (2023). *WIEN2k an Augmented Plane Wave + Local Orbitals Program for Calculating Crystal Properties*, Karlheinz Schwarz, Techn. University at Wien, Austria, ISBN 3-9501031-1-2.
21. Perdew, J.P., Burke, K., Ernzerhof, M. (1996). Generalized gradient approximation made simple. *Physical Review Letters*, *77*, 3865.
22. David, K., Tran, F., Blaha, P. (2012). Improving the modified Becke-Johnson exchange potential. *Physical Review B*, *85*, 155109.
23. Monkhorst, H.J., Pack, J.D. (1976). Special points for Brillouin-zone integrations. *Physical Review B*, *13*, 5188.
24. Okoye, C.M.I. (2003). Theoretical study of the electronic structure, chemical bonding and optical properties of $KNbO_3$ in the paraelectric cubic phase. *Journal of Physics. Condensed Matter*, *15*, 5945.
25. Amin, B., Ahmad, M.I., Maqbool, S., Said, G., Ahmad, R. (2011). Ab-initio study of the bandgap engineering of Al(1-x)Ga(x)N for optoelectronic applications. *Journal of Applied Physics*, *109*, 023109.
26. Yun, Y., Yuhong, H., Fei, M., Zongquan, Z., Xiumei, W. (2017). Effects of oxygen vacancy on the mechanical, electronic and optical properties of monoclinic $BiVO_4$. *Journal of Materials Science*, *52*, 8546–8555.
27. Ozaki, S., Adachi, S. (1994). Optical constants of ZnS_xSe_{1-x} ternary alloys. *Journal of Applied Physics*, *75*, 7470.

17 Challenges and Mitigation Strategies for Chromium-Free Electroless Nickel Plating on as-Printed ABS Surfaces Manufactured by 3D Printing

Koustubh Chouhan, Sumanta Mukherjee, and Vikas Kumar

17.1 INTRODUCTION

Metallic coatings on polymer substrates have become essential across a wide range of industries, including automotive, electronics, telecommunications, and consumer products. By combining the lightweightness, low cost, and moldability of polymers with the functional advantages of metals, such as electrical conductivity, corrosion resistance, wear resistance [1], and decorative appeal, these hybrid components satisfy modern demands for enhanced performance and design flexibility. Applications include electromagnetic interference shielding in electronic housings, aesthetically appealing finishes in appliance parts, and mechanically durable surfaces in automotive interior and exterior components. Among the various metallization techniques, electroless nickel (EN) plating has been widely adopted for its ability to produce uniform coatings on complex shapes without requiring an external power source [2]. Acrylonitrile butadiene styrene (ABS) has long been the polymer of choice for metallization due to its balanced mechanical strength, thermal stability, and ease of processing [3]. The chemical structure of ABS, which includes polar nitrile groups, contributes to better adhesion of metal coatings after suitable surface activation. Its widespread use in automotive trims, sanitary fittings, and electronic casings highlights its industrial importance. However, the successful metallization of ABS typically depends on effective surface roughening and activation to generate catalytic sites necessary for nickel deposition [4].

DOI: 10.1201/9781003653226-17

Traditional EN processes on ABS involve chemical etching, usually with hexavalent chromium-based solutions, followed by sensitization and activation steps using palladium chloride. While effective, the use of hexavalent chromium poses serious environmental and health hazards, leading to increasing regulatory restrictions worldwide [5]. This has driven researchers to explore chromium-free alternatives, including potassium permanganate-based etching, palladium-free catalysts, biodegradable adhesion promoters, and hybrid methods combining chemical and physical activation. Despite progress, achieving consistent adhesion and coating quality comparable to chromium-based processes remains technically challenging.

The rise of additive manufacturing, particularly fused deposition modeling (FDM), adds new complexity to this challenge. 3D printing enables rapid prototyping and small-batch production of complex, custom geometries that are impractical or costly to manufacture by traditional methods. Industries now seek to leverage 3D printing not only for design flexibility but also for lightweight structures and functional parts that can be enhanced with metal coatings [6]. The ability to metallize intricate 3D-printed components could open new applications in electronics, robotics, and automotive design, where weight reduction and customized shapes are critical.

FDM-printed ABS parts, however, introduce unique obstacles to reliable metallization. The layer-by-layer deposition creates visible layer lines and anisotropic mechanical properties that affect coating adhesion [7]. Surface porosity and microvoids inherent to the process can trap chemicals during pretreatment, compromising the uniformity of metal deposition. Furthermore, complex internal features and recessed areas in 3D-printed designs challenge conventional chemical activation and nickel-plating techniques. Mechanical sanding or abrasive pre-treatment, traditionally used to increase surface roughness and improve adhesion, is often impractical for detailed geometries and contradicts the promise of additive manufacturing for producing ready-to-coat parts.

The growing industrial demand for sustainable, high-quality metallic coatings on complex 3D-printed ABS components has motivated research into chromium-free and environmentally friendly surface activation methods. These new approaches aim to maintain or improve coating adhesion and uniformity while eliminating hazardous chemicals and manual sanding steps [8].

Metallic coatings on polymers have been studied for decades, driven by the need to combine the lightweight and design flexibility of plastics with the functional and decorative benefits of metals. EN plating remains the most widely used technique, valued for its ability to coat complex shapes uniformly and its favorable mechanical, electrical, and corrosion-resistant properties [9, 10]. The traditional process begins with chemical etching, usually employing hexavalent chromium-based solutions that selectively attack the butadiene domains in ABS, increasing surface roughness and creating sites for catalyst adsorption. This step is followed by sensitization and activation, typically using palladium chloride, after which autocatalytic nickel deposition completes the coating [5].

The toxicity and regulatory restrictions associated with hexavalent chromium have prompted intensive research into alternative surface activation methods. Rajendran *et al.* [1] proposed potassium permanganate-based etching as a less hazardous substitute, achieving comparable increases in surface roughness. Other researchers, including those mentioned above, explored palladium-free activators to eliminate expensive and toxic noble metals, while Melentiev *et al.* [8] investigated biodegradable adhesion promoters to further improve environmental compatibility. Complementary methods such as UV/ozone treatment and plasma activation, studied by Restrepo *et al.* [11], respectively, increase the surface energy of ABS and introduce reactive groups that facilitate better catalyst attachment. Zhang *et al.* [3] demonstrated that sol–gel coatings enhance wettability and provide additional bonding sites, while Charbonnier *et al.* [12] introduced nano-silica primers to improve mechanical interlocking. Ohtsu *et al.* [13] applied graphene oxide interlayers to reduce delamination and enhance long-term coating durability.

Process optimization also plays a vital role in achieving high-quality coatings. Fayyad *et al.* [14] showed that adjusting bath pH, temperature, and agitation can significantly influence deposition rate, adhesion strength, and uniformity. Bonin *et al.* [15] demonstrated the benefits of ultrasonic agitation for improving nickel deposition in recessed areas, whereas Sudagar *et al.* [16] proposed multi-axis rotation of parts during plating to enhance coverage on complex geometries. Additional studies investigated hybrid surface activation strategies that combine physical and chemical treatments to achieve higher catalyst density and better adhesion.

The adoption of 3D printing, especially FDM, has further complicated the metallization process. FDM-printed ABS parts often display distinct surface roughness and anisotropic layer lines that hinder uniform catalyst deposition and metal coverage. Jones *et al.* [17] reported that filament orientation affects coating uniformity, while Zhang *et al.* [3] found that chemical smoothing reduces surface roughness and promotes better nickel deposition. Ref. [17] explored solvent swelling as a means to increase micro-roughness and mechanical anchoring. Despite these methods, untreated FDM-printed ABS parts typically show poor adhesion in standard mechanical tests, as documented by Sood *et al.* [18], which was later confirmed by Rousseau *et al.* [19].

It is evident from the existing literature that while advances in [20] chromium-free etching, alternative catalysts, and hybrid activation methods have significantly improved the environmental profile of EN plating, significant challenges remain in adapting these processes to complex 3D-printed ABS geometries. Existing methods often require trade-offs between adhesion strength, coating uniformity, environmental sustainability, and scalability. Therefore, this study aims to evaluate various chromium-free EN coating processes specifically tailored for intricate 3D-printed ABS components without the use of sanding. The study focuses on identifying the key challenges associated with adhesion, surface activation, and coating uniformity, while exploring environmentally sustainable alternatives that can deliver functional and durable metal finishes on complex polymer geometries.

17.2 MATERIALS AND METHODS

17.2.1 3D Printing of ABS

ABS substrates of dimension 10 mm × 15 mm × 2 mm were fabricated using a Bambu Lab P1S 3D printer, which employs FDM technology. The printing parameters were selected to ensure uniform surface quality and dimensional stability,

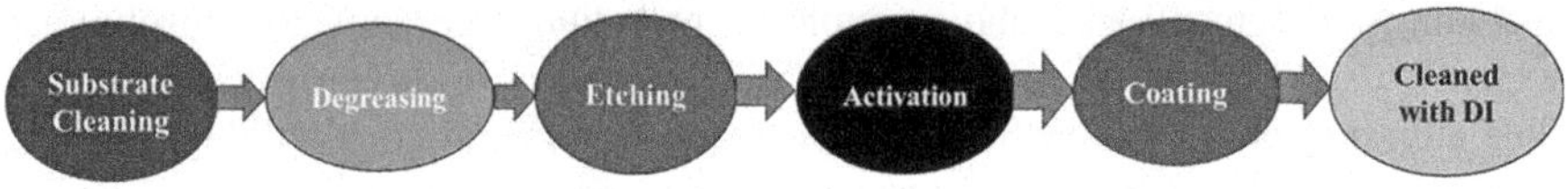

FIGURE 17.1 Steps followed for cleaning and preparation of the 3D-printed ABS surfaces for electroless coating.

critical for subsequent chemical treatment and metallization. The layer height was set at 0.2 mm, providing a balance between surface finish and printing time. The nozzle temperature was maintained at around 230°C, optimized for ABS extrusion, while the print bed was heated to 100°C to minimize warping and enhance adhesion of the first layer. The printing speed was set to approximately 60 mm/s, and an infill density of 100% was used to produce fully dense substrates with sufficient mechanical integrity for subsequent processing.

Following fabrication, the ABS substrates underwent a series of preparatory and metallization steps, as outlined in Figure 17.1, to enable uniform nickel deposition.

17.2.2 Surface Cleaning and Preparation

After printing, the substrates were cleaned thoroughly using deionized (DI) water to remove any dust and loosely bound particles. Degreasing was then performed to eliminate organic contaminants, such as residual oils from handling, which could inhibit chemical reactions during etching and plating. For this, the substrates were immersed in a 100 ml NaOH alcoholic solution (alcohol: water = 1:1) at 45°C for 15 minutes. Afterward, they were placed in a 30°C water bath for 1 minute. This initial cleaning stage ensured that the substrate surfaces were properly conditioned for effective chemical modification.

17.2.3 Etching Process

Since mechanical surface treatment (i.e., manual sanding and polishing) was not performed, chemical etching was critical for ensuring the effective adhesion of the nickel coating. The etching stage was crucial for creating a surface that was conducive to EN plating while eliminating the need for mechanical sanding. Five etchants, designated E1–E5, were selected based on their chemical compositions (Table 17.1) to investigate their effectiveness in enhancing surface roughness.

Etchant E1 consisted of hydrogen peroxide (24 ml) and sulfuric acid (20 ml) per 100 ml of DI water, applied for a treatment time of 30 minutes. Etchant E2 combined hydrogen peroxide (1.53 ml), sulfuric acid (27 ml), and ammonium fluoride (0.3 g) per 100 ml of DI water. Etchant E3 used nitric acid (30 ml) per 100 ml of DI water to chemically roughen the ABS surface. Etchant E4 employed 40 ml of acetone vapor as an organic etching medium, facilitating surface softening and modification without strong mineral acids. Etchant E5 included hydrogen peroxide (20 ml) and sulfuric acid (80 ml) per 100 ml of DI water, and it was also applied for 30 minutes.

TABLE 17.1
Etchant Chemical Compositions

Etchant Code	Chemical	Quantity
E1	H_2O_2	24 ml
	H_2SO_4	20 ml
E2	H_2O_2	1.53 ml
	H_2SO_4	27 ml
	NH_4F	0.3 g
E3	HNO_3	30 ml
E4	Acetone (Vapour)	40 ml
E5	H_2O_2	20 ml
	H_2SO_4	80 ml

17.2.4 Activation of Substrate Surfaces

Following etching, the ABS substrates were chemically activated to deposit catalytic particles necessary for electroless plating. Three types of activation solutions were used, as described in Table 17.2. Activator A1 contained palladium chloride (0.1 g), stannous chloride (2.2 g), and hydrochloric acid (6 ml) per 100 ml of DI water; this solution facilitates the reduction of palladium ions and their deposition on the polymer surface through the presence of stannous ions [21]. This acidic environment ensures that palladium ions remain soluble and available for adsorption onto the etched ABS surface. The palladium ions act as the essential catalytic sites required for the initiation of EN deposition. Activator A2 consisted solely of hydrochloric acid (12 ml) per 100 ml of DI water. Additionally, a dispersion bath was prepared in a larger volume of 1000 ml DI water, containing palladium chloride (0.45 g) and hydrochloric acid (9.5 ml). This dispersion bath helps achieve uniform distribution of catalytic palladium particles across the etched ABS surface, enhancing the subsequent metal deposition process.

17.2.5 Preparation of Coating Bath

The final step in the process was EN plating, where the catalytically activated ABS substrates were immersed in specially formulated plating baths to deposit a uniform

TABLE 17.2
Activator Chemical Composition

Activator	Composition
A1	PdCl2 (0.1 g), SnCl2 (2.2 g), HCl (6 ml) in 100 ml of DI water
A2	HCl (12 ml) in 100 ml of DI water
Dispersion	PdCl2 (0.45 g), HCl (9.5 ml)

TABLE 17.3
Chemical Composition of Baths

Bath	Composition (per 100 ml DI Water)
B1	$NiCl_2$ (2.5 g), $NaPO_2H_2$ (2.4 g), $Na_3C_6H_5O_7$ (4.5 g), NH_4Cl (3 g)
B2	$NiSO_4$ (2.5 g), $NaPO_2H_2$ (3 g), $Na_3C_6H_5O_7$ (2 g), $SC(NH_2)_2$ (0.7 ml), NH_3 (1.5 ml)

nickel layer without the need for an external electric current. Two plating bath compositions were used, as summarized in Table 17.3. Bath B1 consisted of nickel chloride (2.5 g), sodium hypophosphite (2.4 g), sodium citrate (4.5 g), and ammonium chloride (3 g) per 100 ml of DI water, which helps control pH and ionic strength. Bath B2 included nickel sulfate (2.5 g) instead of nickel chloride, combined with sodium hypophosphite (3 g), sodium citrate (2 g), thiourea (0.7 ml) acting as a stabilizer, and ammonia solution (1.5 ml) to adjust the pH. These plating baths were selected to achieve consistent and adherent nickel coatings on the chemically activated ABS substrates. Sodium citrate acts as a complexing agent, while thiourea acts as a stabilizer that improves the refinement of the deposit's microstructure. The presence of sodium hypophosphite as a reducing agent ensures controlled deposition kinetics. Similar bath compositions have been optimized in previous studies to enhance coating uniformity and adhesion on complex geometries [17].

17.3 RESULTS AND DISCUSSION

17.3.1 Process Evaluation and Coating Development

A series of experimental trials were conducted to evaluate the effect of different etching solutions, activators, and plating bath compositions on the EN coating of 3D-printed ABS substrates. The influence of these process variations is summarized in Tables 17.4 and 17.5.

Table 17.4 presents the process parameters and corresponding outcomes for samples S1 through S9. These samples were etched with solution E1 at 35°C, followed by activation with solution A at 60°C and EN deposition using bath B1 at 80°C. The table includes detailed process durations for each step, ranging from 5 to 60 minutes. The observations associated with these samples indicate that most failed to develop continuous or well-adhered coatings. Samples S1, S4, and S8 were recorded as having no coating at all, while samples such as S2, S3, S6, S7, and S9 exhibited non-uniform coating. Only sample S5 showed partial success, forming a coating that lacked adequate adhesion and was classified as "CL" (coating with low adhesion).

Figure 17.2 illustrates these outcomes visually. Figure 17.2(a) shows the untreated 3D-printed ABS surface, highlighting the characteristic ridged texture and layer lines inherent to the FDM process. Figure 17.2(b) presents sample S1, which did not develop any visible nickel deposition; the surface remained matte and unaltered, indicating that activation and catalytic initiation were unsuccessful.

TABLE 17.4
Influence of Etching Solution E1, Activation Step (A), and Electroless Nickel Plating Bath B1 on Coating Quality of 3D-Printed ABS Substrates. Process Durations (in minutes) and Bath Temperatures are Provided for Each Sample (S1–S9)

Process	Bath Temp. (°C)	S1	S2	S3	S4	S5	S6	S7	S8	S9
E1	35	30	15	15	30	30	30	45	40	40
A	60	15	15	15	10	10	15	15	15	15
B	80	10	5	60	60	50	60	60	30	60
Observation		NO	NU	NU	NU	CL	NU	NU	NU	NU

Note: CL—coating with low adhesion; NO—no coating; NU—non-uniform coating.

TABLE 17.5
Effect of Alternative Etching Solutions E2, E3, and E4, Combined with Activation and Plating Bath B1, on Coating Performance for Samples S10–S18

Process	Bath Temp. (°C)	S10	S11	S12	S13	S14	S15	S16	S17	S18
E2	35	30	30	30	30					
E3						40	40			
E4									10	
A	60	15	15	15	15	15	15	15	15	15
B	80	5	60	30	30	30	30			
Observation		NU	NU	CL	SD	NO	NU	NU	NU	SD

Note: CL—coating with low adhesion; NO—no coating; NU—non-uniform coating; SD—sample degraded during treatment/coating.

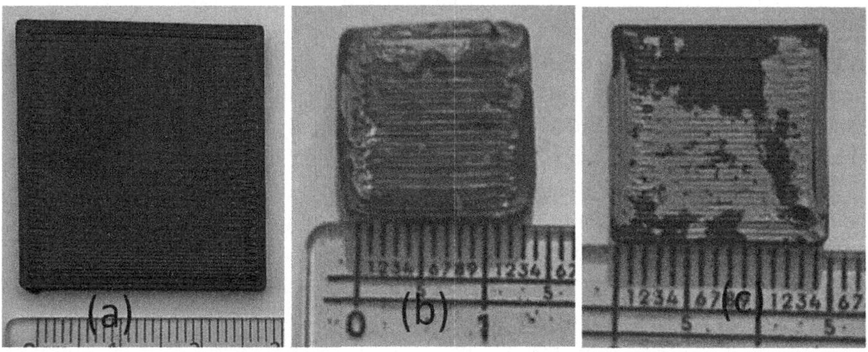

FIGURE 17.2 Surface condition of 3D-printed ABS before and after electroless nickel coating: (a) Untreated ABS substrate; (b) Sample S1; and (c) Sample S3.

FIGURE 17.3 (a) Sample S10—non-uniform coating, (b) S12—coating with low adhesion, and (c) S13—sample degraded during pre-treatment/coating.

Figure 17.2(c) displays sample S3, where a thin and uneven nickel layer was deposited. The coating exhibited a patchy morphology, and large portions of the ABS surface remained exposed, reflecting the "non-uniform" classification in the table. These visual results corroborate the observation that relying solely on chemical etching and activation, without mechanical abrasion or enhanced surface preparation, resulted in poor coating performance.

Table 17.5 presents the outcomes for samples S10 to S18, in which alternative etchants—specifically E2, E3, and E4—were introduced while maintaining the same activation (A at 60°C) and plating bath conditions (B1 at 80°C). The data show mixed results. Samples S10 and S11 were categorized as having non-uniform coatings, while sample S12 developed a low-adhesion coating. Sample S13 was classified as degraded during the process, likely due to over-etching or chemical instability. Further along the sequence, sample S14 showed no coating, and samples S15 through S17 exhibited non-uniform deposition. Additionally, sample S18 displayed surface degradation.

These outcomes are represented in Figure 17.3. Figure 17.3(a) shows sample S10 with patchy nickel coverage. The metallic layer appeared discontinuous, and the overall surface showed a mottled texture, suggesting inconsistent nucleation and growth. Figure 17.3(b) displays sample S12, where the coating was present but appeared thin and detached in some regions. The edges of the coating exhibited noticeable lifting, consistent with poor interfacial bonding. Figure 17.3(c) presents sample S13, which exhibited clear signs of physical damage. The surface was eroded, and the substrate showed visible discoloration and roughening, consistent with the "sample degraded" classification. These images confirm that etchant variations alone were insufficient to produce reliable coatings on unsanded ABS, and in some cases, they increased the likelihood of damaging the substrate.

Table 17.6 presents the results of a modified process applied to samples S19 through S24. These trials introduced additional complexity, including neutralization at 35°C, dual activation using A1 and A2, and a dispersion step (D) prior to deposition. Coating was performed in either bath B1 or B2 at 80°C, with time

TABLE 17.6
Impact of Extended Process Steps Including Etching with E1, E3, E5, and E6, Neutralization, Dual Activation (A1 and A2), Dispersion (D), and Plating Baths B1 and B2 on Samples S19–S24

Process	Bath Temp (°C)	S19	S20	S21	S22	S23	S24
E1	35		40	40	40		
E3	35						60
E5	35	40					
E6	35					60	
Neutralization	35	5					
A1	60				10	12	10
A2	60	5	15	15			
D					5	5	5
Coating (B1)	80				30	12	35
Coating (B2)	80	30min	30min	40s			
Observation		**NO**	**UC**	**B2 decomposed**	**NU**	**CB**	**CB**

Note: CB—coating with blisters; NO—no coating; NU—non-uniform coating; UC—uniform coating.

durations ranging from 40 seconds to 35 minutes. The table records that sample S20 achieved a uniform coating, while sample S21 experienced bath decomposition. Samples S22, S23, and S24 were noted for non-uniform coatings or coatings with blisters.

Figure 17.4 shows sample results from this group. Figure 17.4(a) presents sample S14, which corresponds to a "no coating" result. The image confirms that the ABS surface remained visually unchanged, with no reflective metallic layer. Figure 17.4 (b) shows sample S16 with a noticeable irregular coating. The surface featured a combination of bare and coated regions, consistent with a breakdown in catalytic

FIGURE 17.4 (a) Sample S14—no coating, (b) S16—non-uniform coating, and (c) S18—sample degraded during pre-treatment/coating.

initiation across the substrate. Figure 17.4(c) illustrates sample S18, which suffered severe surface degradation. The surface exhibited signs of erosion, structural failure, and peeling, supporting the classification of sample degradation. The figure confirms that even with the addition of neutralization, dual activation, and dispersion, challenges remained in achieving stable and consistent coatings.

Table 17.7 includes samples S25 through S29, which were treated using combinations of etchants E5, E6, and E7, followed by activation with A1, dispersion (D), and in one case, pH adjustment using ammonia buffer. The nickel coating was deposited using bath B1. The ammonia buffer was applied only to sample S25. Observations indicate that sample S26 achieved a uniform coating, whereas the other samples developed coatings with blisters. The presence of blistering indicated poor gas escape or localized stress under the deposited layer.

Figure 17.5 presents representative samples from this group. Figure 17.5(a) displays sample S21, where nickel was deposited unevenly, and the coating appeared partially delaminated. Figure 17.5(b) shows sample S23, where circular or bubble-like deformations were present beneath the metallic layer, indicative of blister formation. Figure 17.5(c) presents sample S24 with a similar but more widely distributed blister pattern. These visual results match the observations listed in the table and suggest that blistering remained a consistent issue despite the introduction of pH control and improved dispersion.

The final set of results appears in Table 17.8 for samples S30 through S36. These samples underwent multiple etching cycles, consistent activation with A1, and dispersion with variable coating durations. All samples were subsequently plated in bath B1. The observations show that samples S30 to S33 developed coatings with blisters, while samples S34, S35, and S36 produced uniform coatings.

TABLE 17.7
Results for Samples S25–S29 Processed with Multiple Etchants (E5, E6, and E7), Activation (A1), Dispersion, and Optional pH Adjustment Using Ammonia Buffer

Process	**Bath Temp. (°C)**	**S25**	**S26**	**S27**	**S28**	**S29**
E1	35					
E5	35				30	45
E6	35	60	60			
E7	35			50		
A1	60	10	10	10	10	10
D	60	5	5	5	5	5
Coating (B1)	80	15	15	15	30	40
NH_3 buffer		Yes	No	No	No	
Observation		**NU**	**UC**	**CB**	**CB**	**CB**

Note: CB—coating with blisters; NU—non-uniform coating; UC—uniform coating.

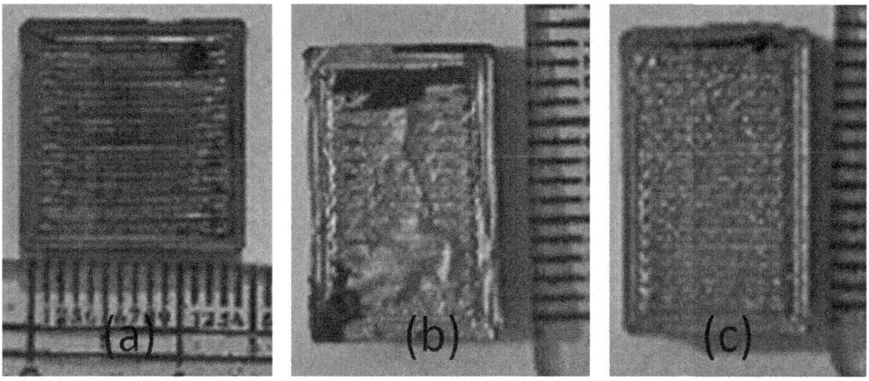

FIGURE 17.5 (a) Sample S1, (b) S23, and (c) S24.

TABLE 17.8
Final Series of Experiments for Samples S30–S36 Involving Repeated Etching Steps, Activation (A1), Dispersion (D), and Plating in Bath B1

Process	Bath Temp. (°C)	S30	S31	S32	S33	S34	S35	S36
E1	35		40					60
E5						40		
E6		40		40	40		60	
A1	60	10	10	10	10	10	10	10
D		5	5	5	8	5	5	5
Coating (B1)	80	15	15	30	15	15	15	12
Observation		CB	**CB**	**CB**	**CB**	UC	UC	**UC**

Note: CB—coating with blisters; UC—uniform coating.

Figure 17.6 presents these final samples. Figure 17.6(a) shows sample S31 with multiple blistering sites disrupting the coating. The nickel layer appeared thick in some areas and lifted in others, pointing to internal stress or gas entrapment during deposition. Figure 17.6(b) presents sample S34 with a smoother, more uniform coating. The metallic surface was continuous and reflective, with fewer observable flaws. Figure 17.6(c) shows sample S35, which also achieved a consistent nickel layer, with no exposed ABS visible. Figure 17.6(d) illustrates sample S36, which exhibited the most uniform and well-adhered coating among the tested samples. The surface appeared smooth and completely covered, suggesting improved process control.

A possible explanation behind such observations is that the inherently smooth and low-energy surface of the FDM-printed ABS limited chemical activation and

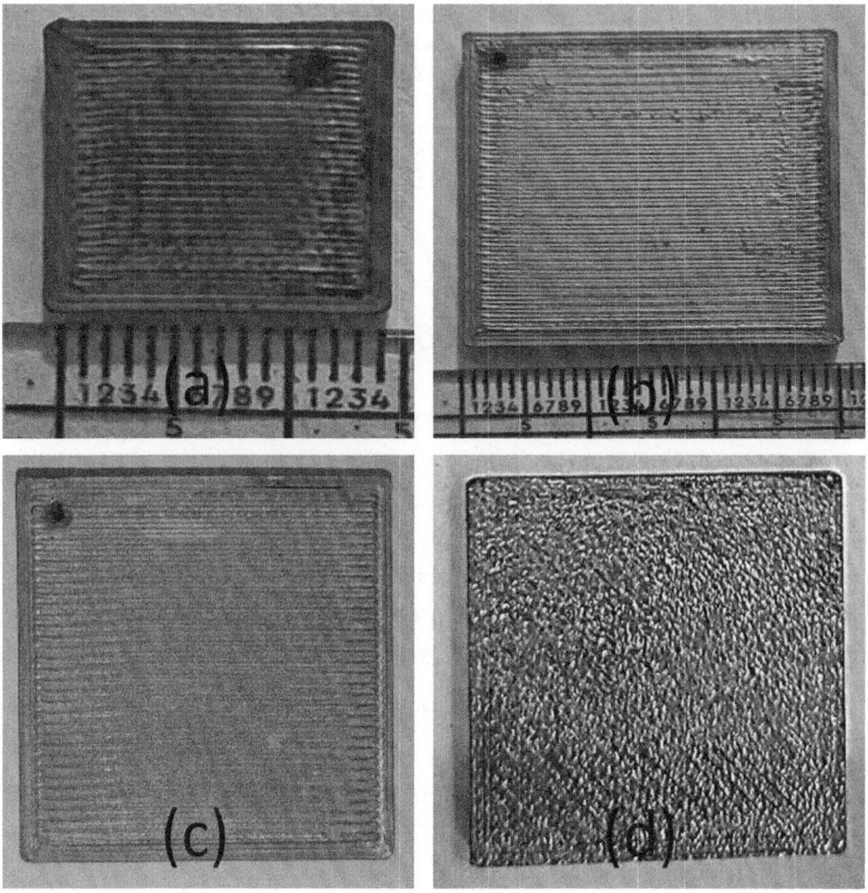

FIGURE 17.6 (a) Sample S31, (b) S34, (c) S35, and (d) S36.

catalytic nucleation. Without mechanical sanding, there was insufficient roughness for anchoring catalyst particles, and the etching process alone was not enough to introduce a dense, reactive network of polar functional groups. Local variations in etching depth and activator distribution could also have contributed to inconsistent nickel deposition.

These results highlight that mere selection of an activator or bath composition is insufficient without considering the role of optimized surface preparation. Similar findings were reported by Choi et al., who emphasized the significance of nano-silica primers in enhancing adhesion [15]. Sudagar *et al.* also underlined the role of surface interlayers in achieving better coating performance [16]. The influence of bath composition on coating uniformity as seen here aligns with the studies by Jones *et al.* [17] and process monitoring methods suggested by Sood *et al.* [18].

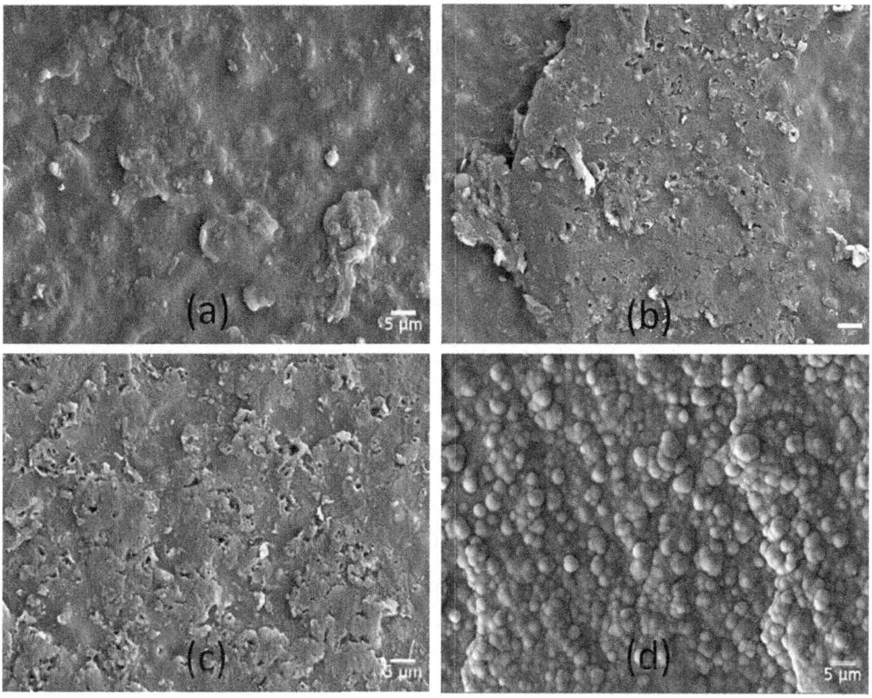

FIGURE 17.7 FESEM analysis of the sample surface, which was obtained from 3D printed (a) Normal ABS, (b) Etched using Etchant E1, (c) Activated using Activator A, and (d) Coated using Bath B.

17.3.2 Surface Morphology and Microstructure

FESEM images (Figure 17.7) depict the surface transformation: from the smooth, layered as-printed ABS (a), to micro-pitted surfaces after etching (b), followed by scattered palladium catalyst clusters after activation (c), and finally, partial nickel coverage with nodules after plating (d).

The as-printed ABS surface (Figure 17.7(a)) reveals a smooth, unmodified morphology, characterized by the layer lines of the FDM process. The lack of microscale texture, polar functional groups, and mechanical interlocks results in inherently low surface energy. This surface offers poor wettability and provides minimal anchoring points for catalytic or metallic layers. Such smoothness, without surface roughening (sanding), significantly impairs both the initiation of metallization and final adhesion quality. Chemical etching with Piranha solution generates visible micro-pitting and increases surface porosity (Figure 17.7(b)). This oxidative treatment enhances wettability and provides sites for catalyst adsorption. However, non-uniform etching leads to inconsistent surface activation and may initiate micro-defects. Importantly, chemical etching alone does not achieve the same depth or consistency of roughness as sanding, resulting in only partial compensation for mechanical anchoring, and failing to ensure robust coating adhesion. The activated surface (Figure 17.7(c)) shows palladium nanoparticle clusters scattered across the etched terrain. While these serve as catalytic sites for nickel deposition, their uneven distribution is symptomatic of variable etch quality and poor surface reactivity. In the absence of prior sanding, the physical anchorage for these particles is further diminished, promoting weak nickel nucleation and contributing to subsequent adhesion failures. The final surface (Figure 17.7(d)) shows dense, cauliflower-like nickel nodules indicative of localized successful deposition. However, large patches reveal incomplete or discontinuous coverage, attributed to insufficient activation and poor adhesion. Importantly, the lack of sanding impedes mechanical interlocking, resulting in delamination-prone interfaces and structurally weak coatings.

The smooth FDM layers reduce adhesion due to fewer anchoring sites. Chemical etching partially compensates by increasing surface roughness and polarity, but uneven etching limits uniform activation. Consequently, the final nickel layer shows regions of good deposition interspersed with bare or poorly covered areas.

17.3.3 Chemical Composition of Coatings

The quality and elemental composition of electroless Ni–P coatings applied to the 3D-printed ABS polymer substrates were evaluated using both elemental mapping (EDX) and energy dispersive spectroscopy (EDS) techniques. These complementary methods provide insights into not only the elemental composition but also the spatial distribution of elements across the coated surface [22]. The EDX mapping in Figure 17.8 reveals a well-dispersed and nearly uniform distribution of nickel (green) and phosphorus (red) over the ABS substrate. The predominance of nickel signals (77%) suggests effective catalytic activation and deposition, while the 23% phosphorus indicates the typical characteristics of electroless Ni–P alloy coatings. A uniform green overlay demonstrates good coating adherence and consistent coverage on the 3D surface, even though the underlying substrate geometry introduces certain challenges due to the absence of sanding.

Further supporting this, the EDS spectrum (Figure 17.9) and quantitative analysis (Table 17.9) show a nickel content of 91.31 wt% and a phosphorus content of 8.69 wt%, aligning closely with the expected composition for medium-phosphorus electroless coatings. The sharp and well-defined peaks for Ni in the EDS spectrum also affirm the elemental purity and confirm successful deposition. A comparison

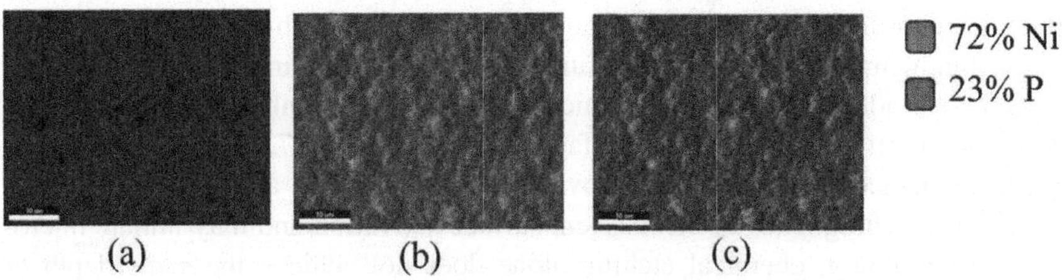

FIGURE 17.8 Elemental mapping for Ni–P coated 3D printed ABS polymer components.

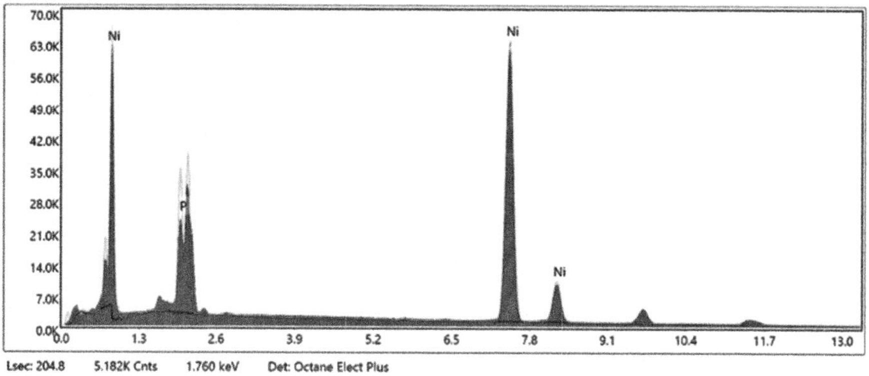

FIGURE 17.9 Percentage (wt.%) of nickel and phosphorus.

TABLE 17.9
Wt% of Ni and P.

Element	Weight %
P	8.69
Ni	91.31

between EDX and EDS data highlights a slight variation in elemental percentages (77% Ni in mapping vs. 91.31% in EDS), likely due to differences in sampling area and analysis depth. While EDS quantifies bulk surface composition at specific localized points, EDX offers a broader spatial perspective. This discrepancy underscores the importance of employing both techniques for a comprehensive understanding.

Despite the promising uniformity in elemental distribution, minor variations and sparse unmapped zones in the EDX image suggest limitations in complete surface wetting or inadequate access to recessed regions within the 3D-printed layers. These factors can cause inconsistent thickness and possibly weak adhesion zones, especially in highly intricate geometries.

17.3.4 Effect of Surface Pre-Treatment on the Substrates

The Fourier Transform Infrared (FTIR) spectra (Figure 17.10) highlight the chemical and elemental transformations that occur due to surface etching and activation. A broad and intensified peak near 3400 cm^{-1} corresponds to –OH stretching vibrations, indicating the formation of hydroxyl groups on the polymer surface due to oxidative cleavage [11]. These polar functionalities are essential, as they improve surface hydrophilicity, which facilitates better adhesion of the Ni–P layer. In addition, the appearance of a peak around 1700 cm^{-1} is attributed to C=O

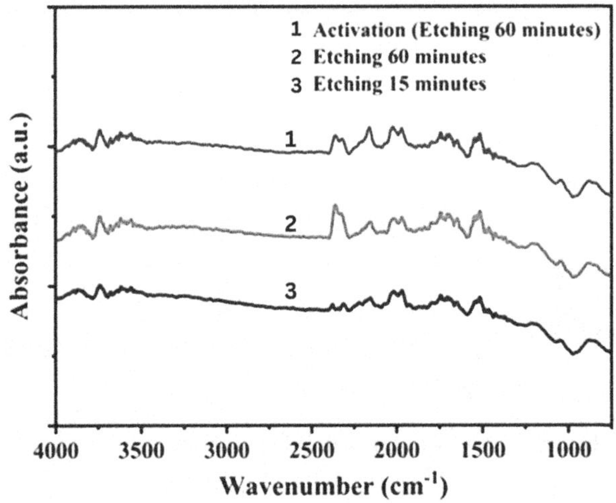

FIGURE 17.10 FTIR spectra of activated, 60 min etching and 15 min etching.

stretching, further confirming oxidation reactions likely resulting from strong acidic treatment [23]. Such groups contribute to increased surface energy and act as nucleation centers for metallic deposition. The persistent aromatic C=C stretching peak at ~1600 cm^{-1} suggests that the styrene-based aromatic ring structure of the ABS remains chemically intact, ensuring the mechanical integrity of the substrate [11]. Bands observed between 1450 and 1380 cm^{-1} associated with CH_2/CH_3 bending and those at 1100–1000 cm^{-1}, potentially indicating phosphate or ether linkages, support the presence of surface-bound moieties introduced during the activation step [8].

Complementing these observations, the EDX analysis confirms the successful deposition of a Ni–P alloy with 91.31 wt% Ni and 8.69 wt% P. This composition falls within the medium phosphorus range, known for its balanced hardness and corrosion resistance properties. The relatively high nickel concentration suggests efficient catalytic nucleation, made possible by the chemically active sites introduced via etching. The synergy between chemical surface activation (evident from FTIR) and compositional uniformity (verified via EDX) confirms the effective preparation of the ABS substrate for electroless coating. Thus, the integration of hydroxyl, carbonyl, and phosphate groups plays a critical role in enhancing coating adherence and functionality.

17.3.5 Surface Topography

The surface morphology of 3D-printed ABS components was investigated using an optical profilometer to understand the challenges associated with chromium-free EN coating. The four stages of surface evolution (Figure 17.11), that is, (a) Without Etching, (b) After Etching using E1 Etchant, (c) Activated, and (d) Coated, are depicted through 3D profilometric surface maps.

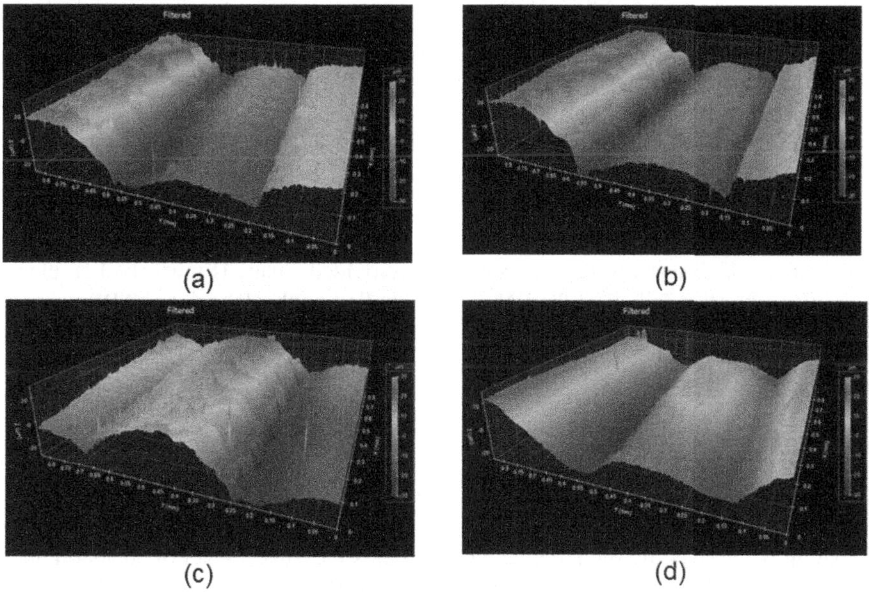

FIGURE 17.11 Topography of the 3D printed ABS polymer component surface obtained by optical profilometer: (a) Without etching, (b) After etching using E1 etchant, (c) Activated, (d) Coated.

The as-printed surface displayed distinct stratification resulting from the FDM process (Figure 17.11(a)). With an Sa value of 10.6 μm and an Sq value of 12.1 μm, this surface was highly irregular, showing prominent ridges and valleys. These deep crevices compromised wetting and made the surface inherently difficult for chemical treatments to penetrate [24]. Optical profilometry revealed large topographical variations of up to ±30 μm, which confirmed the need for an effective surface activation protocol.

After etching with the Piranha solution ($H_2SO_4 + H_2O_2$), the surface became more textured and reactive (Figure 17.11(b)). Although the surface appeared visually rougher due to micro-pitting, the Sa slightly decreased to 9.1 μm and Sq to 11.6 μm, indicating the creation of a more uniformly roughened surface rather than sharp peaks. The Piranha solution oxidized the ABS surface, increased its surface energy, and promoted chemical anchoring, which was essential for palladium nanoparticle seeding. However, uneven etching in confined geometries remained a challenge.

The activated surface (Figure 17.11(c)) showed well-distributed catalytic regions interspersed with inactivated patches. The surface roughness further decreased to Sa 7.9 μm and Sq 9.5 μm, indicating a smoothing effect resulting from chemical interactions and palladium deposition. While this facilitated localized reactivity, the inconsistency across complex 3D contours was problematic because smoother regions might have lacked sufficient catalyst for uniform nickel nucleation.

The coated surface (Figure 17.11(d)) exhibited a semi-continuous metallic film while retaining residual topography. Post-deposition values of Sa 8.5 μm and

Sq 10.5 µm suggested partial leveling of surface irregularities as nickel filled some voids. Nevertheless, certain areas showed nickel deficiency, likely caused by insufficient activation or excessive surface porosity. This outcome underscored the difficulty of achieving fully uniform coatings on complex, unsanded ABS geometries using a chromium-free process.

17.3.6 Coating Adhesion

The cross-hatch test results, as seen in the provided image (Figure 17.12), clearly showed extensive failure of the EN coating applied on the 3D-printed ABS component. The adhesive tape retained a significant portion of the metallic layer, and the substrate underneath exhibited large areas where the coating had been stripped away. The damage was primarily concentrated along the scribed hatch lines, although the failure extended beyond the immediate intersections. This indicated very weak adhesion between the nickel layer and the ABS surface. Based on the visual condition of both the tape and the test specimen, this result corresponded to a 0B rating according to ASTM D3359 standards. In this classification, more than 65% of the coating is removed by the tape, which represents the lowest level of adhesion under this method. The detachment of the coating in sheets or large flakes suggested that the bond between the metallic layer and the plastic was either inadequately formed or was too fragile to withstand even moderate mechanical stress.

The failure was evidently adhesive in nature. There was no visible evidence of cohesive fracture within the nickel coating itself; instead, the separation occurred cleanly at the interface between the nickel layer and the ABS substrate. This type of failure pointed toward insufficient chemical or mechanical anchoring of the coating to the surface. Since this ABS substrate was not sanded before coating, the unmodified surface topography could have limited the effectiveness of the bond. The visual texture of the exposed ABS after coating removal still showed the characteristic layered ridges of the 3D printing process, which might have prevented uniform chemical activation or complete wetting by the activator solution.

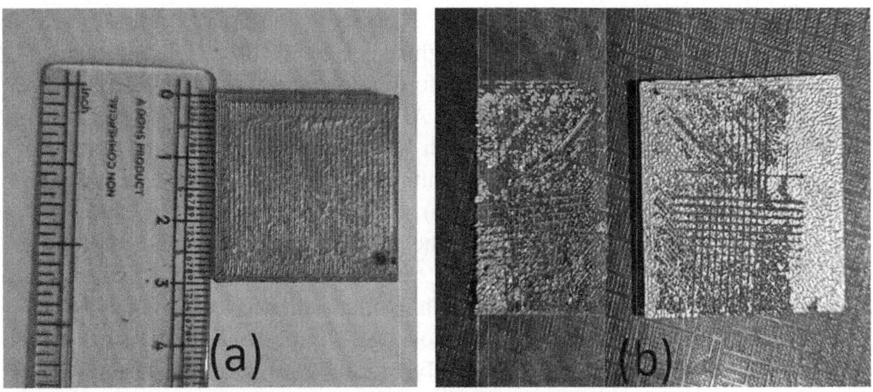

FIGURE 17.12 (a) Before the cross-hatch test and (b) After the cross-hatch test.

The pattern of detachment was not random but showed consistent removal across most of the cross-hatch area, which confirmed that the problem lay with fundamental adhesion rather than isolated process defects. The surface preparation, the performance of the activator, and the compatibility between the ABS substrate and the nickel bath were all likely contributing factors to this failure. The existing surface roughness of the unsanded 3D-printed ABS, combined with possibly suboptimal activator performance, may have resulted in inadequate palladium deposition, thereby limiting nucleation for the subsequent nickel layer.

According to [4] ASTM D3359, a 0B rating is assigned when more than 65% of the coating within the grid area is removed after tape pull. In the observed sample, a significant portion of the EN coating detached not only along the scribed cuts but also across broader areas, leaving the ABS substrate exposed. The adhesive tape retained a substantial amount of the coating, visually confirming that adhesion failed over most of the tested area. This failure was severe and represented the lowest grade in the ASTM D3359 standard, indicating that the bond between the coating and the ABS substrate was extremely weak. For reference, a 5B rating corresponds to 0% removal (perfect adhesion), 4B to ≤5% removal, 3B to 5–15% removal, 2B to 15–35% removal, 1B to 35–65% removal, and 0B to more than 65% removal.

The result indicated complete adhesive failure, where the coating separated cleanly at the interface under mechanical stress without significant resistance. This confirmed major challenges in achieving strong adhesion in chromium-free EN coatings applied to unsanded, intricate 3D-printed ABS surfaces.

17.4 CONCLUSION

This study investigated the limitations and failure mechanisms associated with the application of chromium-free EN coatings on geometrically complex 3D-printed ABS components produced by FDM, without the use of mechanical surface roughening such as sanding. The experimental results indicated that the absence of mechanical texturing significantly reduced the adhesion strength between the deposited nickel layer and the polymer substrate. Although a palladium chloride-based activator was employed as an alternative to traditional chromium-based surface treatments, it did not consistently generate the catalytic activity required for robust metal deposition. This limitation was reflected in repeated adhesion test failures, specifically 0B ratings as per ASTM D3359 cross-hatch adhesion tests.

Surface characterization techniques, including energy-dispersive X-ray spectroscopy, Fourier-transform infrared spectroscopy, and scanning electron microscopy, confirmed the presence of nickel and phosphorus on the surface and indicated partial success in metal deposition. However, these methods also revealed common defects such as peeling, delamination, and inconsistent coating coverage, particularly in regions with complex or recessed geometries. The findings suggested that the inherent characteristics of FDM-printed ABS, such as pronounced layer lines, internal voids, and uneven surface profiles, limited both the uniformity of chemical activation and the mechanical anchoring needed for reliable adhesion.

In the absence of sufficient surface functionalization or micro-texturing, the deposited nickel layer could not achieve a continuous or durable bond with the

ABS substrate. This limitation arises because the chemical and mechanical anchoring points necessary for reliable adhesion are either missing or unevenly distributed on the smooth, layered surface typical of FDM. Overcoming this challenge requires a more comprehensive and multi-step surface engineering strategy that addresses both chemical reactivity and physical surface morphology.

Future research may explore the application of alternative surface activation techniques such as low-pressure plasma treatment, which can introduce polar functional groups onto the polymer surface and increase its surface energy, thereby improving wettability and the affinity for catalytic particles. Similarly, UV/ozone exposure has the potential to oxidize the surface selectively, creating oxygen-containing functional groups without significant material removal. Combined chemical–physical methods, which might include sequential etching and plasma exposure, could further enhance activation by balancing chemical reactivity with improved surface roughness.

In addition to surface activation, optimization of the EN bath composition presents another important pathway for improving coating performance. Modifying the bath with suitable stabilizers could reduce unwanted decomposition and maintain consistent deposition rates, while the addition of surfactants may help to lower surface tension, leading to better wetting of complex geometries. Complexing agents can regulate the availability of nickel ions, resulting in more uniform nucleation and growth of the metal layer across the entire substrate.

In parallel, there is a need to investigate and develop alternative activator systems that do not rely on palladium or hexavalent chromium compounds, which are associated with high cost and environmental concerns. Research into novel catalytic materials, including less hazardous metal complexes or organic molecules capable of initiating nickel deposition, may provide a more sustainable solution. By incorporating these physicochemical and material innovations, it may be possible to achieve uniform, adherent, and environmentally responsible EN coatings suitable for use on complex, unsanded 3D-printed ABS components.

REFERENCES

[1] T. R. Tamilarasan, R. Rajendran, M. Siva Shankar, U. Sanjith, G. Rajagopal, and J. Sudagar, 2016 "Wear and scratch behaviour of electroless Ni-P-nano-TiO_2: Effect of surfactants," *Wear*, vol. 346–347, pp. 148–157.

[2] D. Ashkenazi, A. Inberg, Y. Shacham-Diamand, and A. Stern, 2021 "Gold, silver, and electrum electroless plating on additively manufactured laser powder-bed fusion SlSi10Mg parts: A review," *Coatings*, vol. 11, p. 422.

[3] H. Zhang, Z. Kang, J. Sang, and H. Hirahara, "Surface metallization of ABS plastics for nickel plating by molecules grafted method," *Surface and Coating Technology*, vol. 340, pp. 8–16, 2018.

[4] A. Equbal, A. Equbal, and A. K. Sood, "Metallization on FDM processed parts using electroless procedure," *Procedia Materials Science*, vol. 6, pp. 1197–1206, 2014.

[5] M. García-Gabaldón, V. Pérez-Herranz, and H. Reyes, "Regeneration of hexavalent chromium from a simulated rinse etching solution using an electrochemical reactor with two compartments separated by a ceramic membrane," *International Journal of Electrochemical Science*, vol. 6, pp. 1493–1507, 2011.

[6] A. Equbal and A. K. Sood, "Investigations on metallization in FDM build ABS part using electroless deposition method," *Journal of Manufacturing Processes*, vol. 19, pp. 22–31, 2015.
[7] A. Equbal and A. Kumar Sood, "Metallization on FDM parts using the chemical deposition technique," *Coatings*, vol. 4, no. 3, pp. 574–586, 2014.
[8] R. Melentiev, R. Tao, and G. Lubineau, "Greener electrochemical plating of ABS polymer with unprecedented adhesion via hierarchical micro–nanotexturing," *Journal of Materials Research and Technology*, vol. 24, pp. 3575–3587, 2023.
[9] I. Baek, C. M. Lim, K. Y. Park, and B. K. Ryu, "Enhanced metal coating adhesion by surface modification of 3D printed PEKKs," *Coatings*, vol. 12, no. 6, 2022.
[10] F. Shamsi, M. Khorasanian, and S. M. Lari Baghal, "Effect of potassium permanganate on corrosion and wear properties of ceramic coatings manufactured on CP-aluminum by plasma electrolytic oxidation," *Surface and Coating Technology*, vol. 346, pp. 63–72, 2018.
[11] S. Restrepo *et al.*, "Antibacterial evaluation of electroless Ni–P coating with ZnO nanoparticles on 3D printed ABS," *International Journal of Advanced Manufacturing Technology*, vol. 128, no. 1–2, pp. 209–220, 2023.
[12] M. Charbonnier, M. Romand, and Y. Goepfert, "Ni direct electroless metallization of polymers by a new palladium-free process," *Surface and Coatings Technology*, vol. 200 (16–17), pp. 5028–5036, 2006.
[13] N. Ohtsu, T. Takahara, M. Hirano, and H. Arai, "Effect of treatment temperature on the biocompatibility and mechanical strength of hydroxyapatite coating formed on titanium using calcium phosphate slurry," *Surface and Coating Technology*, vol. 239, pp. 185–190, 2014.
[14] E. M. Fayyad, A. M. Abdullah, A. M. A. Mohamed, G. Jarjoura, Z. Farhat, and M. K. Hassan, "Effect of electroless bath composition on the mechanical, chemical, and electrochemical properties of new NiP–C 3 N 4 nanocomposite coatings," *Surface and Coating Technology*, vol. 362, pp. 239–251, 2019.
[15] L. Bonin, V. Vitry, and F. Delaunois, "The tin stabilization effect on the microstructure, corrosion and wear resistance of electroless NiB coatings," *Surface and Coating Technology*, vol. 357, pp. 353–363, 2019.
[16] J. Sudagar, J. Lian, and W. Sha, "Electroless nickel, alloy, composite and nano coatings – A critical review," *Journal of Alloys and Compounds*, vol. 571, pp. 183–204, 2013.
[17] C. G. Jones, B. E. Mills, R. K. Nishimoto, and D. B. Robinson, "Electroless deposition of palladium on macroscopic 3D-printed polymers with dense microlattice architectures for development of multifunctional composite materials," *Journal of the Electrochemical Society*, vol. 164, no. 13, pp. D867–D874, 2017.
[18] A. K. Sood, A. Equbal, and N. K. Dixit, "Electroless plating on plastic," *International Journal of Scientific and Engineering Research*, vol. 4, no. 8, 12–18, 2013.
[19] A. F. Rousseau *et al.*, "Microstructural and tribological characterisation of a nitriding/TiAlN PVD coating duplex treatment applied to M2 High Speed Steel tools," *Surface and Coating Technology*, vol. 272, pp. 403–408, 2015.
[20] R. Suman, D. Nandan, A. Haleem, S. Bahl, and M. Javaid, "Experimental study of electroless plating on acrylonitrile butadiene styrene polymer for obtaining new eco-friendly chromium-free processes," in Amit Chauhan and Shankar Sehgal (eds.), *Materials Today: Proceedings*, Chandigarh, India: Elsevier Ltd, 2020, p. 843.
[21] M. Petrova, M. Georgieva, D. Lazarova, D. Dobrev, and T. Pavlov, "Electroless metallisation of ABS polymer samples produced by different technologies," *Transactions of the Institute of Metal Finishing*, vol. 99, no. 4, pp. 188–193, 2021.

[22] A. Garcia *et al.*, "ABS polymer electroless plating through a one-step poly(acrylic acid) covalent grafting," *ACS Applied Materials & Interfaces*, vol. 2, no. 4, pp. 1177–1183, 2010.

[23] Antonia Stoyanova and Reneta Boukoureshtlieva. "National conference "sofia electrochemical days 2012" (SED2012)," *Bulgarian Chemical Communications*, vol. 45, p. 5, 2013.

[24] V. Kumar, B. B. Mandal, S. Das, B. Oraon, and S. Mukherjee, "Tailoring electroless Ni-W-P polyalloy coatings: Unveiling the synergistic impact of cerium addition and annealing on surface functional properties," *Surface and Coating Technology*, vol. 485, 2024, pp. 148–157.

18 Predictive Modeling and Optimization of Additive Manufacturing and Laser Cladding Processes Using Machine Learning and Artificial Intelligence

Shakti Kumar, Priya Kesari, Bappa Das, Bishub Choudhury, Manowar Hussain, and K. Pavan Chandu

18.1 INTRODUCTION

Laser cladding and additive manufacturing (AM), also known as 3D printing, have emerged as revolutionary technologies in contemporary production. These technologies offer unmatched capabilities in producing intricate geometries, reducing material waste, and facilitating on demand production and quick prototyping. In the past two decades, these technologies have significantly changed, evolving from prototyping tools to reliable manufacturing platforms utilized in the energy, medicinal, automotive, and aerospace industries (Figure 18.1). Laser cladding is a surface engineering technique that deposits material onto a substrate using a powerful laser. It is equally important for improving surface characteristics, fixing expensive parts, and creating functionally graded materials.

To reach the full potential of AM and laser cladding, their processes must be optimized. Melting, solidification, fluid flow, heat transfer, and phase changes are some of the intricate thermophysical phenomena involved in these processes. These processes are all heavily impacted by a number of interdependent parameters, including laser power, scan speed, hatch spacing, powder feed rate, and ambient conditions. Due to the need for simplifying assumptions and the constraints of processing resources, traditional empirical and physics-based modeling approaches frequently fail to adequately capture these complex relationships. Consequently, there is a growing interest in creating predictive modeling frameworks that are able to adjust to changing production environments and manage huge, high-dimensional

DOI: 10.1201/9781003653226-18

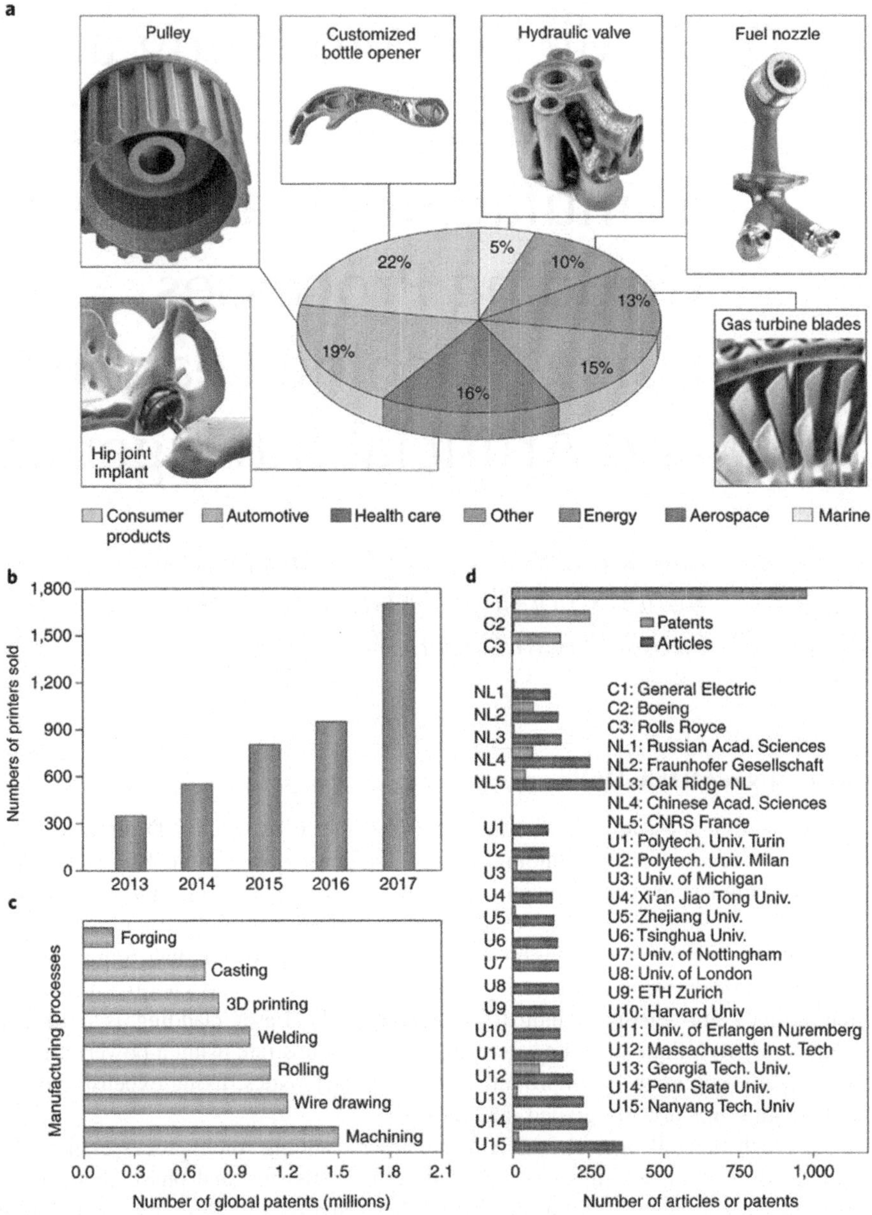

FIGURE 18.1 Applications of metal printing in various industries [1].

datasets. Modeling, controlling, and optimizing AM and laser cladding processes have undergone a paradigm shift with the introduction of machine learning (ML) and artificial intelligence (AI). By leveraging real-time and historical data, ML/AI algorithms may uncover hidden patterns, forecast process results, spot abnormalities, and recommend ideal process parameters. This capability greatly improves production productivity, part quality, and process robustness. It also opens the door to intelligent, self-governing production systems that can monitor themselves and adjust their control.

Recent research demonstrates several uses of AI and ML in AM and laser cladding. To describe the correlations between input parameters and output metrics like surface roughness, density, and tensile strength, regression-based techniques, including Linear Regression, Polynomial Regression, and Support Vector Regression (SVR), have been employed [2, 3]. In analyzing feature importance and the categorization of defect categories in AM components, Decision Trees and Random Forests (RFs) have shown good performance [4]. Artificial Neural Networks (ANNs) have been widely used in the prediction of melt pool shape, porosity development, and mechanical properties in Laser Powder Bed Fusion (LPBF) and Directed Energy Deposition (DED) processes because of their capacity to simulate intricate nonlinear interactions [1, 5]. Bead shape and surface quality are impacted when laser cladding is applied to slanted surfaces. Cladding geometry is precisely estimated by XGBoost-based prediction models, improving process efficiency and optimization. For example, Scime and Beuth's study used Convolutional Neural Networks (CNNs) to accurately identify and categorize flaws in LPBF layer-wise images [6]. Recurrent Neural Networks (RNNs) have also been applied to in-situ sensor data for time-series prediction of temperature histories [7]. For data pre-processing, noise reduction, and locating underlying structures in sensor datasets, unsupervised learning methods like Principal Component Analysis (PCA) and K-means clustering are used [8]. Despite being relatively new in this field, reinforcement learning (RL), in which the learning agent iteratively refines its approach based on input from the environment, has promise for real-time control of AM systems [9]. In order to optimize processes and tune parameters, ML-based surrogate models are frequently combined with optimization techniques such as Bayesian Optimization, Particle Swarm Optimization (PSO), and Genetic Algorithm (GA) [10, 11]. These hybrid technologies speed up development cycles by drastically lowering the number of necessary simulations or experiments. The literature also highlights the usage of digital twins, which are ML/AI-enhanced virtual copies of manufacturing systems, for predictive maintenance and real-time monitoring. These systems use a variety of data sources, including sensors, simulations, and historical documents, to update the model continually and forecast future conditions [12, 13]. Smart factories and Industry 4.0 projects are made possible in large part by digital twins. Another crucial area of study involves using ML/AI to forecast and control microstructures. Studies have shown that deep learning can correlate grain size, orientation, and phase distributions with process parameters [14, 15]. By manipulating process parameters, designers may control the microstructure and customize mechanical qualities. In laser cladding, ML techniques have been used to anticipate hardness profiles, minimize dilution, and optimize clad geometry. For example, bead height and width as a function of laser power, scan speed, and powder flow rate have been predicted using ANNs and SVR models [16]. ML has also been applied to real-time image-based feedback for the detection and classification of faults like lack-of-fusion, porosity, and cracks [17]. One significant development is the use of multi-modal data sources, which integrate force, acoustic, thermal, and visual data to provide thorough process monitoring. These data-driven approaches enhance the accuracy of predictions and offer a comprehensive view of the procedure [18]. Even with these developments, there are still a number of difficulties. Both the quantity and quality of data are crucial; ML/AI models require big, properly labeled datasets for training, which are often scarce in industrial contexts. To tackle this problem, methods for domain adaptation and transfer learning are being investigated, which use knowledge from one domain to another with little retraining [19]. Furthermore, cross-platform integration and the scalability of AI solutions are hampered by the absence of standardization in data formats and protocols [20]. Significant problems also surround the interpretability and explainability of ML models, especially in safety-critical sectors like the production of medical devices and aircraft. Efforts are underway to create interpretable AI models that retain predicted accuracy while offering insights that are intelligible to humans [21]. Furthermore, reliable computational architectures and effective algorithms with low-latency inference are necessary for the real-time deployment of ML/AI in AM and laser cladding systems. To handle rising computational needs, cloud-based solutions and edge computing are being explored [22, 23]. Last but not least, there is increasing interest in combining physics-based models with ML techniques to create hybrid models that combine the advantages of mechanistic and data-driven insights. These models are especially helpful when physical rules are well understood but data is limited [24, 25]. To sum up, the combination of ML and AI with AM and laser cladding has the potential to completely transform contemporary manufacturing. To fully reap the benefits of these intelligent systems, future research must address issues with data quality, model interpretability, standardization, and real-time deployment, even though significant advancements have been made in predictive modelling, defect detection, process optimization, and microstructure control.

This book chapter presents a novel method for optimizing inclined laser cladding processes using sophisticated ML models. While conventional research focuses on horizontal cladding, this work specifically tackles the difficulties of predicting bead morphology on inclined surfaces, which is essential for fixing intricate geometries in the heavy, automotive, and aerospace industries. Bead width, height, and peak shift may all be precisely controlled by combining AI algorithms with experimental data. These factors are essential for preserving quality. This chapter represents an important step toward intelligent, adaptive manufacturing systems by enhancing laser cladding's intelligent path planning and process efficiency.

18.2 OVERVIEW OF MACHINE LEARNING AND ARTIFICIAL INTELLIGENCE

In general, ML approaches can be divided into three categories: RL, unsupervised learning, and supervised learning. By using labeled input–output pairings, supervised learning allows the model to understand the relationship between input parameters (like laser power and scanning speed) and outputs (like surface hardness and melt pool size). ANNs, Decision Trees, RFs, and Support Vector Machines (SVMs) are examples of popular supervised algorithms. For instance, ANNs have demonstrated remarkable accuracy in predicting bead shape in laser cladding, while SVM has been used to categorize defect types in components made with LPBF. Multi-layered neural networks that can learn complicated representations from unprocessed input are used in deep learning—a subfield of ML. The ability to detect porosity or cracks in real-time from high-speed camera or thermal imaging data makes CNNs especially helpful for image-based analysis. In order to forecast sequential events, such as thermal fluctuations during scanning or layer-by-layer faults in AM, RNNs and long short-term memory (LSTM) networks are utilized to handle time-series data. Additionally, more and more people are using Generative Adversarial Networks (GANs) to generate synthetic training datasets, which are especially useful in situations when there is a lack of high-quality, labeled data [2, 26–28]. ML and optimization algorithms are commonly used together to attain the best possible process control and quality. GA, PSO, and Bayesian optimization are popular methods. These methods enable multi-objective optimization of competing parameters, such as minimizing porosity and increasing deposition rate (Figure 18.2). For instance, GA has been used to determine the best laser power, feed rate, and track spacing combinations in DED in order to increase mechanical characteristics and minimize dilution [30, 31].

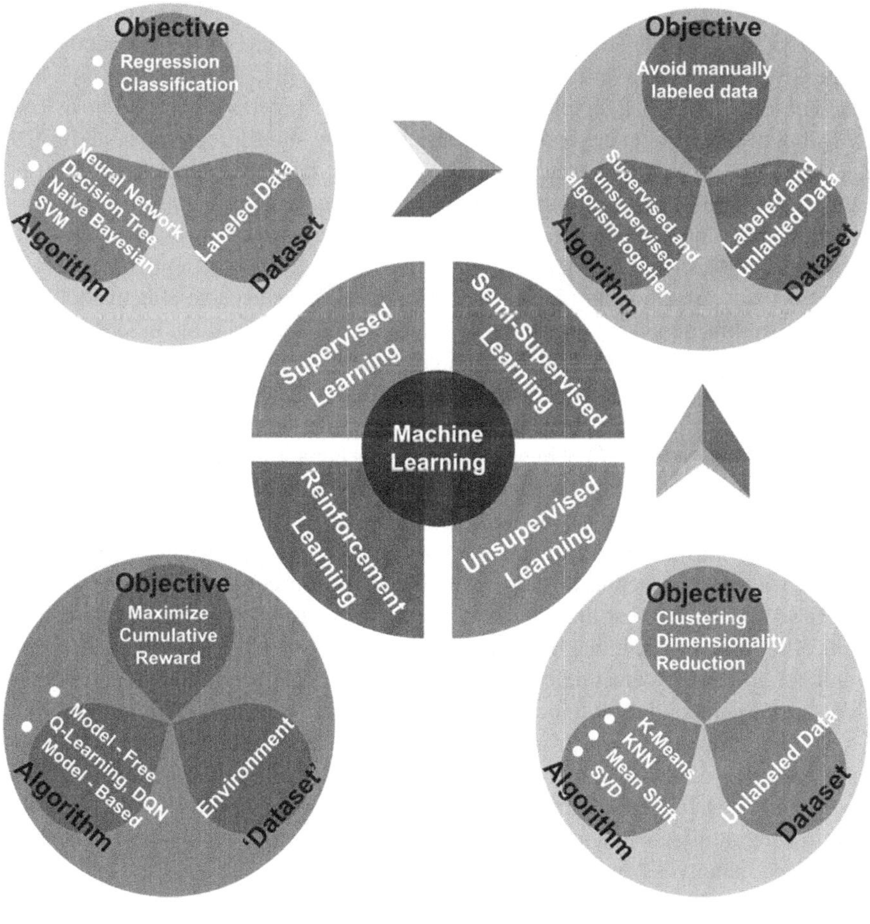

FIGURE 18.2 Schematic procedure of machine learning-aided design of thermal metamaterials. This approach is also applicable to the design of other functional materials [29].

18.3 CLADDING

In laser cladding, a subset of DED, a high-power laser melts and deposits a feedstock material—typically in wire or powder form—onto a substrate to form a metallurgically bonded layer with improved surface properties. This method has several advantages, such as less thermal distortion from targeted heating, stronger bonding, and enhanced resistance to wear and corrosion [32].

Numerous studies have been conducted in the literature on the impact of process factors on the clad quality. The effects of various process variables on the clad geometry and residual stress distribution have been illustrated by a thermo-metallurgical model created for laser cladding [33]. Porosity, cracking, and delamination are examples of frequent flaws that can result from incorrect parameter combinations. These difficulties have the potential to impair the clad layer's mechanical performance, particularly the development of flaws brought on by abrupt changes in temperature or irregular powder feeding. Moreover, process variability is influenced by changes in flow parameters and powder morphology, which makes repeatability a recurring problem. However, even when accurate, traditional physics-based models like computational fluid dynamics (CFD) and finite element approaches are often computationally expensive and require in-depth domain knowledge [34, 35]. ML and AI have become effective tools for predictive modeling and process optimization in laser cladding in order to get beyond these restrictions. SVMs, RF regression, and ANNs are examples of supervised learning techniques that researchers have used to determine correlations between input parameters and output measures like hardness, dilution rate, and clad height. ANN models were employed to estimate the bead geometry in P420 steel powder over an AISI1018 substrate and employed, achieving high accuracy, as shown in Figure 18.3. AI and ML have become effective tools for process optimization and predictive modeling in laser cladding in order to get beyond these restrictions. To determine correlations between input parameters and output measures, including clad height, dilution rate, and hardness, researchers have used supervised learning methods like SVMs, RF regression, and ANNs. For example, Zhang et al. [16] achieved over 95% accuracy in their prediction of bead shape in laser cladding of Inconel 718 using ANN models. Additionally, surface hardness can be estimated using the Gaussian process regression (GPR) model, which takes into account different scanning speeds and powder flow rates. When paired with sensor feedback, these models not only minimize experimental effort but also allow for real-time process control. In-situ monitoring, facilitated by optical sensors, thermal imaging, and acoustic emission analysis, has been integrated with ML models to detect defects on the fly. CNNs have been employed to analyze high-speed camera data and identify porosity formation during cladding [36–38].

Hybrid approaches, which combine data-driven ML models with physics-based simulations, are becoming more popular. These hybrid digital twins enable forecasts to be both scalable and interpretable. Additionally, multi-objective aims, including minimizing dilution while maximizing deposition efficiency and hardness, have been optimized using GAs and RL [39].

A physics-informed Temporal Convolutional Network (TCN) and a high-fidelity thermal-fluid model were created by the laser cladding system to precisely forecast

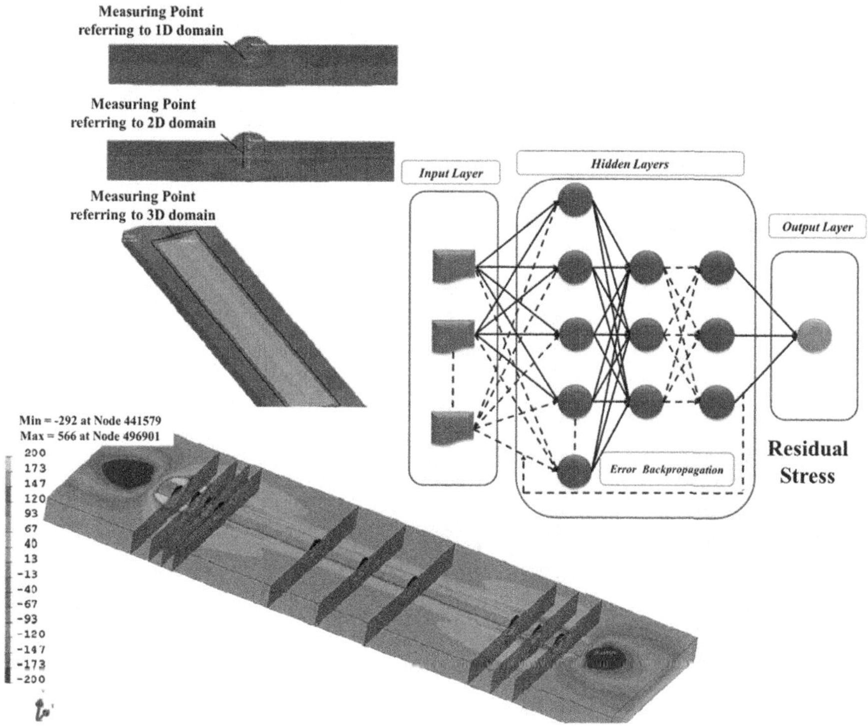

FIGURE 18.3 Single track bead and noted the variations of the residual stress for different slice planes orthogonal to the bead 18.3 (a) 1D-3D domain and ANN model to predict the residual stresses [36].

the depth of the molten pool in fiber laser cladding of 316L steel, as shown in Figure 18.4. By incorporating physical elements such as Marangoni convection, the model outperformed RNN, LSTM, and GRU, achieving up to an 82.7% reduction in MAE, indicating enhanced prediction accuracy and process comprehension [40].

Apart from the advancements, the industrial deployment of AI-enabled laser wire cladding still faces challenges related to dataset quality, real-time computational demands, and a lack of standardized protocols. Therefore, ongoing research is focusing on developing lightweight ML models, transfer learning strategies for cross-material applications, and federated learning to share knowledge across different machines and setups without compromising proprietary data, as shown in Figure 18.5 [38]. In conclusion, laser cladding stands as a critical technique within advanced manufacturing, and its fusion with AI and ML methodologies is catalyzing a shift toward intelligent, adaptive, and efficient production systems. Through predictive modeling and real-time optimization, manufacturers can ensure superior part quality, minimize waste, and unlock new capabilities in material design and surface engineering.

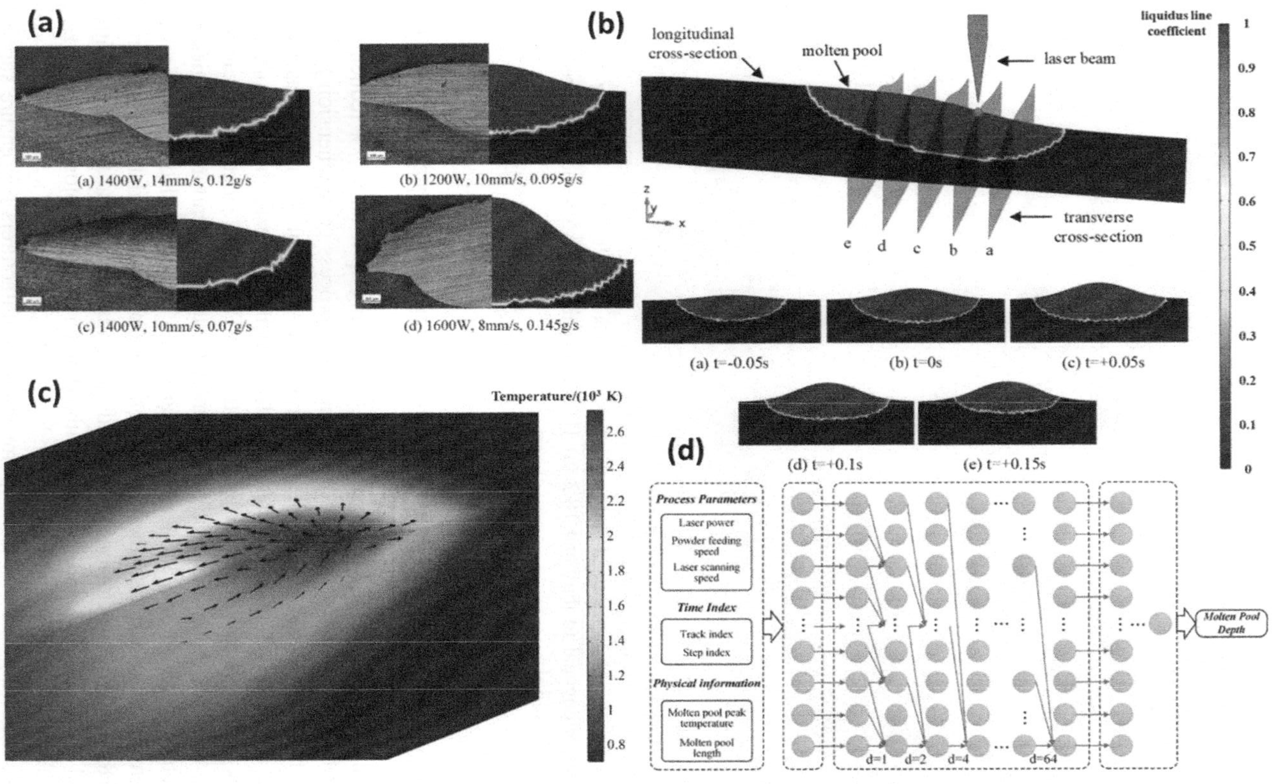

FIGURE 18.4 Process for forecasting the depth of a molten pool using physics-informed machine learning and thermal-fluid modeling. Experimental laser cladding; thermal-fluid modeling; temperature and flow field analysis; geometry and dimensionless analysis; the physics-informed TCN model's architecture; and the prediction of molten pool depth [40].

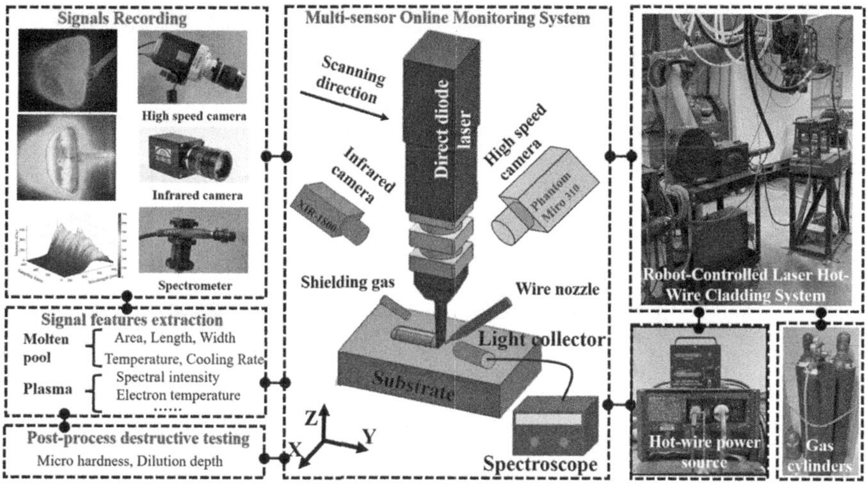

FIGURE 18.5 The robotic arm-controlled laser hot-wires the cladding system with a multi-sensor monitoring subsystem is shown schematically [41].

18.4 DATA ACQUISITION AND PRE-PROCESSING IN AM AND LASER CLADDING

In the fields of laser cladding and AM, data collection and pre-processing are essential processes that facilitate the successful use of ML and AI models. Thorough prediction and optimization systems that can manage the intricacy of these high-energy production processes require accurate and varied data sources, from simulation results to experimental observations. Sensor-based monitoring devices such as pyrometers, photodiodes, infrared (IR) thermography, high-speed cameras, acoustic emission sensors, and optical microscopes are the main tools used to collect experimental data. Throughout the building process, these technologies offer comprehensive temporal and geographic information about surface geometry, cooling rates, temperature fields, melt pool dynamics, and defect evolution (Figure 18.6). In LPBF systems, for example, thermal cameras and coaxial vision systems have been used to monitor melt pool size, track stability, and layer uniformity in real-time (Figure 18.6). Multi-modal sensors measure bead height, laser power absorption, and powder feed rate in situ in laser cladding and DED [42].

Numerical modeling methods like Finite Element Analysis (FEA), CFD, and Phase-Field Modeling are used to create simulation-based data that complements experimental data. Predicting mechanical and thermal behavior, including residual stress and deformation, is made possible using FEA. Melt pool dynamics, gas–metal interaction in the laser–material interaction zone, Marangoni convection, and keyhole creation are all commonly studied using CFD. Meanwhile, phase-field models replicate microstructural changes, including dendritic growth or grain coarsening, which is essential for predicting structure–property relationships in components that are additively built, as shown in Figure 18.7(a–d). When experimental data is

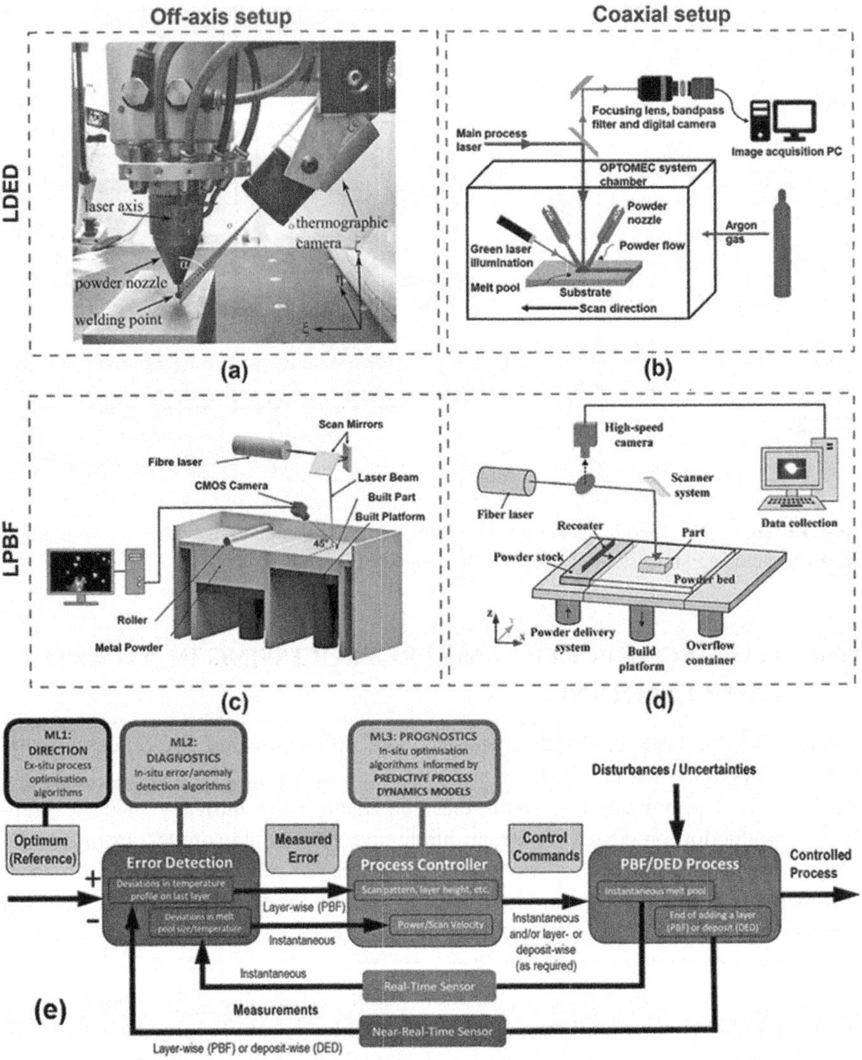

FIGURE 18.6 Comparison and visualization of off-axis and coaxial camera setups: (a) Off-axis camera mounting in LDED. (b) Coaxial camera setup in LDED. (c) Off-axis camera configuration in LPBF [132]. (d) Coaxial camera setup in LPBF (e) ML in implementing real-time and near-real-time diagnostic-prognostic CLC of AM processes [37].

insufficient, these simulations can be used as data augmentation tools and surrogate models [45–47].

To provide ML algorithms with clean, high-quality input, pre-processing is required after data collection. Feature extraction, which involves identifying significant parameters from unprocessed data, is among the initial stages. Heat gradients, surface roughness, porosity indicators, melt pool width, and height are typical

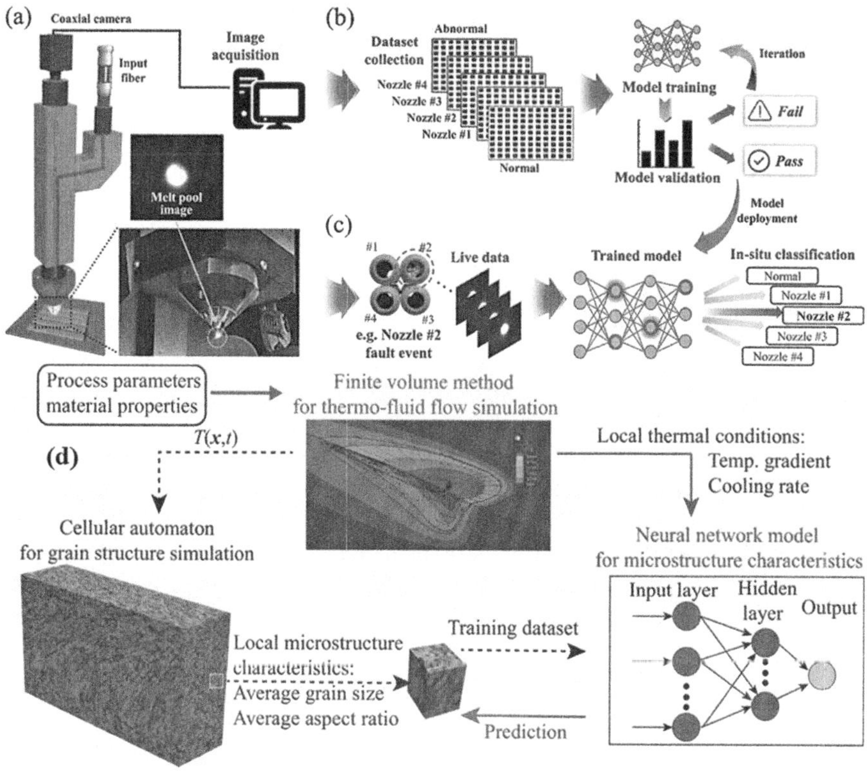

FIGURE 18.7 Schematic of predictively additively manufactured grain structure with a machine learning method and predicting grain structure characteristics in directed energy deposition [43, 44].

characteristics of AM and laser cladding. These features can be derived from acoustic signatures, geometric scans, or thermal profiles [48]. Problems with data quality, such as noise, missing values, and class imbalance, provide a significant pre-processing problem. If noise in thermographic or acoustic data arising from ambient conditions or sensor limitations is not handled, it can seriously impair model accuracy. To smooth and denoise signals, filtering techniques, including wavelet transforms, Kalman filtering, and Savitzky-Golay filters, are frequently used [49]. Model-based or statistical approaches can be used to impute missing data, which often arises from recording errors or sensor dropout. Another frequent problem for classification models is data imbalance (e.g., few defect events vs. many normal instances). Cost-sensitive learning algorithms, under-sampling, or oversampling (like SMOTE) are other solutions [50]. To summarize, the combination of simulation and experimental data, along with meticulous pre-processing, lays the groundwork for superior predictive analytics in AM and laser cladding. These procedures allow ML models to accurately anticipate geometry, microstructure, flaws, and mechanical performance by organizing raw data into representations that are information-rich, noise-resistant, and well-structured. At this point, maintaining data accuracy is crucial to maximizing the promise of AI-powered manufacturing systems.

18.5 PREDICTIVE MODELING USING ML/AI

ML and AI-based predictive modeling has become a game-changing method for overcoming the challenges of laser cladding and AM. These processes involve thermomechanical, multivariable, and highly nonlinear phenomena that are difficult to represent using traditional physics-based models alone. ML/AI tools are being increasingly used to forecast important process outcomes like melt pool characteristics, defect formation, microstructural features, and mechanical properties with high accuracy and generalizability by utilizing massive datasets produced by in-situ monitoring, simulations, and experimental post-processing. Since melt pool dimensions—such as width, depth, height, and area—have a direct impact on surface finish, interlayer bonding, and construction quality, their prediction is a major area of focus. Sensor-derived features, such as thermal signals, laser power, and scan speed, have been successfully used to train supervised ML algorithms like RFs, Gradient Boosting Machines (GBM), and ANNs to predict the temperature field and melt pool geometry for AM processes, as illustrated in Figure 18.8. These models offer quick prediction skills that can help with optimization and control in real time [52, 53].

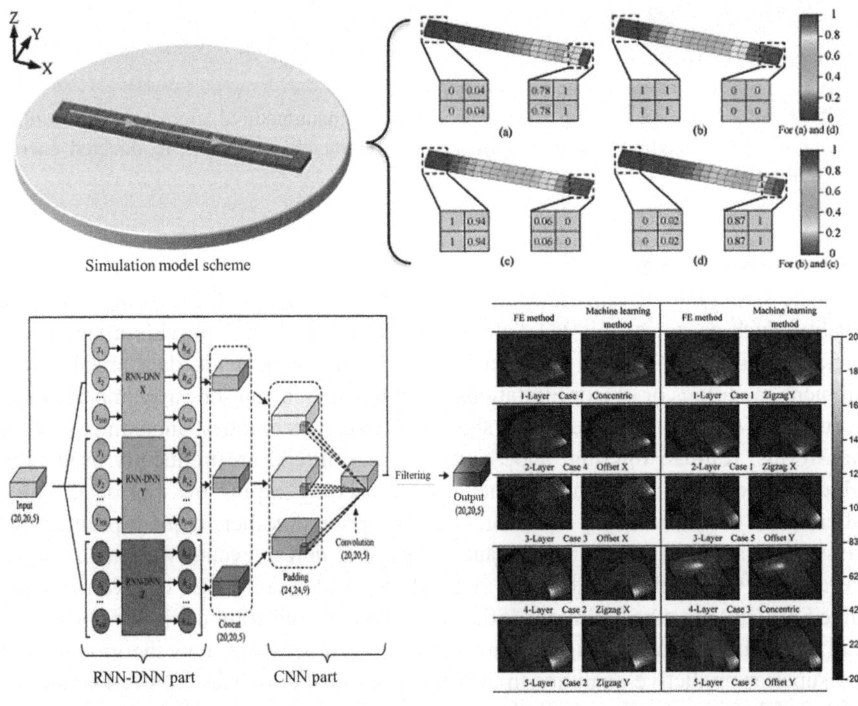

FIGURE 18.8 RNN-DNN and CNN models are used to forecast the 3D printing process's temperature field [51].

Another crucial application is defect prediction. Based on process parameters and in-process monitoring data, ML models have been created to predict porosity, lack of fusion, and cracking. For example, to identify and locate holes and fractures in LPBF of alloys such as Ti-6Al-4V and IN718 [54], CNNs have been trained on thermographic pictures and acoustic emissions. Unusual patterns that precede defect formation have also been found using unsupervised clustering techniques like k-means and DBSCAN [42]. This method improves process robustness by facilitating both early identification and root-cause analysis of failures.

In the field of materials science, modeling microstructure and phase evolution has advanced significantly through the application of AI and ML approaches. With models trained on simulation and experimental data, it is now possible to infer phase transformation, grain shape, and dendritic growth patterns from process histories. Hybrid ML-physics models and physics-informed neural networks (PINNs) have demonstrated potential in forecasting phase fractions and grain size distributions, especially in high-strength alloys like IN718 and maraging steels [5, 6]. Through controlled processing, these predictions are crucial for customizing the required material properties.

Hardness and yield strength in laser-clad steels and Ni-based alloys have been predicted using SVR and Deep Neural Networks (DNNs) [47]. In safety-critical applications like biomedical implants and aircraft, Bayesian neural networks are becoming increasingly popular because they can quantify uncertainty in addition to making predictions [55]. Deep learning models have been created in the LPBF of IN718 to link layer-wise thermal history and processing circumstances to residual stresses and crack susceptibility [56]. Decision trees and ANN-based models have effectively predicted the clad geometry and process parameters in laser cladding of steels [57]. Microstructure informed fatigue resistance estimates for Ni-based alloys have been made possible by hybrid data-driven models that combine thermodynamic calculations and experimental data [58]. ML and AI-based predictive modeling greatly improve our comprehension and efficiency of intricate AM and laser cladding operations. These technologies bring in a new era of intelligent manufacturing by facilitating real-time process control, defect mitigation, and material design in addition to reducing reliance on trial-and-error experimentation.

18.6 INTELLIGENT MONITORING AND CONTROL

The use of 3D point clouds as the feedback data was investigated for an adaptive dimension correction method based on sensors. To detect the intermediate layers' on-machine laser scanning and adjust for surface deviations during processing, a laser displacement sensor was incorporated into an LAAM system. Lastly, an investigation was carried out on in-situ process monitoring and control, as shown in Figure 18.9. A closed-loop controller that regulates the laser intensity was educated with the measured melt pool width. With the help of an auto-tuning unit that optimizes controller parameters automatically, the controller's innovative multitasking architecture allows for flexible process control across a wide range of part forms, materials, toolpaths, and process parameters. Implementing this monitoring and control system can enhance the deposited material's geometric correctness and process stability [59].

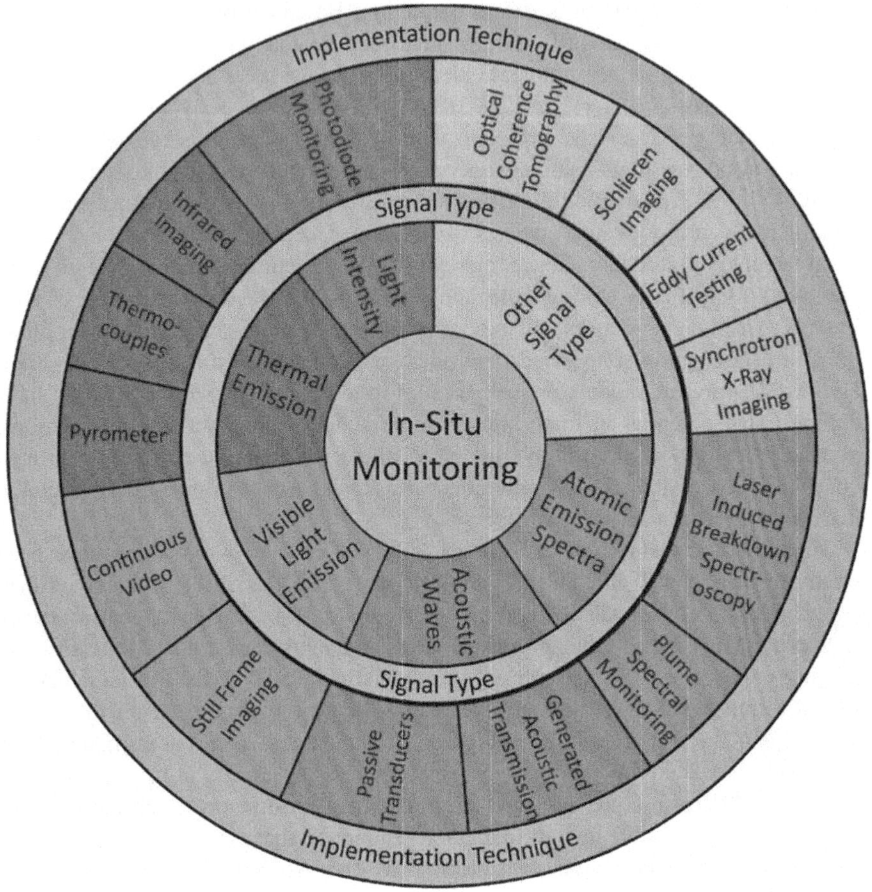

FIGURE 18.9 Laser-based metal AM's radial map of in-situ monitoring technologies. The methods used to monitor the process are depicted in the outer circle, while the inner circle depicts the sort of signal to be watched [59].

Process control has advanced significantly with the incorporation of these sensor modalities into a digital twin (Figure 18.10). A digital twin is a virtual, real-time model of the actual manufacturing process, continuously updated using simulation feedback and sensor data. Robustness and reproducibility can be improved by using AI-driven prediction models in the digital twin to automatically detect and fix deviations from ideal conditions. For example, the integration of ML with finite element (FE) and CFD models enables the prediction of residual stress distribution and thermal history, offering a feedback mechanism to modify parameters such as scan strategy or laser power. ML-predicted feedback control is the foundation of closed-loop manufacturing systems. Closed-loop controllers in LPBF and laser cladding use real-time predictions from ML algorithms to dynamically modify processing parameters in reaction to deviations, guaranteeing uniform build quality

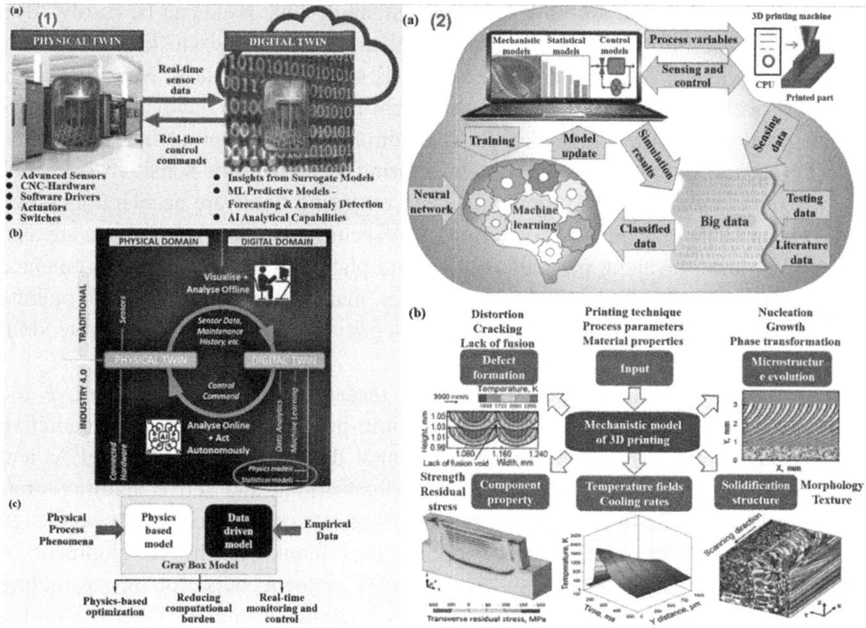

FIGURE 18.10 Schematic maps of (a) DT and (b) the mechanistic model of 3D printing [60].

throughout layers and components. Adaptive control of melt pool temperature and form has also been achieved through the use of RL techniques, showing promise for fully autonomous parameter tuning in complex geometries [60].

18.7 CHALLENGES AND RESEARCH OPPORTUNITIES

Significant progress has been made, but there are still a number of obstacles preventing the mainstream industrial use of ML and AI in AM and laser cladding. Training data quantity and quality are a big challenge. To ensure reliable generalization, ML models need vast, varied, and high-quality datasets; nevertheless, creating such data is often costly, time-consuming, and susceptible to variations in equipment, materials, and environmental factors. Additionally, there are difficulties with standardization and model interoperability due to data heterogeneity, which comes from various sensors (optical, acoustic, and thermal), simulation models, and construction platforms.

The interpretability of AI models represents yet another significant constraint. CNNs and GANs are examples of deep learning architectures that have shown remarkable prediction power; however, they often operate as "black boxes," offering little information about the physical processes they model. This erodes confidence and prevents adoption in sectors that depend on safety, like biomedical manufacturing and aerospace. The capacity to function consistently across various machines, materials, or geometries without requiring a lot of retraining is another issue with many of the models that are now in use. Promisingly, this issue can be resolved by creating physics-informed ML frameworks that integrate physical limitations and domain knowledge. Deployment in real time is still a difficult task. ML algorithms that work well in offline training settings might not be able to handle the computational speed and resilience needed for high-throughput production lines' real-time monitoring and control. In smart manufacturing systems, time-sensitive decision-making relies on the edge AI techniques and emerging hardware accelerators (such as GPUs and TPUs). Concerns regarding cybersecurity and data governance are also becoming more prevalent, particularly as cloud platforms and digital twins enhance connectivity in AM systems. For intelligent manufacturing systems to operate dependably, it is essential to guarantee data integrity, intellectual property protection, and resilience against cyberattacks [61, 62].

The future holds a number of exciting research prospects. To improve the accuracy of process–property predictions, multi-modal and multi-scale predictive models that include simulated and experimental data (e.g., from CFD, FEA, and phase-field models) can be developed. Transfer learning and active learning strategies can increase platform adaptability and lower data requirements. Process design and virtual certification have new opportunities thanks to the development of generative models (e.g., GANs and variation AL autoencoders) for microstructure synthesis and defect repair and RL for adaptive parameter tuning. Aligned with the Industry 5.0 vision, a self-aware manufacturing system is conceived as one that can sense, predict, decide, and adjust on its own in response to real-time feedback.

18.8 CONCLUSION

- A paradigm shift in the design, monitoring, and optimization of advanced manufacturing processes is represented by the combination of ML and AI with AM and laser cladding. These technologies make use of data-driven models to provide predictive insights into intricate thermal-fluid behavior, microstructure evolution, and defect generation mechanisms that were previously challenging to simulate only with physics-based methods.
- Together, the predictive power of supervised learning, the power of deep learning for pattern recognition, and the adaptive optimization potential of evolutionary algorithms and RL have created new opportunities for intelligent production.
- This chapter has provided a thorough summary of the state of ML/AI applications in laser cladding and AM. It has highlighted how sophisticated algorithms, capable of predicting melt pool properties, porosity, residual stresses, and mechanical performance, are fed by contemporary data collection methods, including sensors, thermographic imaging, and high-fidelity simulations.
- Additionally, the chapter highlighted the critical role of feature selection, dimensionality reduction, and pre-processing in ensuring accurate model predictions. With growing emphasis on digital twins and closed-loop control, the integration of real-time monitoring combined with AI models is opening the door for autonomous and adaptive production systems that support Industry 4.0 and 5.0 goals.

- However, issues, including cybersecurity, generalizability across machines and materials, interpretability of models, and data scarcity, still pose a serious threat despite significant advancements. Future studies should concentrate on creating physics-aware, standardized AI frameworks; strong cross-platform transfer learning models; and hybrid strategies that combine mechanistic and data-driven insights.
- Materials science, computer vision, control engineering, and manufacturing technology must collaborate in a transdisciplinary manner to realize truly intelligent AM systems. The future of high-performance, dependable, and sustainable AM and laser cladding systems depends on ML and AI.

REFERENCES

1. Mukherjee, T., & DebRoy, T. (2018). Scientific, technological and economic issues in metal printing and their solutions. *Nature Materials*, 17(9), 934–943.
2. DebRoy, T., Wei, H. L., Zuback, J. S., Mukherjee, T., Elmer, J. W., Milewski, J. O., . . . & Zhang, W. (2018). Additive manufacturing of metallic components–Process, structure and properties. *Progress in Materials Science*, 92, 112–224. https://doi.org/10.1016/j.pmatsci.2017.10.001
3. Frazier, W. E. (2014). Metal additive manufacturing: A review. *Journal of Materials Engineering and Performance*, 23(6), 1917–1928. https://doi.org/10.1007/s11665-014-0958-z
4. Herzog, D., Seyda, V., Wycisk, E., & Emmelmann, C. (2016). Additive manufacturing of metals. *Acta Materialia*, 117, 371–392. https://doi.org/10.1016/j.actamat.2016.07.019
5. Thompson, M. K., Moroni, G., Vaneker, T., Fadel, G., Campbell, R. I., Gibson, I., . . . & Bernard, A. (2016). Design for additive manufacturing: Trends, opportunities, considerations, and constraints. *CIRP Annals*, 65(2), 737–760. https://doi.org/10.1016/j.cirp.2016.05.004
6. Scime, L., & Beuth, J. (2019). Anomaly detection and classification in a laser powder bed additive manufacturing process using a trained computer vision algorithm. *Additive Manufacturing*, 19, 114–126. https://doi.org/10.1016/j.addma.2017.11.009
7. King, W. E., Barth, H. D., Castillo, V. M., Gallegos, G. F., Gibbs, J. W., Hahn, D. E., . . . & Smith, R. J. (2015). Observation of keyhole-mode laser melting in laser powder-bed fusion additive manufacturing. *Journal of Materials Processing Technology*, 214(12), 2915–2925. https://doi.org/10.1016/j.jmatprotec.2014.12.015
8. Tapia, G., & Elwany, A. (2014). A review on process monitoring and control in metal-based additive manufacturing. *Journal of Manufacturing Science and Engineering*, 136(6), 060801. https://doi.org/10.1115/1.4028540
9. Qian, B., Zhang, D., Han, J., Wang, G., & Bai, Y. (2019). Machine learning in additive manufacturing: State-of-the-art and perspectives. *Additive Manufacturing*, 28, 565–586. https://doi.org/10.1016/j.addma.2019.05.006
10. Li, C., Liu, Z. Y., Fang, X. Y., & Guo, Y. B. (2017). Residual stress in metal additive manufacturing. *Procedia CIRP*, 71, 348–353. https://doi.org/10.1016/j.procir.2017.04.050
11. Zhang, Y., & Bernard, A. (2020). Machine learning for additive manufacturing: A review. *Virtual and Physical Prototyping*, 15(3), 276–294.
12. Wu, D., Jennings, C., Terpenny, J., Gao, R. X., & Kumara, S. (2017). A comparative study on machine learning algorithms for smart manufacturing: Tool wear prediction using random forests. *Journal of Manufacturing Science and Engineering*, 139(7), 071018. https://doi.org/10.1115/1.4036350

13. Leung, C. L. A., Marussi, S., Atwood, R. C., Towrie, M., Withers, P. J., & Lee, P. D. (2018). In situ X-ray imaging of defect and molten pool dynamics in laser additive manufacturing. *Nature Communications*, 9(1), 1–9. https://doi.org/10.1038/s41467-018-03734-7
14. Yao, X., Yang, Z., & Zhang, J. (2020). Review of machine learning applications in additive manufacturing. *Materials and Design*, 189, 108515. https://doi.org/10.1016/j.matdes.2020.108515
15. Murzin, S. P. (2024). Artificial intelligence-driven innovations in laser processing of metallic materials. *Metals*, 14(12), 1458.
16. Zhang, Y., Bai, P., Li, Z., & Zhang, J. (2025). Rapid optimization of iron-based alloy laser cladding process based on orthogonal experiment and machine learning for Q345. *Optics & Laser Technology*, 182, 112086.
17. Li, C. (2023). Research progress and prospect of laser cladding technology. *Journal of Physics: Conference Series* 2419, 012049.
18. Huang, J. S., Liew, J. X., Ademiloye, A. S., & Liew, K. M. (2021). Artificial intelligence in materials modeling and design. *Archives of Computational Methods in Engineering*, 28(5), 3399–3413.
19. Gu, D., Meiners, W., Wissenbach, K., & Poprawe, R. (2012). Laser additive manufacturing of metallic components: Materials, processes and mechanisms. *International Materials Reviews*, 57(3), 133–164. https://doi.org/10.1179/1743280411Y.0000000014
20. Choudhary, K., DeCost, B., Chen, C., Jain, A., Tavazza, F., Ward, L., . . . & Agrawal, A. (2022). Recent advances and applications of machine learning in experimental materials science. *npj Computational Materials*, 8(1), 1–26. https://doi.org/10.1038/s41524-021-00657-8
21. Toprak, İ. B. (2025). Prediction and optimization of hardness in AlSi10Mg alloy produced by laser powder bed fusion using statistical and machine learning approaches. *Scientific Reports*, 15(1), 17966.
22. Xu, X., Mi, G., Xiong, L., Jiang, P., Shao, X., & Wang, C. (2018). Morphologies, microstructures and properties of TiC particle reinforced Inconel 625 coatings obtained by laser cladding with wire. *Journal of Alloys and Compounds*, 740, 16–27.
23. Deepak, G., Parthiban, M., Nath, S. S., Alfurhood, B. S., Mouleswararao, B., & Kishore, V. R. (2024). Ai-enhanced thermal modeling for integrated process-product-system optimization in zero-defect manufacturing chains. *Thermal Science and Engineering Progress*, 55, 102945.
24. Niaki, M. K., & Nonino, F. (2020). Impact of additive manufacturing on supply chain performance: A literature review. *The International Journal of Production Economics*, 226, 107621. https://doi.org/10.1016/j.ijpe.2020.107621
25. Tan, X., Kok, Y., Tan, Y. J., Descoins, M., Mangelinck, D., & Tor, S. B. (2015). Graded microstructure and mechanical properties of additive manufactured Ti–6Al–4V via electron beam melting. *Acta Materialia*, 97, 1–16. https://doi.org/10.1016/j.actamat.2015.06.027
26. Jordan, M. I., & Mitchell, T. M. (2015). Machine learning: Trends, perspectives, and prospects. *Science*, 349(6245), 255–260.
27. LeCun, Y., Bengio, Y., & Hinton, G. (2015). Deep learning. *Nature*, 521(7553), 436–444.
28. Sahar, T., Rauf, M., Murtaza, A., Khan, L. A., Ayub, H., Jameel, S. M., & Ahad, I. U. (2023). Anomaly detection in laser powder bed fusion using machine learning: A review. *Results in Engineering*, 17, 100803.
29. Zhu, C., Bamidele, E. A., Shen, X., Zhu, G., & Li, B. (2024). Machine learning aided design and optimization of thermal metamaterials. *Chemical Reviews*, 124(7), 4258–4331.
30. Lee, K. H., Lee, H. W., & Yun, G. J. (2023). A defect detection framework using three-dimensional convolutional neural network (3D-CNN) with in-situ monitoring data in laser powder bed fusion process. *Optics & Laser Technology*, 165, 109571.

31. Era, I. Z., Farahani, M. A., Wuest, T., & Liu, Z. (2023). Machine learning in directed energy deposition (DED) additive manufacturing: A state-of-the-art review. *Manufacturing Letters*, 35, 689–700.
32. Fauchais, P., & Vardelle, A. (2012). Thermal sprayed coatings used against corrosion and corrosive wear. *Advanced Plasma Spray Applications*, 10, 34448.
33. Zhong, X., Li, G., Wang, M., Pan, W., & Wang, Y. (2025). Advances in research on numerical simulations of residual stress in laser cladding additive-subtractive hybrid manufacturing. *The International Journal of Advanced Manufacturing Technology*, 138, 815–854.
34. Zeng, C., Tian, W., Liao, W. H., & Hua, L. (2016). Microstructure and porosity evaluation in laser-cladding deposited Ni-based coatings. *Surface and Coatings Technology*, 294, 122–130.
35. Wang, H., Cao, Y., Huang, Z., Liu, Y., Hu, P., Luo, X., . . . & Sun, Y. (2024). Recent advances on machine learning for computational fluid dynamics: A survey. arXiv preprint arXiv:2408.12171.
36. Mohajernia, B., Mirazimzadeh, S. E., Pasha, A., & Urbanic, R. J. (2022). Machine learning approaches for predicting geometric and mechanical characteristics for single P420 laser beads clad onto an AISI 1018 substrate. *The International Journal of Advanced Manufacturing Technology*, 118(11), 3691–3710.
37. Gunasegaram, D. R., Barnard, A. S., Matthews, M. J., Jared, B. H., Andreaco, A. M., Bartsch, K., & Murphy, A. B. (2024). Machine learning-assisted in-situ adaptive strategies for the control of defects and anomalies in metal additive manufacturing. *Additive Manufacturing*, 81, 104013.
38. Kumar, S., Gopi, T., Harikeerthana, N., Gupta, M. K., Gaur, V., Krolczyk, G. M., & Wu, C. (2023). Machine learning techniques in additive manufacturing: A state of the art review on design, processes and production control. *Journal of Intelligent Manufacturing*, 34(1), 21–55.
39. Shi, S., Liu, X., Wang, Z., Chang, H., Wu, Y., Yang, R., & Zhai, Z. (2024). An intelligent process parameters optimization approach for directed energy deposition of nickel-based alloys using deep reinforcement learning. *Journal of Manufacturing Processes*, 120, 1130–1140.
40. Hu, K., Wang, Y., Li, F., Zhou, Y., & Li, W. (2024). Thermal-fluid modeling and physics-informed machine learning for predicting molten pool depth in single-layer multi-track fiber laser cladding. *The International Journal of Advanced Manufacturing Technology*, 135(7), 3591–3613.
41. Yao, M., Kong, F., & Tong, W. (2023). A 3D finite element analysis of thermally induced residual stress distribution in stainless steel coatings on a mild steel by laser hot wire cladding. *The International Journal of Advanced Manufacturing Technology*, 126(1), 759–776.
42. Wang, J., Fan, Z., & Liang, Z. (2021). Anomaly detection using unsupervised learning for quality control in additive manufacturing. *Additive Manufacturing Letters*, 1, 100008.
43. Kats, D., Wang, Z., Gan, Z., Liu, W. K., Wagner, G. J., & Lian, Y. (2022). A physics-informed machine learning method for predicting grain structure characteristics in directed energy deposition. *Computational Materials Science*, 202, 110958.
44. Bandyopadhyay, A., & Traxel, K. D. (2018). Invited review article: Metal-additive manufacturing—Modeling strategies for application-optimized designs. *Additive Manufacturing*, 22, 758–774.
45. Yamanaka, T., Arima, T., & Tanaka, M. (2021). Microstructure evolution prediction using machine learning in metal AM. *Materials Science and Engineering A*, 801, 140369.
46. Li, Z., Zhang, L., & Lin, X. (2022). Physics-informed deep learning for grain morphology prediction in LPBF. *Computational Materials Science*, 206, 111216.

47. Singh, P., & Ghosh, A. (2020). Machine learning-based prediction of mechanical properties in laser-clad coatings. *Surface and Coatings Technology*, 402, 126341.
48. Le-Hong, T., Lin, P. C., Chen, J. Z., Pham, T. D. Q., & Van Tran, X. (2023). Data-driven models for predictions of geometric characteristics of bead fabricated by selective laser melting. *Journal of Intelligent Manufacturing*, 34(3), 1241–1257
49. Zhao, N., Wei, J., Long, Z., Yang, C., Bi, J., Wan, Z., & Dong, S. (2023). An integrated method for tunnel health monitoring data analysis and early warning: Savitzky–Golay smoothing and wavelet transform denoising processing. *Sensors*, 23(17), 7460.
50. Gibert, K., Sànchez-Marrè, M., & Izquierdo, J. (2016). A survey on pre-processing techniques: Relevant issues in the context of environmental data mining. *AI Communications*, 29(6), 627–663
51. Ng, W. L., Goh, G. L., Goh, G. D., Ten, J. S. J., & Yeong, W. Y. (2024). Progress and opportunities for machine learning in materials and processes of additive manufacturing. *Advanced Materials*, 36(34), 2310006.
52. Scime, L., & Beuth, J. (2019). A multi-scale convolutional neural network for autonomous anomaly detection and classification in a laser powder bed fusion additive manufacturing process. *Additive Manufacturing*, 24, 273–286. https://doi.org/10.1016/j.addma.2018.10.033
53. Khanzadeh, M., Bian, L., & Tapia, G. (2018). Data-driven prediction of melt pool size in metal additive manufacturing. *Journal of Manufacturing Science and Engineering*, 140(10), 101010. https://doi.org/10.1115/1.4040610
54. Zhang, Y., Qian, Y., Wu, D., & Zhang, Y. (2020). In-situ defect detection for laser powder bed fusion using machine learning and imaging. *Journal of Materials Processing Technology*, 285, 116807.
55. Wigh, D. S., Goodman, D., & Neumann, T. (2022). Uncertainty-aware AI models for mechanical property prediction in AM. *Nature Computational Science*, 2(3), 147–155.
56. Ren, Z., & Zhang, W. (2021). Deep learning for residual stress prediction in LPBF of Inconel 718. *Journal of Materials Processing Technology*, 296, 117196.
57. Bahl, S., Ganeriwala, R. K., & Chawla, V. (2021). ANN modeling for laser cladding of H13 tool steel. *Journal of Manufacturing Processes*, 67, 403–412.
58. He, P., Hu, Y., & Wang, Y. (2023). Microstructure-based fatigue prediction in Ni alloys via machine learning. *Materials & Design*, 225, 111371.
59. Herzog, T., Brandt, M., Trinchi, A., Sola, A., & Molotnikov, A. (2024). Process monitoring and machine learning for defect detection in laser-based metal additive manufacturing. *Journal of Intelligent Manufacturing*, 35(4), 1407–1437.
60. Shen, T., & Li, B. (2024). Digital twins in additive manufacturing: A state-of-the-art review. *The International Journal of Advanced Manufacturing Technology*, 131(1), 63–92.
61. Wang, Z., Yang, W., Liu, Q., Zhao, Y., Liu, P., Wu, D., . . . & Chen, L. (2022). Data-driven modeling of process, structure and property in additive manufacturing: A review and future directions. *Journal of Manufacturing Processes*, 77, 13–31.
62. Zheng, P., Wang, H., Sang, Z., Zhong, R. Y., Liu, Y., Liu, C., . . . & Xu, X. (2018). Smart manufacturing systems for Industry 4.0: Conceptual framework, scenarios, and future perspectives. Frontiers of *Mechanical Engineering*, 13(2), 137–150.

19 Predictive Modeling of Electroless Polyalloy Coating Properties Using Artificial Neural Networks

Vikash Kumar, Buddhadeb Oraon, and Sumanta Mukherjee

19.1 INTRODUCTION

Electroless polyalloy coatings represent critical advancements in surface engineering technologies, providing enhanced corrosion resistance, wear performance, hardness, and multifunctionality on a broad spectrum of metallic substrates. Unlike conventional electroplating methods that rely on externally supplied electrical energy, electroless plating deposits alloys uniformly across the substrate surface through an autocatalytic chemical reduction process. This approach enables the formation of conformal pinhole-free films, which are essential for protecting complex geometries and components used in advanced engineering sectors such as aerospace, automotive, electronics, and chemical processing [1–3].

The unique properties of electroless polyalloy coatings stem from their complex multicomponent chemistries, often comprising nickel, phosphorus, silicon, boron, tungsten, aluminum, and titanium in various combinations [4, 5]. These alloying elements interact synergistically to tailor the coating microstructure, surface morphology, and functional attributes. For instance, the incorporation of tungsten is known to improve tribological properties, whereas copper is vital in applications demanding high electrical conductivity and in creating antimicrobial surfaces, which are key to minimizing healthcare-associated infections [6–8]. The resulting microstructures frequently exhibit fractal-like self-similarity and multifunctional tribological behavior, enabling customized solutions for high-performance applications. However, these complex ternary or quaternary alloy systems introduce a highly nonlinear and coupled parameter space involving bath chemistry, pH, temperature, stabilizer concentrations, and deposition time, which govern critical properties, such as plating rate, hardness, wear resistance, and corrosion behavior.

DOI: 10.1201/9781003653226-19

Experimentally investigating these vast multidimensional variables through traditional approaches is time-consuming, expensive, and often inefficient due to the intricacy of underlying chemical kinetics and material interactions. Nonlinear dependencies and multivariate interplays challenge conventional empirical design and trial-and-error optimization strategies, impeding rapid advancements in coating formulations and process parameters. Consequently, there is a pressing need for computational predictive models that can simulate and reliably forecast coating properties prior to experimental fabrication, thereby accelerating material design cycles, reducing development costs, and enabling real-time process control and quality assurance [9, 10].

Artificial Neural Networks (ANNs) have emerged as powerful computational tools capable of modeling the intricate nonlinear relationships inherent in electroless polyalloy systems. Inspired by biological neural architectures, ANNs consist of interconnected nodes, or neurons, organized in layered structures that perform weighted summations and nonlinear transformations. This architecture enables them to approximate complex functional mappings from multiple process input variables, such as chemical composition, temperature, pH, and deposition time, to output coating properties, including microhardness, wear performance, and corrosion resistance. Their flexibility surpasses that of traditional regression models and complements other machine learning techniques such as support vector machines (SVMs), fuzzy inference systems, and gradient boosting (e.g., XGBoost), enabling researchers to capture highly nonlinear and multicollinear phenomena in coating behavior [9, 11].

The successful application of ANNs to electroless coating modeling relies heavily on comprehensive high-fidelity experimental datasets. Experimental design techniques such as Taguchi's orthogonal arrays allow systematic variation of key bath parameters while minimizing the required experimental runs [12] and maximizing the information gained for ANN training. Hybrid modeling approaches that incorporate metaheuristic optimizers (e.g., genetic algorithms [13] and artificial hummingbird algorithms [14]) enhance ANN training by optimizing network weights and hyperparameters and dynamically escaping local minima to improve predictive accuracy and model robustness. Moreover, physics-informed neural networks embed governing physical laws and mechanistic constraints into ANN frameworks to improve interpretability, data efficiency, and extrapolative capability, thus bridging the gap between data-driven and physics-based modeling paradigms [15].

Despite these advancements, challenges persist. Data scarcity, especially for novel, rare, or multiscale alloy systems, limits the generalizability and cross-validation of the model. The variability and noise inherent to experimental measurements introduce uncertainty, which can lead to overfitting and diminish model reliability. Recent developments in uncertainty quantification, such as Bayesian Neural Networks [16], address these limitations by providing probabilistic confidence intervals around predictions, which are crucial for risk-informed decision-making in coating design and maintenance planning. The integration of self-supervised learning and natural language processing techniques for systematic dataset augmentation from literature corpora promises to alleviate data paucity challenges and enable broader model applicability [17].

Industry 4.0 and the emergence of smart manufacturing further motivate the integration of ANN-driven predictive models into digital coating production lines. AI-based frameworks facilitate accelerated R&D cycles, autonomous process optimization, and predictive maintenance by enabling rapid, real-time adjustments based on coating performance forecasts, thereby significantly enhancing throughput, consistency, and cost-effectiveness. Standardization of experimental protocols, data formats, and collaborative data sharing will be critical to fully harnessing the power of these digital tools for industrial impact [18, 19].

This chapter presents an in-depth study on the predictive modeling and optimization of electroless Ni–W–P–Ce coatings using ANNs and the Taguchi design methodology. It begins by outlining the fundamentals of electroless polyalloy deposition and the critical process parameters, specifically the concentrations of nickel sulfate, sodium hypophosphite, sodium tungstate, and ceric sulfate, that influence the coating performance. This chapter details the application of the Taguchi L9 orthogonal array for experimental design, supported by analysis of variance (ANOVA) to assess parameter significance. This chapter demonstrates how intelligent modeling and statistical optimization can effectively guide the development of high-performance protective coatings.

19.2 MATERIALS AND METHODOLOGY

19.2.1 Substrate and Coating Bath Preparation

Low-carbon steel (AISI 1010) substrates with dimensions of $15 \times 10 \times 1$ mm^3 were used. The substrates were mechanically polished using silicon carbide sandpapers with progressively finer grits ranging from 400 to 1200 grit. Prior to coating, the substrates were activated by immersion in a 30% HCl solution for 30 s [20], followed by thorough rinsing with distilled water. The coating bath was prepared using the chemicals and conditions detailed in Table 19.1, and all chemicals were procured from Merck Life Science Pvt. Ltd. A magnetic stirrer equipped with a hot plate was used to maintain uniform heating of the bath. The substrates were suspended in a solution using a cotton thread. After the coating process, the samples were rinsed with warm distilled water, dried, and stored for subsequent characterization.

TABLE 19.1
Bath Composition and Conditions

Chemicals	Amount (g/L)	Operating Conditions
Nickel sulfate	20–30	pH: 7.5–8
Sodium hypophosphite	14–20	Temp: 85 ± 2°C
Sodium tungstate	15–25	Time: 60 mins
Tri-sodium citrate	20	
Ceric sulfate	1–2	
Thiourea	0.001	

19.2.2 Experimental Design Using Taguchi Method

The experiments were structured using a Taguchi L9 orthogonal array, considering the concentrations of the four key coating components as the primary factors, each varying at three distinct levels, as listed in Table 19.2. These factor levels were selected on the basis of several preliminary (pilot) experiments. Minitab 2022 software was used to perform the Taguchi analysis, with microhardness (HV) chosen as the response variable. To maximize the microhardness, the signal-to-noise (S/N) ratio (dB) was calculated using the "larger-the-better" criterion, as defined by Equation (19.1), where n represents the total number of experiments and y_i represents the measured response value for the ith experiment [21]. Subsequently, a confirmation test was conducted using the optimal parameter combination to validate the results derived from the Taguchi method.

$$S/N = -10 \times \log_{10}\left(\frac{1}{n}\sum_{i=1}^{n}\frac{1}{y_i^2}\right) \tag{19.1}$$

19.2.3 Surface Characterization

Vickers microhardness measurements were performed using a UHL VMHT microhardness tester. For each coated sample, indentations were made at five different locations, and the average of these values was taken as the representative microhardness. The testing parameters were standardized with an indentation load of 50 gf, a dwell time of 12 s, and an indentation speed of 25 μm/s.

The surface morphology of the optimally coated samples was examined using a field-emission scanning electron microscope (FEI Inspect F50). Elemental analysis was conducted by energy-dispersive X-ray spectroscopy (EDS) integrated with field emission scanning electron microscope (FESEM). The crystallographic structure of the coatings was characterized using X-ray diffraction (XRD, Rigaku Miniflex-600). Additionally, surface texture analysis was performed using a non-contact 3D optical profilometer (Taylor Hobson).

TABLE 19.2
Control Factors and Parameter Levels

			Parameter Levels		
Control Factors	**Designation**	**Unit**	**−1**	**0**	**1**
Nickel sulfate	A	g/L	20	25[a]	30
Sodium Hypophosphite	B	g/L	14	17[a]	20
Sodium Tungstate	C	g/L	15	20[a]	25
Ceric Sulfate	D	g/L	1	1.5[a]	2

[a] Mid-level combination as the initial parameter.

19.2.4 ANN Model Development

A feed-forward neural network (FFNN) [22] was developed in MATLAB to predict the microhardness based on four input parameters from a Taguchi L9 design. The network architecture (shown in Figure 19.1) consists of an input layer with four neurons, two hidden layers with ten and six neurons, and an output layer with one

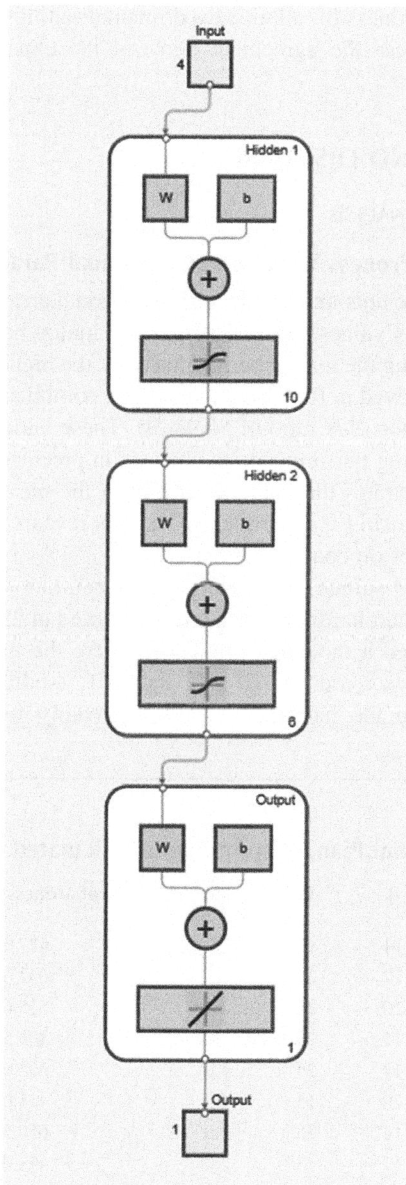

FIGURE 19.1 Feed-forward neural network architecture for microhardness prediction.

neuron. The model was trained using the Levenberg–Marquardt algorithm (trainlm) with early stopping (maximum of 20 validation failures) and a limit of 1000 epochs. The input and output data were normalized using mapminmax, and 50 random splits (three training, three validation, and three test samples) were used to ensure robustness. The best-performing model, selected based on the highest overall correlation coefficient (R), was retrained and evaluated using the mean squared error (MSE), root mean square error (RMSE), and mean absolute error (MAE). Finally, the ANN predicted the microhardness for all nine experimental settings, and a comparison plot was generated to assess the agreement between the experimental and predicted values.

19.3 RESULTS AND DISCUSSION

19.3.1 Taguchi Analysis

19.3.1.1 Effect of Process Variables and Optimal Parameter Selection

Table 19.3 presents the uncoded Taguchi L9 orthogonal array design, corresponding average microhardness values, and calculated S/N ratios based on the "larger-the-better" criterion. Among the nine experimental runs, the highest microhardness value of 706.6 HV was achieved in Run 9 with parameter combination A3B3C2D1, which also recorded the highest S/N ratio of 56.98 dB. These variations clearly reflect the sensitivity of the coating performance to changes in precursor concentrations.

Figure 19.2(a) illustrates the main effect plot of the mean microhardness values obtained from the Taguchi L9 experiments. The plot reveals the average influence of each process parameter on coating hardness.

For factor A (nickel sulfate), increasing the concentration from 20 g/L to 30 g/L generally improved microhardness, with a dip observed at 25 g/L. Factor B (sodium hypophosphite) showed a moderate influence, where the microhardness peaked at 17 g/L and slightly decreased at 20 g/L. Factor C (sodium tungstate) exhibited a positive trend, with the hardness increasing steadily from 15 g/L to 25 g/L.

TABLE 19.3
Uncoded Experimental Plan, Response, and Calculated S/N Ratios

Run No.	A	B	C	D	Microhardness (HV 0.05)	S/N Ratio
1	20	14	15	1.0	627.0	55.95
2	20	17	20	1.5	632.0	56.01
3	20	20	25	2.0	535.6	54.58
4	25	14	20	2.0	405.5	52.16
5	25	17	25	1.0	693.0	56.81
6	25	20	15	1.5	451.0	53.08
7	30	14	25	1.5	650.3	56.26
8	30	17	15	2.0	471.0	53.46
9	30	20	20	1.0	706.6	56.98

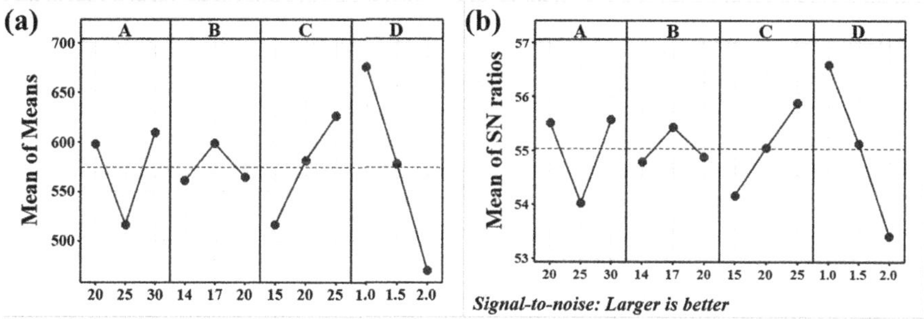

FIGURE 19.2 (a) Main effect plot of means; (b) Main effect plot of S/N ratios.

The microhardness of an electroless coating increases with a higher sodium tungstate concentration because it promotes greater tungsten incorporation while reducing the phosphorus content in the deposit. Lower phosphorus levels favor the formation of crystalline or nanocrystalline structures instead of amorphous ones, which enhances the hardness of the coating by improving its structural integrity and resistance to deformation [23]. The most prominent effect was observed for factor D (ceric sulfate). The microhardness was the highest at 1.0 g/L, then declined sharply at 1.5 g/L and continued to decrease further at 2.0 g/L. A small amount of ceric sulfate enhanced the microhardness of the electroless coating by refining the grain structure during deposition, resulting in a denser and more uniform microstructure. Cerium ions hinder grain growth, promote grain boundary strengthening, and introduce dispersion and solid solution strengthening effects, all of which impede dislocation movement and increase the hardness. However, this effect is concentration-dependent; only low levels of ceric sulfate yield these benefits, whereas excess amounts may lead to coarser grains or undesirable phases that reduce the coating quality [24, 25].

The S/N ratio plot (Figure 19.2(b)) further supports these findings, emphasizing the robustness and consistency of the same optimal parameter combination (A3B2C3D1). A steady rise in S/N ratios with increasing sodium tungstate levels indicates its positive effect on coating stability, while a steep drop beyond 1.0 g/L of ceric sulfate indicates performance degradation at higher concentrations. Together, these plots confirm that the careful tuning of the input concentrations significantly improves both the hardness and reliability of the electroless Ni–W–P–Ce coatings. Overall, the mean effect analysis indicated that the optimal hardness was achieved at A3B2C3D1 (30 g/L $NiSO_4$, 17 g/L NaH_2PO_2, 25 g/L Na_2WO_4, and 1.0 g/L $Ce(SO_4)_2$), which is consistent with the best-performing experimental run.

Table 19.4 presents the mean S/N ratios for the microhardness corresponding to each level of the four process parameters. The difference between the maximum and minimum mean S/N ratios (max–min) indicates the relative influence of each parameter on the coating hardness. Ceric sulfate (D) exhibited the highest max–min value of 3.18, making it the most influential parameter, followed by sodium tungstate (C) with a difference of 1.72. Nickel sulfate (A) showed a moderate effect (1.55), whereas sodium hypophosphite (B) had the least influence, with a max–min value of 0.64. These rankings highlight that the ceric sulfate concentration has the

TABLE 19.4
Mean Signal-to-Noise Ratio for Microhardness

		Mean S/N Ratio				
Designation	**Parameters**	**Level 1**	**Level 2**	**Level 3**	**Max-Min**	**Rank**
A	Nickel sulfate	55.51	54.02	55.57	1.55	3
B	Sodium hypophosphite	54.79	55.43	54.88	0.64	4
C	Sodium tungstate	54.16	55.05	55.88	1.72	2
D	Ceric sulfate	56.58	55.12	53.4	3.18	1

strongest impact on microhardness, emphasizing the importance of optimizing its level to enhance the coating performance.

19.3.1.2 ANOVA Results and Confirmation Testing

Table 19.5 presents the results of the ANOVA performed to determine the relative contribution of each process parameter to the variation in coating microhardness. Among the four parameters, ceric sulfate (D) exhibited the highest percentage contribution (63.4%), indicating that it was the most influential factor affecting the microhardness. Sodium tungstate (C) and nickel sulfate (A) contributed 18.46% and 15.51%, respectively, indicating moderate influence. In contrast, sodium hypophosphite (B) had the least impact, contributing only 2.63%. These results are consistent with the findings from the S/N ratio and main effects plots, reaffirming that optimizing the concentration of ceric sulfate is critical for achieving higher hardness in electroless Ni–W–P–Ce coatings.

However, the error term in this ANOVA analysis had zero degrees of freedom (DOF), making it impossible to calculate the mean square or variance of the error term and, hence, the F-ratio. The F-ratio is defined as the variance attributed to an individual factor divided by the variance of the error term. When the DOF error is zero, this calculation cannot be performed directly. To address this, pooling has been applied [21], which involves ignoring or combining the least significant factor(s) to

TABLE 19.5
ANOVA for Microhardness

Parameters	DOF	Sum of Squares	Mean Square	F-Ratio	% Contribution
A	2	15410	7705	–	15.51
B	2	2610	1305	–	2.63
C	2	18341	9171	–	18.46
D	2	62978	31489	–	63.4
Error	0	–	–	–	–
Total	8	99339			100

TABLE 19.6
Pooled ANOVA for Microhardness (Pooled Factor is Shown in { })

Parameters	DOF	Sum of Squares	Mean Square	F-Ratio	P-Value	% Contribution
A	2	15410	7705	5.90	0.14	15.51
{B}	{2}	{2610}	–	–	–	–
C	2	18341	9171	7.03	0.12	18.46
D	2	62978	31489	24.13	0.04[a]	63.4
Pooled error	2	2610	1305	–	–	2.63
Total	8	99339				100

Note: Confidence level = 95%.
[a] Significant (P-value < 0.05).

create a pooled error term with a nonzero DOF. This allows the estimation of the error variance and calculation of the F-ratio for significance testing.

Table 19.6 shows the pooled ANOVA results for microhardness, where sodium hypophosphite (factor B) was treated as the least significant and pooled with the error term to address the issue of zero DOF in the original ANOVA. This adjustment enabled the calculation of the MSE and F-ratios for statistical significance testing. Among the parameters, ceric sulfate (D) was found to be statistically significant at the 95% confidence level, with a P-value of 0.04 and the highest contribution (63.4%). Although nickel sulfate (A) and sodium tungstate (C) exhibited a moderate influence, with contributions of 15.51% and 18.46%, respectively, their P-values (0.14 and 0.12) indicated a lack of statistical significance. These findings reinforce that ceric sulfate concentration plays the most dominant role in enhancing the coating microhardness, which is consistent with the trends observed in the main effects and S/N ratio plots.

The confirmation test results comparing the initial and optimum parameter combinations for microhardness are listed in Table 19.7. The initial combination (A2B2C2D2), selected using the mid-level values in Table 19.2, served as the baseline. The optimum combination (A3B2C3D1) was determined using Taguchi analysis. Experimental validation of both settings was performed, and microhardness

TABLE 19.7
Confirmation Test Results for Microhardness

Combination Type	Parameter Setting	Microhardness (HV 0.05)	S/N Ratio (dB)
Initial combination	A2B2C2D2	487.2	53.75
Optimum combination	A3B2C3D1	744.6	57.43
Improved S/N ratio			3.68
Increase in microhardness		34.57%	

values were recorded. The optimum combination yielded a significantly higher microhardness (744.6 HV compared to 487.2 HV) than the initial setting, representing a 34.57% improvement. Correspondingly, the S/N ratio increased from 53.75 dB to 57.43 dB, indicating reduced variability and enhanced consistency. These results confirm the effectiveness of the Taguchi method in optimizing the process parameters to improve the coating performance.

19.3.2 ANN Prediction

The regression analysis (Figure 19.3) highlights the strong agreement between the predicted and actual microhardness values across all the data subsets. The high correlation coefficients in the training and validation sets indicate that the neural network effectively learned the underlying patterns and generalized well to the unseen data. Although the test set showed a slightly lower correlation (R =

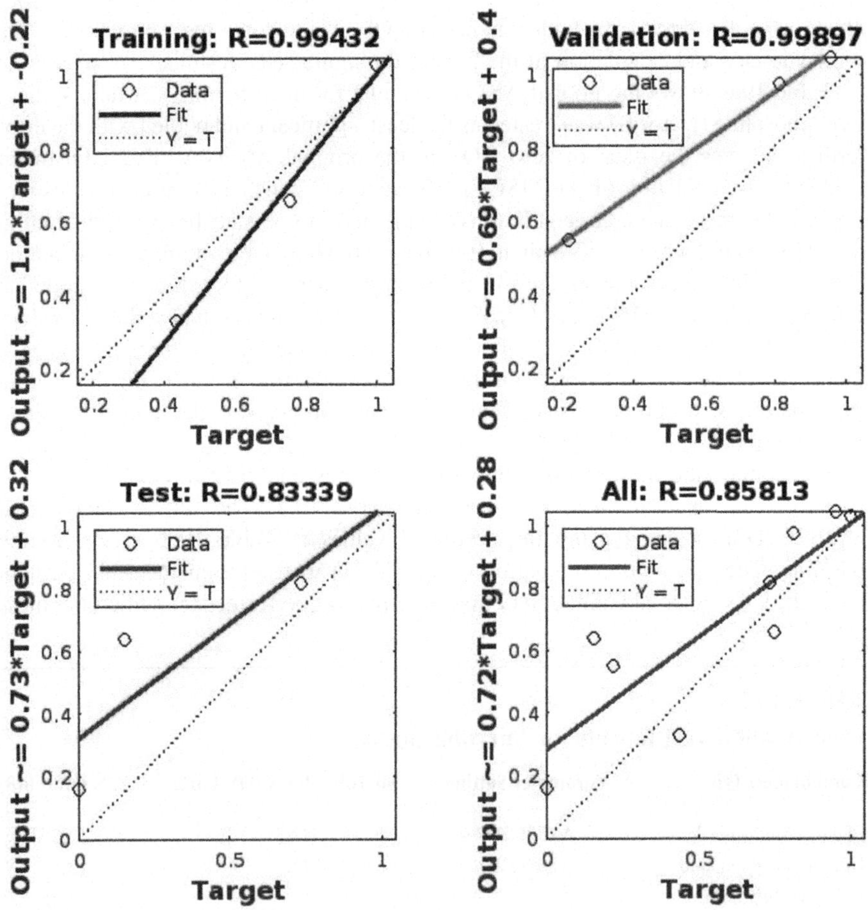

FIGURE 19.3 Regression plots for training, validation, testing, and overall datasets.

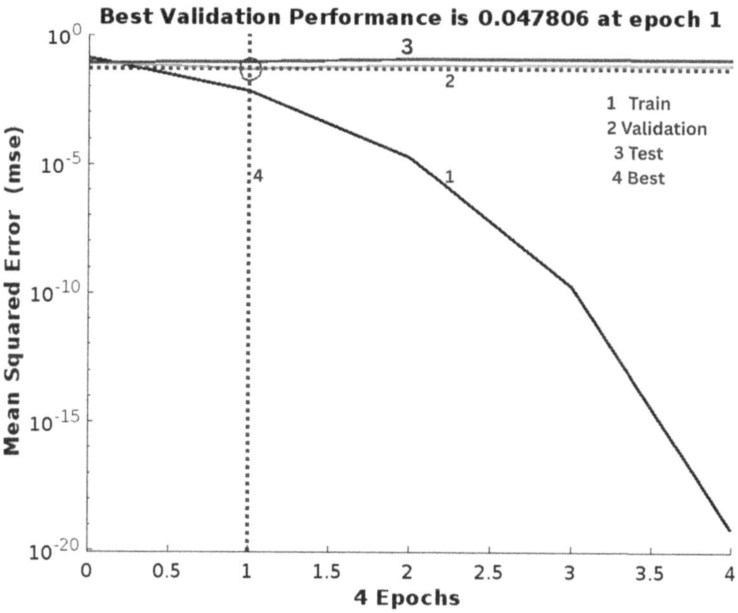

FIGURE 19.4 Mean squared error (MSE) performance plot.

0.83339), the overall R-value of 0.85813 demonstrates the robustness of the model and its ability to capture the nonlinear relationship between process parameters and microhardness with reasonable accuracy, despite the limited dataset size.

The performance trend over the epochs is presented in Figure 19.4, where the best validation performance (MSE = 0.047806) was achieved at epoch 1. Although the training error continued to decrease, the validation and test errors remained relatively constant, suggesting the onset of overfitting beyond this point. This early stopping behavior, although limiting further learning, helped to maintain the model's generalization ability, as evidenced by the regression results. The network successfully extracted key features from the data within a few iterations, thereby indicating rapid and efficient learning.

Further insights into the training dynamics are shown in Figure 19.5, which depicts the evolution of the gradient, the learning rate parameter (mu), and validation checks. The gradient decreased steadily to a minimal value of 3.59×10^{-10} by epoch 4, suggesting convergence toward an optimal solution. Simultaneously, the mu value dropped consistently, indicating stable learning progress. The validation check counter reached its limit of three at epoch 4, prompting the early termination of training to prevent overfitting. These observations confirm that the training process is well-regulated and that the final model achieves a good balance between accuracy and generalization.

Table 19.8 presents a comparison between the experimental and ANN-predicted microhardness values for all nine Taguchi L9 experimental runs, along with the corresponding absolute error percentages. The results demonstrated that the ANN

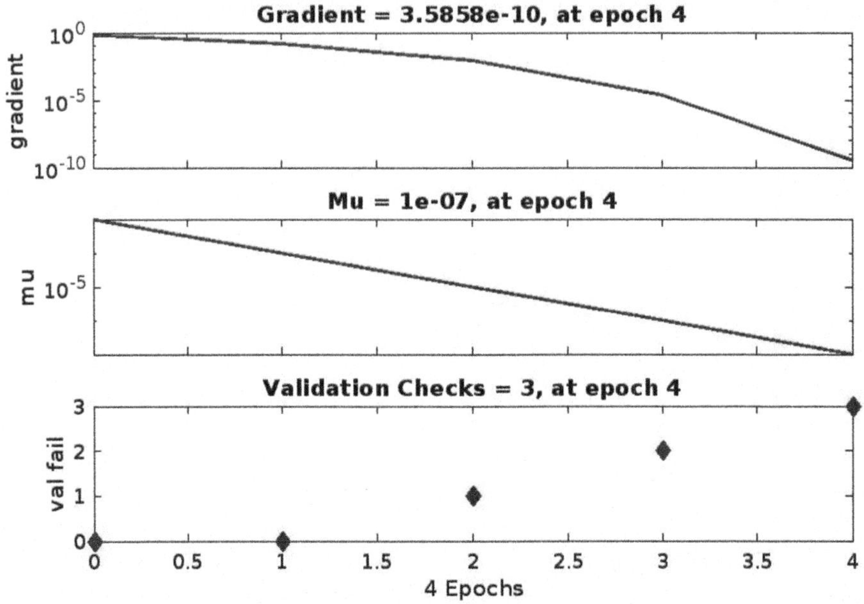

FIGURE 19.5 Training state: gradient, learning rate (mu), and validation checks over epochs.

TABLE 19.8
Experimental versus ANN-Predicted Microhardness with Absolute Error (%)

					Microhardness (HV)		
Run No.	A	B	C	D	Experimental	ANN Predicted	Absolute Error (%)
1	20	14	15	1.0	627.0	651.66	3.93
2	20	17	20	1.5	632.0	603.51	4.51
3	20	20	25	2.0	535.6	504.16	5.87
4	25	14	20	2.0	405.5	453.12	11.74
5	25	17	25	1.0	693.0	720.58	3.98
6	25	20	15	1.5	451.0	597.71	32.53
7	30	14	25	1.5	650.3	698.48	7.41
8	30	17	15	2.0	471.0	570.6	21.15
9	30	20	20	1.0	706.6	714.95	1.18

model was able to closely approximate the actual microhardness in most cases, with absolute errors generally below 10% for the majority of runs. The lowest error was observed in Run 9 (1.18%), indicating excellent predictive accuracy. However, higher deviations were observed in Runs 6 and 8, with absolute errors of 32.53% and 21.15%, respectively, which may be attributed to the limited training data or complex local variations that were not fully captured by the model. Overall, the ANN

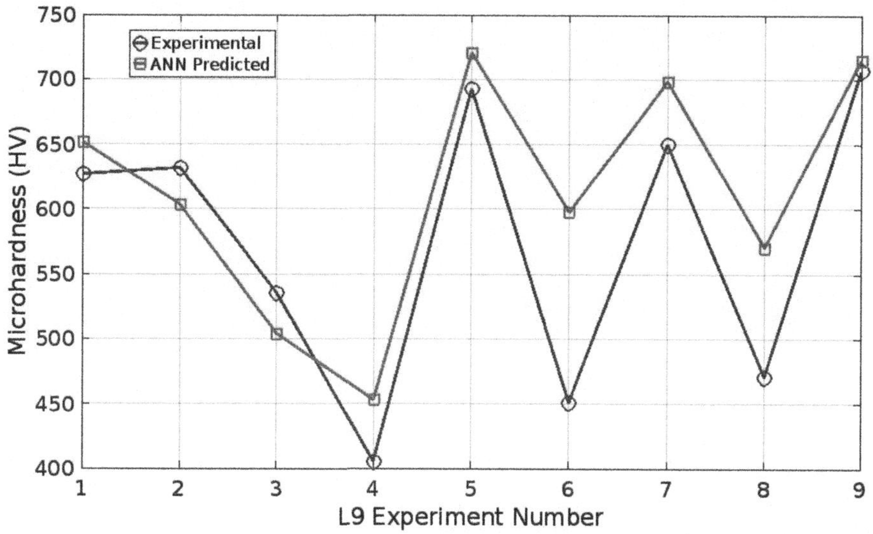

FIGURE 19.6 Comparison of experimental and ANN-predicted microhardness values.

predictions aligned well with the experimental results, supporting the model's effectiveness in capturing the underlying process behavior.

The comparison plot in Figure 19.6 illustrates the experimental and ANN-predicted microhardness values for each of the nine Taguchi L9 experimental runs. The predicted values closely followed the trend of the experimental data, demonstrating the capability of the model to accurately capture the variations in microhardness across different process parameter combinations. While slight deviations were observed in some runs—particularly Runs 6 and 8—the overall agreement remained strong, further validating the predictive accuracy and generalization ability of the developed ANN model.

19.4 CHARACTERIZATION OF THE OPTIMIZED COATING SURFACE

Surface characterization was performed only for the electroless Ni–W–P–Ce coating produced using the optimum parameter combination (A3B2C3D1) identified through Taguchi analysis. This sample is referred to as the "optimal coating" in this section.

The surface morphology of the optimal coating at two different magnifications is shown in Figure 19.7. The surface appeared dense and uniform and free from visible cracks or pores. Distinct, uniformly sized coating nodules were observed, indicating consistent deposition. Some black spots were present, which may be attributed to carbon contamination due to delayed FESEM characterization caused by late slot allocation. The overall coating structure exhibited a characteristic cauliflower-like morphology [26].

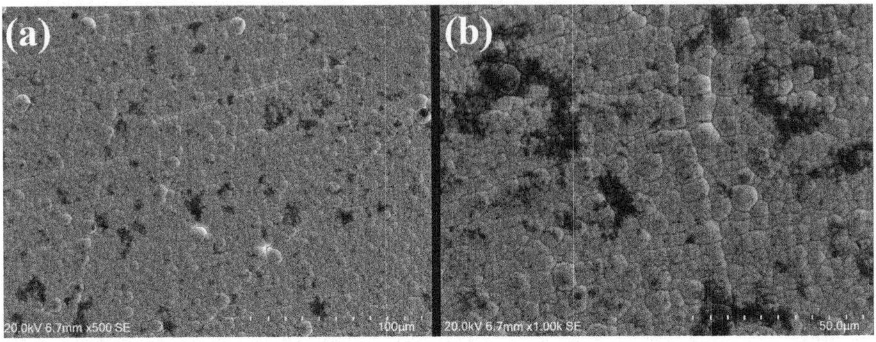

FIGURE 19.7 Surface morphology of the optimal Ni–W–P–Ce coating.

TABLE 19.9
EDX Results for the Ni–W–P–Ce Coating

Element (wt.%)	Nickel, Ni	Tungsten, W	Phosphorus, P	Cerium, Ce
Weight %	86.02	5.68	6.51	1.78
Atomic %	84.73	1.78	12.16	1.32

The elemental compositions of the Ni–W–P–Ce coatings are presented in Table 19.9. The coating primarily consisted of nickel (86.02 wt.%), with moderate amounts of phosphorous (6.51 wt.%) and tungsten (5.68 wt.%), along with a small quantity of cerium (1.78 wt.%). Based on the phosphorus content, the coating falls under the category of medium phosphorus electroless coatings, which are known for their ductility, resistance to cracking under mechanical loading, and the presence of an initial amorphous Ni–P phase along with fcc–Ni crystallites [27]. These structural characteristics contribute to enhanced toughness and mechanical integrity. The small amount of Ce detected indicates effective incorporation during co-deposition, which may enhance surface properties such as hardness and corrosion resistance.

Elemental mapping of the optimal Ni–W–P–Ce coating, shown in Figure 19.8, confirms the uniform distribution of all constituent elements across the coating surface. Nickel was observed as the dominant phase, whereas tungsten and phosphorus were evenly co-deposited throughout the matrix. Cerium, although present in smaller amounts, was well-dispersed, indicating successful incorporation during the electroless deposition process. The uniform elemental distribution supports the homogeneity of the coating and is consistent with the compact and crack-free surface morphology observed previously.

The 3D surface profile of the optimal coating, captured using an optical profilometer, is shown in Figure 19.9. The surface topography appeared uniform with evenly distributed microfeatures, indicating consistent and homogeneous coating deposition. The measured surface roughness parameters include an average

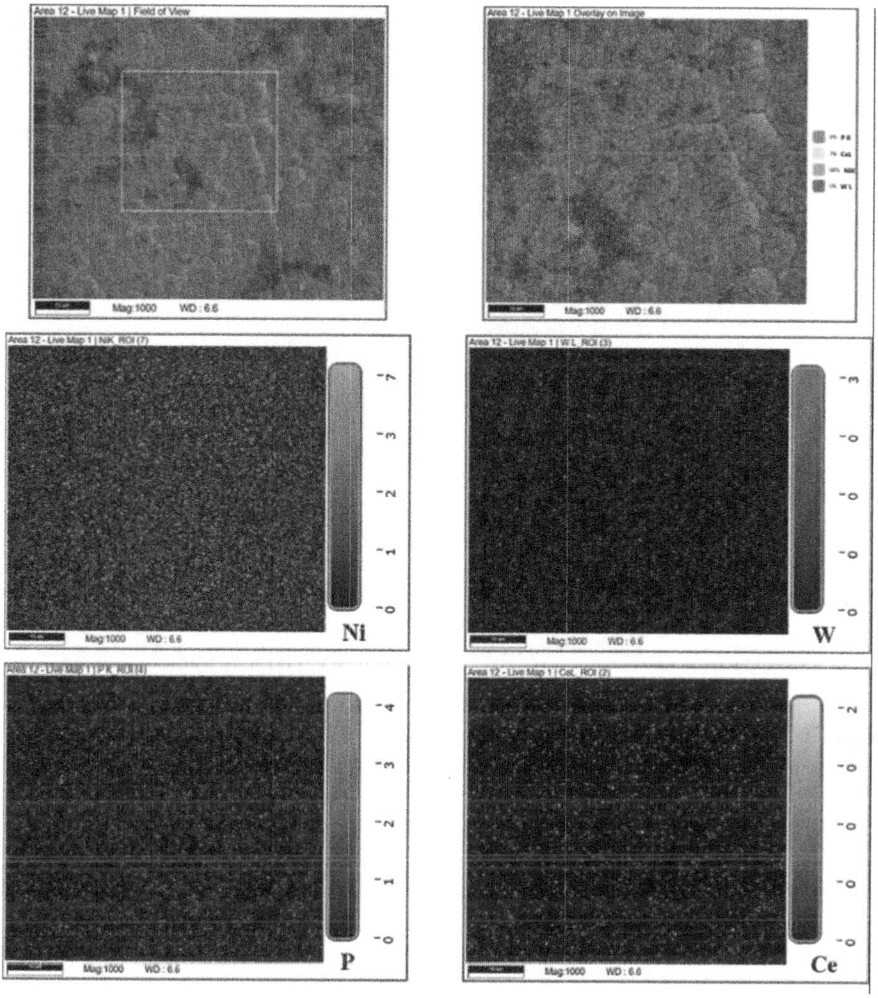

FIGURE 19.8 Elemental mapping of the optimal Ni–W–P–Ce coating.

roughness (Sa) of 0.66 μm and a root mean square roughness (Sq) of 1.15 μm, reflecting a moderately textured surface. These roughness values indicate the effective incorporation of alloying and reinforcing elements, which contribute to the surface refinement.

The XRD pattern of the optimal Ni–W–P–Ce coating (Figure 19.10) revealed a strong and sharp peak at approximately 44°, corresponding to the (111) plane of face-centered cubic (fcc) nickel, indicating a predominant crystalline orientation. The smaller peak observed near 51.7° corresponds to the (220) plane of Ni [28]. The presence of these distinct peaks suggested a nanocrystalline structure. The average crystallite size was calculated to be 6.5 nm using the Debye–Scherrer equation [29], confirming the fine-grained nature of the coating.

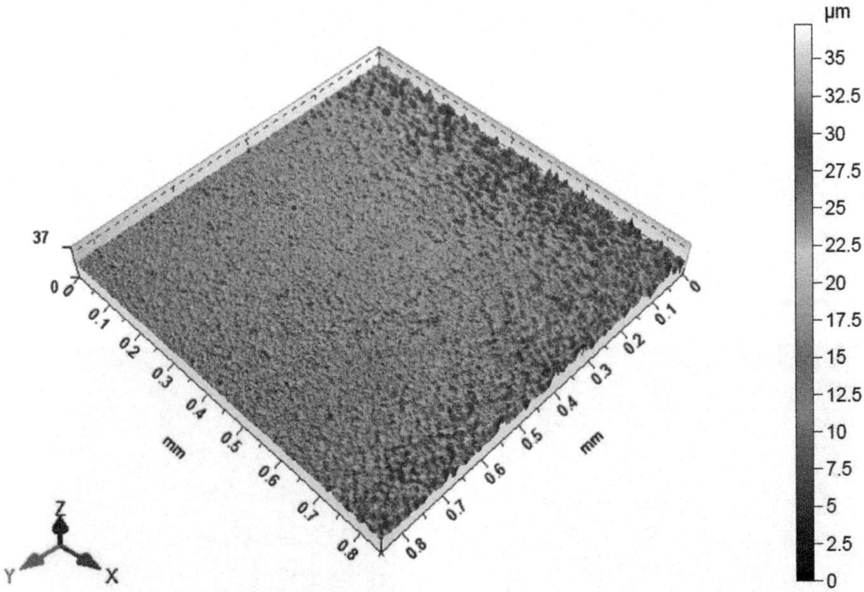

FIGURE 19.9 3D surface topography of the optimal electroless Ni–W–P–Ce coating.

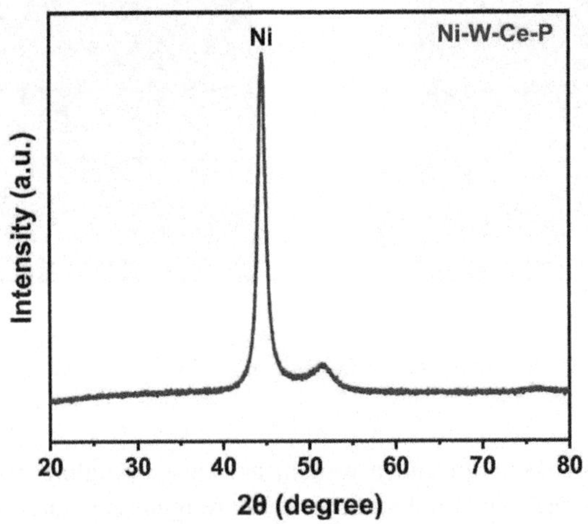

FIGURE 19.10 XRD pattern of the optimized Ni-W-P-Ce coating.

19.5 CONCLUSIONS

This study successfully developed and optimized an electroless Ni–W–P–Ce coating using a synergistic approach that combined the Taguchi design of experiments, ANOVA, and ANN modeling. The Taguchi L9 orthogonal array identified an optimal combination of 30 g/L nickel sulfate, 17 g/L sodium hypophosphite, 25 g/L sodium tungstate, and 1 g/L ceric sulfate, leading to a 34.57% increase in microhardness over the initial setting. The ANOVA results highlighted ceric sulfate as the most influential factor (63.4%), followed by sodium tungstate and nickel sulfate. The ANN model developed for predicting microhardness showed a strong predictive capability, with an overall correlation coefficient (R) of 0.85813, despite the limited dataset. Surface characterization of the optimized coating revealed a dense, crack-free morphology with a cauliflower-like structure, while XRD analysis confirmed a nanocrystalline phase with an average crystallite size of 6.5 nm. EDX analysis indicated a medium phosphorus content (6.51 wt.%) along with the effective incorporation of tungsten (5.68 wt.%) and cerium (1.78 wt.%), contributing to enhanced mechanical performance. This integrated methodology demonstrates a robust framework for the design and optimization of advanced electroless coatings and can be effectively extended to other multifunctional surface engineering applications.

REFERENCES

1. Kumar, V., Mandal, B. B., Das, S., Oraon, B., & Mukherjee, S. (2024). Tailoring electroless Ni-WP polyalloy coatings: Unveiling the synergistic impact of cerium addition and annealing on surface functional properties. *Surface and Coatings Technology*, 485, 130874.
2. Kumar, V., Shrivastava, A., Chakraborty, S. S., & Mukherjee, S. (2023). Directed energy deposition of Ni-based hard-facing alloy. In *Laser Applications in Manufacturing* (pp. 81–98). Boca Raton, FL: CRC Press.
3. Rauta, C. S., Majumdar, G., & Sarkar, S. (2022). Parametric optimization of microhardness of electroless Ni-Zn-Cu-P coating using Taguchi design and artificial neural network. *JOM*, 74(12), 4564–4574.
4. Ekmekci, D., & Bülbül, F. (2015). Preparation and characterization of electroless Ni–B/nano-SiO_2, Al_2O_3, TiO_2 and CuO composite coatings. *Bulletin of Materials Science*, 38, 761–768.
5. Nemane, V., & Chatterjee, S. (2020). Scratch and sliding wear testing of electroless Ni–B–W coating. *Journal of Tribology*, 142(2), 021705.
6. Barman, M., Barman, T. K., & Sahoo, P. (2024). Tribological behavior of electroless Ni-B ternary and quaternary coatings with inclusion of W and Mo. *Physica Scripta*, 99(2), 025924.
7. Varol, T., Hacısalihoğlu, İ., Kaya, G., Güler, O., Yıldız, F., Aksa, H. C., & Akçay, S. B. (2021). The effect of selective laser melting process on the microstructure, density, and electrical conductivity of silver-coated copper cores. *Journal of Materials Engineering and Performance*, 30, 5216–5226.
8. Bharadishettar, N., Bhat K. U., & Bhat Panemangalore, D. (2021). Coating technologies for copper based antimicrobial active surfaces: A perspective review. *Metals*, 11(5), 711.
9. Zhai, S., Sugio, K., & Sasaki, G. (2025). Prediction and optimization of microhardness of electroless Ni–P alloy coating based on machine learning. *Journal of the Electrochemical Society*, 172(3), 032501.
10. Mukhopadhyay, A., Duari, S., Barman, T. K., & Sahoo, P. (2016). Wear analysis of electroless Ni-P coating under lubricated condition using fuzzy logic. *Portugaliae Electrochimica Acta*, 34(1), 63–83.
11. Kumar, A., Jana, T., Ghuku, S., & Mukhopadhyay, A. (2025). Fractal, tribo-mechanical and corrosion characterization of hydrophobic and lead free electroless Ni-B mono and bi-layer coating developed using artificial intelligence. *Surface Topography: Metrology and Properties*, 13, 015015.
12. Mandal, R., Ghosh, A., Sarkar, S., De, J., Mandal, T., Sen, R. S., & Majumdar, G. (2024). Parametric optimization to maximize microhardness of electroless Ni-Sn-P coating using

Taguchi and evolutionary approaches. *Journal of The Institution of Engineers (India): Series C*, 105(5), 1217–1231.

13. De, J., Banerjee, T., Sen, R. S., Oraon, B., & Majumdar, G. (2016). Multi-objective optimization of electroless ternary Nickel–Cobalt–Phosphorous coating using non-dominant sorting genetic algorithm-II. *Engineering Science and Technology, an International Journal*, 19(3), 1526–1533.
14. Sadoun, A. M., Najjar, I. M. R., Fathy, A., Abd Elaziz, M., Al-Qaness, M. A., Abdallah, A. W., & Elmahdy, M. (2023). An enhanced Dendritic Neural Algorithm to predict the wear behavior of alumina coated silver reinforced copper nanocomposites. *Alexandria Engineering Journal*, 65, 809–823.
15. Song, R., Eo, S., Lee, M., & Lee, J. (2022). Physics-informed machine learning for optimizing the coating conditions of blade coating. *Physics of Fluids*, 34(8), 082112.
16. Paturi, U. M. R., Reddy, N. S., Cheruku, S., Narala, S. K. R., Cho, K. K., & Reddy, M. M. (2021). Estimation of coating thickness in electrostatic spray deposition by machine learning and response surface methodology. *Surface and Coatings Technology*, 422, 127559.
17. Fu, N., Wei, L., & Hu, J. (2024). Physics-guided dual self-supervised learning for structure-based material property prediction. *The Journal of Physical Chemistry Letters*, 15(10), 2841–2850.
18. Verma, J., & Khanna, A. S. (2023). Digital advancements in smart materials design and multifunctional coating manufacturing. *Physics Open*, 14, 100133.
19. Rannetbauer, W., Hambrock, C., Hubmer, S., & Ramlau, R. (2024). Enhancing predictive quality in HVOF coating technology: A comparative analysis of machine learning techniques. *Procedia Computer Science*, 232, 1377–1387.
20. Kumar, V., Mandal, B. B., & Oraon, B. (2022). Statistical analysis of activator and poly-alloy electroless coating on plain carbon steel by full factorial design. *NeuroQuantology*, 20(10), 9396–9405.
21. Mandal, B. B., Kumar, V., & Oraon, B. (2024). Optimization, prediction, and characterization of electroless Ni-WP-nanoTiO_2 composite coatings on pipeline steel. *Arabian Journal for Science and Engineering*, 49(2), 1643–1657.
22. Suryawanshi, S., Bhosale, D. G., Vasudev, H., & Prabhu, T. R. (2024). Back propagation model for prediction of deposition parameters in plasma sprayed WC-based coatings. *International Journal on Interactive Design and Manufacturing (IJIDeM)*, 19(3), 1837–1848.
23. Yanhai, C., Shuai, C., Qingqiang, H., Dongtai, H., & Zhengtong, H. (2016). Effect of tungsten addition on the anti-fouling property of the electroless Ni-WP deposits. *Rare Metal Materials and Engineering*, 45(8), 1931–1937.
24. Valova, E., Armyanov, S., Hristova, G., Vassilev, T., Steenhaut, O., Dille, J., . . . & Vandendael, I. (2016). Electroless deposited NiCeP coatings. *Surface and Coatings Technology*, 304, 468–475.
25. Kumar, V., Mandal, B. B., & Oraon, B. (2022). Synthesis and characterization of electroless Ni–W-Ce-P coating on AISI 1010 substrate. *Materials Today: Proceedings*, 59, 1183–1188.
26. Mohamed Anuar, R. A., Ismail, A., Ismail, N., Taib, I., Abd Latif, N. S., Hamdan, I. F., & A Razak, (2025). *Optimisation of Electroless Nickel Electroless Palladium Immersion Silver (ENEPImAg) Surface Finish Parameters Using the Taguchi Method for Enhanced Solder Joint Reliability*. Soldering & Surface Mount Technology.
27. Sharma, T., Bera, H., Brown, D. A., Schulze, A., & Brüning, R. (2023). Thermal evolution, phase composition and fracture toughness of electroless Ni-P, Ni-WP and Ni-Mo-WP films for solderable surfaces on copper. *Surface and Coatings Technology*, 467, 129722.

28. Salicio-Paz, A., Grande, H., Pellicer, E., Sort, J., Fornell, J., Offoiach, R., . . . & García-Lecina, E. (2019). Monolayered versus multilayered electroless NiP coatings: Impact of the plating approach on the microstructure, mechanical and corrosion properties of the coatings. *Surface and Coatings Technology*, 368, 138–146.
29. Unni, M., & Jothi, S. (2025). Ni-Co-W-Zr (P) quinary-based medium entropy alloy achieved by an electrochemical route and its properties. *Journal of the Electrochemical Society*, 172(1), 012502.

Index

Pages in *italics* refer to figures and pages in **bold** refer to tables.

U

V

W

X

Z

For Product Safety Concerns and Information please contact our EU representative GPSR@taylorandfrancis.com Taylor & Francis Verlag GmbH, Kaufingerstraße 24, 80331 München, Germany

Printed and bound by CPI Group (UK) Ltd, Croydon, CR0 4YY
14/09/2026
02216256-0014